NEW TRENDS IN DIFFERENTIAL EQUATIONS, CONTROL THEORY AND OPTIMIZATION

NEW TRENDS IN DIFFERENTIAL EQUATIONS, CONTROL THEORY AND OPTIMIZATION

Proceedings of the 8th Congress of Romanian Mathematicians

"Al I Cuza" University of Iaşi, Romania 26 June – 5 July 2015

Edited by

VIOREL BARBU
CĂTĂLIN LEFTER
IOAN I VRABIE

"Al I Cuza" University of Iaşi, Romania &
Romanian Academy, Romania

World Scientific

NEW JERSEY • LONDON • SINGAPORE • BEIJING • SHANGHAI • HONG KONG • TAIPEI • CHENNAI • TOKYO

Published by

World Scientific Publishing Co. Pte. Ltd.

5 Toh Tuck Link, Singapore 596224

USA office: 27 Warren Street, Suite 401-402, Hackensack, NJ 07601

UK office: 57 Shelton Street, Covent Garden, London WC2H 9HE

Library of Congress Cataloging-in-Publication Data

Names: Congress of Romanian Mathematicians (8th : 2015 : Iaşi, Romania) |
Barbu, Viorel, editor. | Lefter, Cătălin, editor. | Vrabie, I. I. (Ioan I.), 1951– editor.

Title: New trends in differential equations, control theory, and optimization : proceedings of the 8th Congress of Romanian Mathematicians : 8th Congress of Romanian Mathematicians, "Al I Cuza" University of Iaşi, Romania, 26 June–5 July 2015 / edited by Viorel Barbu (Romanian Academy, Romania & "Al I Cuza" University of Iaşi, Romania), Cătălin Lefter ("Al I Cuza" University of Iaşi, Romania & Romanian Academy, Romania), Ioan I. Vrabie ("Al I Cuza" University of Iaşi, Romania & Romanian Academy, Romania).

Description: New Jersey : World Scientific, 2016. |
Includes bibliographical references and author index.

Identifiers: LCCN 2016022162 | ISBN 9789813142855 (hardcover : alk. paper)

Subjects: LCSH: Differential equations--Congresses. | Control theory--Congresses. |
Mathematical optimization--Congresses.

Classification: LCC QA370 .C678 2015 | DDC 515/.35--dc23

LC record available at https://lccn.loc.gov/2016022162

British Library Cataloguing-in-Publication Data

A catalogue record for this book is available from the British Library.

Printed in Singapore

Preface

The 8th *Congress of Romanian Mathematicians* was held at the "Alexandru Ioan Cuza" University of Iaşi, during the period June 26 – July 5, 2015. The Congress was organized by the Section of Mathematical Sciences of the Romanian Academy, the "Simion Stoilow" Institute of Mathematics of the Romanian Academy, the Faculty of Mathematics of "Alexandru Ioan Cuza" University of Iaşi, the Faculty of Mathematics and Computer Science of the University of Bucharest, the "Octav Mayer" Institute of Mathematics of the Romanian Academy, Iaşi, and the Romanian Mathematical Society.

The website of the Congress is http://www.math.uaic.ro/cmr2015/. We address a special thank to Dr. Marius Apetrii who developed the web environment as a support for this event.

The Congress was devoted to recent advances in various branches of Pure and Applied Mathematics and brought together more than 400 mathematicians from all over the world.

This volume is composed of 23 invited papers in Differential Equations, Control Theory and Optimization, written by top specialists in their fields, participants in the Congress sections *Ordinary and Partial Differential Equations, Variational Methods, Optimal Control, Optimization and Games Theory*, and *Mechanics, Numerical Analysis, Mathematical Models in Sciences*, and covering important present trends in the above mentioned disciplines.

We express our gratitude to all contributors as well as to the referees for their support in reviewing the papers. We address our warmest thanks to World Scientific for accepting to publish these proceedings and especially to Mrs. E. H. Chionh, the corresponding Editor, for her fruitful and very pleasant cooperation offered during the whole period of preparation of the manuscript. Finally, we would like to thank Mrs. Carmen Savin for her very careful work in arranging the volume in its final form.

Iaşi, April 23, 2016 — Viorel Barbu, Cătălin–George Lefter, Ioan I. Vrabie

Organizing Committee of the Congress

Viorel Barbu[a,f]

Marius Iosifescu[a]

Solomon Marcus[a]

Ioan Tomescu[a]

Lucian Beznea[b]

Dan Timotin[b]

Gabriela Marinoschi[c]

Victor Tigoiu[d]

Cătălin–George Lefter[e,f]

Radu Gologan[g]

Volume Editors

Viorel Barbu[a,f]

Cătălin–George Lefter[e,f]

Ioan I. Vrabie[e,f]

---

[a] *Romanian Academy*
[b] *"Simion Stoilow" Institute of Mathematics of the Romanian Academy*
[c] *"Gheorghe Mihoc–Caius Iacob" Institute of Mathematical Statistics and Applied Mathematics of the Romanian Academy*
[d] *University of Bucharest*
[e] *Faculty of Mathematics of "Alexandru Ioan Cuza" University of Iaşi*
[f] *"Octav Mayer" Institute of Mathematics of the Romanian Academy, Iaşi branch*
[g] *The Romanian Mathematical Society*

Contents

Dirichlet problems with mean curvature operator in Minkowski space

Cristian Bereanu

Faculty of Mathematics and Computer Science, University of Bucharest
14, Str. Academiei, 70109-Bucharest, Romania
and Institute of Mathematics "Simion Stoilow" of the Romanian Academy
E-mail: cristian.bereanu@imar.ro

Petru Jebelean* and Călin Şerban**

Department of Mathematics, West University of Timişoara
4, Blvd. V. Pârvan 300223-Timişoara, Romania
*E-mail: *petru.jebelean@e-uvt.ro, **cserban2005@yahoo.com*

We survey some existence and multiplicity results for Dirichlet problems with mean curvature operator in Minkowski space. We deal with positive radial solutions when the problem has radial symmetry, as well as nontrivial constant sign solutions for problems in general bounded domains. Discontinuous Dirichet problems are also considered. Our approaches combine variational techniques, lower and upper solutions method and topological degree arguments.

Keywords: Mean curvature operator; Dirichlet problem; Critical point; Radial solutions; Lower and upper solutions; Leray-Schauder degree.

1. Introduction

In this paper we survey some recent existence and multiplicity results for Dirichlet problems involving the mean curvature operator in Minkowski space:

$$\mathcal{M}(u) := \operatorname{div}\left(\frac{\nabla u}{\sqrt{1-|\nabla u|^2}}\right).$$

Here and below, $|\cdot|$ stands for the Euclidean norm in $\mathbb{R}^N$ $(N \geq 2)$. A first existence result for a Dirichlet problem involving the operator $\mathcal{M}$ is due to F.J. Flaherty[14] and it is motivated by the seminal paper of S.-Y. Cheng and S.-T. Yau[8]. Extensions of the results from Ref. 14 were obtained by R. Bartnik and L. Simon[1] and C. Gerhardt[15]. For a discussion concerning earlier achievements on this topic we refer the reader to Refs. 1,15 and the references therein. Then, in the last decade, various topological and variational methods were combined to study Dirichlet problems with operator $\mathcal{M}$ (Refs. 2–6,10–12). Essentially, this paper presents ideas and results from Refs. 2–6.

First, we are concerned with positive solutions for the problem

$$\mathcal{M}(v) + \lambda\left[\varphi(|x|)v^q\right] = 0 \quad \text{in} \quad \mathcal{B}(R), \quad v = 0 \quad \text{on} \quad \partial\mathcal{B}(R), \tag{1}$$

where $\lambda > 0$ is a parameter, $q > 1$, $\varphi : [0, \infty) \to \mathbb{R}$ is continuous, strictly positive on $(0, \infty)$, with $\varphi(0) = 0$ and $\mathcal{B}(R) = \{x \in \mathbb{R}^N : |x| < R\}$. We show in Theorem 4.1 that there exists $\Lambda > 0$ such that problem (1) has zero, at least one or at least two positive radial solution according to $\lambda \in (0, \Lambda)$, $\lambda = \Lambda$ or $\lambda > \Lambda$.

Next we deal with the Dirichlet problem

$$\mathcal{M}(u) = f(x, u) \quad \text{in} \quad \Omega, \quad u|_{\partial\Omega} = 0, \tag{2}$$

where Ω is a general bounded domain in $\mathbb{R}^N$ $(N \geq 2)$ with boundary $\partial\Omega$ of class C^2, and $f : \Omega \times \mathbb{R} \to \mathbb{R}$ is a Carathéodory function, i.e.,

(i) $f(x, \cdot) : \mathbb{R} \to \mathbb{R}$ is continuous for a.e. $x \in \Omega$,
(ii) $f(\cdot, s) : \Omega \to \mathbb{R}$ is measurable for all $s \in \mathbb{R}$.

By a *solution* of (2) we mean a function $u \in W^{2,p}(\Omega)$ with some $p > N$, such that $\|\nabla u\|_\infty < 1$, which satisfies the equation a.e. in Ω and vanishes on $\partial\Omega$. Note that, since $W^{2,p}(\Omega) \subset C^1(\overline{\Omega})$, the evaluation at $\partial\Omega$ is understood in the usual sense. Here $\|\cdot\|_\infty$ stands for the usual sup-norm on $C(\overline{\Omega})$.

We show (Theorem 5.2) that under the only hypothesis:

(L^∞) *for each* $\rho > 0$ *there is some* $\alpha_\rho \in L^\infty(\Omega)$ *such that* $|f(x, s)| \leq \alpha_\rho(x)$ *for a.e.* $x \in \Omega$, $\forall\, s \in \mathbb{R}$ *with* $|s| \leq \rho$.

problem (2) is solvable – this is an "universal" existence result originated in Ref. 1. Then, we obtain (Theorem 6.3) the existence of four nontrivial solutions of constant sign for the one-parameter Dirichlet problem

$$\mathcal{M}(u) = \mu(x)|u|^{q-1}u - \lambda g(x, u) \quad \text{in} \quad \Omega, \quad u|_{\partial\Omega} = 0, \tag{3}$$

under the following hypothesis on the data:
(H) *$\mu \in L^\infty(\Omega)$ is non-negative a.e. in Ω, $q > 0$ is fixed and $g : \Omega \times \mathbb{R} \to \mathbb{R}$ is a Carathéodory function satisfying the growth condition (L^∞) and*

$$g(x, 0) = 0 \quad \textit{for a.e.} \quad x \in \Omega. \tag{4}$$

No symmetry assumptions are made on $g(x,\cdot)$. These results are slightly more general than the ones given in Ref. 2.

In the final section we deal with the discontinuous Dirichlet problem

$$\mathcal{M}(u) \in \left[\underline{f}(x,u), \overline{f}(x,u)\right] \quad \text{in} \quad \Omega, \qquad u|_{\partial\Omega} = 0, \tag{5}$$

where $f : \Omega \times \mathbb{R} \to \mathbb{R}$ is now a measurable function satisfying

$$|f(x,s)| \leq \mathcal{C}(1 + |s|^{q-1}), \quad \text{a.e.} \quad x \in \Omega \quad \text{and all} \quad s \in \mathbb{R}, \tag{6}$$

with some $q \in (1,\infty)$ and $\mathcal{C}$ a positive constant. As usual, for a.e. $x \in \Omega$ and all $s \in \mathbb{R}$, we have denoted $\underline{f}(x,s) := \lim_{\delta \searrow 0} \operatorname{essinf}\{f(x,t) : |t-s| < \delta\}$ and $\overline{f}(x,s) := \lim_{\delta \searrow 0} \operatorname{esssup}\{f(x,t) : |t-s| < \delta\}$.

Also, we assume in addition that

$$\underline{f} \text{ and } \overline{f} \text{ are } N\text{-measurable} \tag{7}$$

(recall, a function $h : \Omega \times \mathbb{R} \to \mathbb{R}$ is called N-*measurable* if $h(\cdot, v(\cdot)) : \Omega \to \mathbb{R}$ is measurable whenever $v : \Omega \to \mathbb{R}$ is measurable).

By *a solution* of problem (5) we will understand a function $u \in W^{2,p}(\Omega)$ for some $p > N$, such that $\|\nabla u\|_\infty < 1$, which satisfies

$$\mathcal{M}(u)(x) \in \left[\underline{f}(x,u(x)), \overline{f}(x,u(x))\right], \quad \text{a.e.} \quad x \in \Omega$$

and vanishes on $\partial\Omega$. This type of solutions, but for ordinary differential inclusions, was firstly considered by A.F. Filippov [13]. Also, for partial differential inclusions we refer to the pioneering works of I. Massabo and C.A. Stuart [20], J. Rauch [21], C.A. Stuart and J.F. Toland [22]. We prove for problem (5) an "universal" existence result (Theorem 7.1), similar to the one obtained for problem (2).

2. A fixed point approach, lower and upper solutions, degree

When dealing with radial solutions of (1), we are led (setting $r = |x|$ and $v(x) = u(r)$) to consider the one-dimensional problem

$$\left(r^{N-1}\frac{u'}{\sqrt{1-u'^2}}\right)' + r^{N-1}\lambda\varphi(r)u^q = 0, \qquad u'(0) = 0 = u(R), \tag{8}$$

which makes of interest mixed boundary problems of type

$$(r^{N-1}\phi(u'))' + r^{N-1}\ell(r,u) = 0, \qquad u'(0) = 0 = u(R), \tag{9}$$

under the following hypotheses:

(H_ϕ) $\phi : (-a,a) \to \mathbb{R}$ $(0 < a < \infty)$ *is an odd, increasing homeomorphism;*

(H_ℓ) *$\ell : [0,R] \times [0,\infty) \to [0,\infty)$ is continuous and $\ell(r,s) > 0$ for all $(r,s) \in (0,R] \times (0,\infty)$.*

By *a solution* of (9) we mean a function $u \in C^1 := C^1[0,R]$ with $\|u'\|_\infty < a$, such that $\phi(u')$ is differentiable and (9) is satisfied. Here and below we denote by $\|\cdot\|_\infty$ the usual sup-norm on $C := C[0,R]$.

For the beginning we consider problem (9) *with $\ell : [0,R] \times \mathbb{R} \to \mathbb{R}$ an arbitrary continuous function, without necessarily (H_ℓ).* In order to associate to problem (9) a fixed point operator we will use the closed subspace of C^1 defined by $C^1_M = \{u \in C^1 : u'(0) = 0 = u(R)\}$. The open ball of center $u_0 \in C^1_M$ and radius ρ is denoted by $B(u_0,\rho)$ and, for shortness, we put B_ρ instead $B(0,\rho)$.

Next, setting $\sigma(r) = 1/r^{N-1}$ $(r > 0)$, we consider the linear operators

$$S : C \to C, \quad Su(r) = \sigma(r)\int_0^r t^{N-1}u(t)dt \quad (r \in (0,R]), \quad Su(0) = 0;$$

$$K : C \to C^1, \quad Ku(r) = \int_r^R u(t)dt, \quad (r \in [0,R]).$$

It is easy to see that K is a bounded operator and standard arguments, invoking the Arzela-Ascoli theorem, show that S is compact.

An easy computation shows that, for a given $h \in C$, the mixed problem

$$(r^{N-1}\phi(u'))' + r^{N-1}h(r) = 0, \qquad u'(0) = 0 = u(R)$$

has a unique solution u given by $u = K \circ \phi^{-1} \circ S \circ h$. Moreover, the operator $K \circ \phi^{-1} \circ S : C \to C^1$ is compact. Consider now the nonlinear operator

$$N_\ell : C \to C, \quad N_\ell(u) = \ell(\cdot, u(\cdot)).$$

Notice that N_ℓ is continuous and takes bounded sets into bounded sets. Denoting by d_{LS} the Leray-Schauder degree, we have the following

Lemma 2.1. [Ref. 5] *A function u is a solution of (9) if and only if $u \in C^1_M$ is a fixed point of the compact nonlinear operator*

$$\mathcal{N}_\ell : C^1_M \to C^1_M, \qquad \mathcal{N}_\ell = K \circ \phi^{-1} \circ S \circ N_\ell.$$

Moreover, any fixed point u of $\mathcal{N}_\ell$ satisfies

$$\|u'\|_\infty < a, \qquad \|u\|_\infty < aR,$$

and

$$d_{LS}[I - \mathcal{N}_\ell, B_\rho, 0] = 1 \qquad \textit{for all } \rho \geq a(R+1).$$

A *lower solution* of (9) is a function $\alpha \in C^1$ such that $\|\alpha'\|_\infty < a$, $r^{N-1}\phi(\alpha') \in C^1$ and

$$(r^{N-1}\phi(\alpha'(r)))' + r^{N-1}\ell(r, \alpha(r)) \geq 0 \quad (r \in [0, R]), \quad \alpha(R) \leq 0.$$

Similarly, an *upper solution* of (9) is defined by reversing the above inequalities.

Proposition 2.1. [Ref. 5] *(i) If* (9) *has a lower solution* α *and an upper solution* β *such that* $\alpha(r) \leq \beta(r)$ *for all* $r \in [0, R]$, *then* (9) *has a solution* u *such that* $\alpha(r) \leq u(r) \leq \beta(r)$ *for all* $r \in [0, R]$.
(ii) Assume that (9) *has a lower solution* α *and an upper solution* β *such that* $\alpha(r) \leq \beta(r)$ *for all* $r \in [0, R]$, *and consider the set*

$$\Omega_{\alpha,\beta} = \{u \in C^1_M : \alpha \leq u \leq \beta\}.$$

Assume also that problem (9) *has a unique solution* u_0 *in* $\Omega_{\alpha,\beta}$ *and there exists* $\rho_0 > 0$ *such that* $\overline{B}(u_0, \rho_0) \subset \Omega_{\alpha,\beta}$. *Then,*

$$d_{LS}[I - \mathcal{N}_\ell, B(u_0, \rho), 0] = 1 \qquad \textit{for all } \; 0 < \rho \leq \rho_0.$$

Notice that *if* (H_ℓ) *holds true*, a nontrivial solution of the mixed boundary value problem

$$(r^{N-1}\phi(u'))' + r^{N-1}\ell(r, |u|) = 0, \qquad u'(0) = 0 = u(R) \tag{10}$$

is a strictly decreasing positive solution of (9). We shall need the following

Lemma 2.2. [Ref. 5] *Assume* (H_ϕ), (H_ℓ),

$$\lim_{s \to 0^+} \frac{\ell(r, s)}{\phi(s)} = 0 \quad \textit{uniformly for} \quad r \in [0, R]$$

and

$$\liminf_{s \to 0} \frac{\phi(\sigma s)}{\phi(s)} > 0 \qquad \textit{for all } \; \sigma > 0.$$

Then there exists $\rho_0 > 0$ *such that*

$$d_{LS}[I - \mathcal{N}_\ell, B_\rho, 0] = 1 \qquad \textit{for all } \; 0 < \rho \leq \rho_0$$

(here $\mathcal{N}_\ell$ *is the fixed point operator associated to* (10)*).*

3. Solutions as critical points

We are concerned with problem (9) *with* $\ell : [0, R] \times \mathbb{R} \to \mathbb{R}$ *a continuous function*, but assuming that ϕ is of potential type, precisely we ask:

(H_Φ) *ϕ is an increasing homeomorphism, $\phi(0) = 0$, and there exists $\Phi : [-a, a] \to \mathbb{R}$ continuous, of class C^1 on $(-a, a)$, $\Phi(0) = 0$, with $\Phi' = \phi$ on $(-a, a)$.*

Denoting $W^{1,\infty} := W^{1,\infty}(0, R)$, we introduce

$$\mathcal{K}_0 := \{v \in W^{1,\infty} \; : \; \|v'\|_\infty \leq a, \;\; v(R) = 0\}.$$

One has that $\mathcal{K}_0$ is a convex, compact subset of C. Notice also that

$$\|v\|_\infty \leq aR \quad \text{for all} \quad v \in \mathcal{K}_0. \tag{11}$$

Next, the functional $\psi : C \to (-\infty, +\infty]$, defined by

$$\psi(v) = \begin{cases} \displaystyle\int_0^R r^{N-1}\Phi(v')\,dr, & \text{if } v \in \mathcal{K}_0, \\ +\infty, & \text{if } v \in C \setminus \mathcal{K}_0 \end{cases}$$

is proper, convex and lower semicontinuous. Also, ψ is bounded on $\mathcal{K}_0$.

Setting

$$L(r, s) = \int_0^s \ell(r, \xi)d\xi, \qquad (r, s) \in [0, R] \times \mathbb{R},$$

we define $\mathcal{L} : C \to \mathbb{R}$ by

$$\mathcal{L}(v) = -\int_0^R r^{N-1}L(r, v)dr, \qquad v \in C,$$

which is of class C^1 on C. Then, the energy functional $\mathfrak{I} := \psi + \mathcal{L}$ has the structure required by the critical point theory of Szulkin[23]. Accordingly, a function $u \in C$ is *a critical point* of $\mathfrak{I}$ if $u \in \mathcal{K}_0$ and

$$\int_0^R r^{N-1}\left[\Phi(v') - \Phi(u') - \ell(r, u)(v - u)\right] dr \geq 0 \quad \text{for all} \quad v \in \mathcal{K}_0. \tag{12}$$

Lemma 3.1. [Ref. 4] *For every $h \in C$, the problem*

$$(r^{N-1}\phi(u'))' + r^{N-1}h(r) = 0, \qquad u'(0) = 0 = u(R),$$

has a unique solution u_h, which is also the unique solution in $\mathcal{K}_0$ of

$$\int_0^R r^{N-1}\left[\Phi(v') - \Phi(u') - h(v - u)\right] dr \geq 0 \quad \textit{for all} \quad v \in \mathcal{K}_0. \tag{13}$$

Theorem 3.1. *Assume* (H_Φ) *and that* $\ell : [0, R] \times \mathbb{R} \to \mathbb{R}$ *is continuous. Then, each critical point of* $\mathfrak{I}$ *is a solution of* (9). *Moreover,* (9) *has a solution which is a minimum point of* $\mathfrak{I}$ *on* C.

Proof. Let $u \in \mathcal{K}_0$ be a critical point of $\mathfrak{I}$. This means that u solves (13) with $h = \ell(\cdot, u)$ (see (12)). But, then Lemma 3.1 ensures that u is a solution of (9). Next, on $C \setminus \mathcal{K}_0$ one has that $\mathfrak{I}(v) \equiv +\infty$, which implies that

$$\inf_C \mathfrak{I} = \inf_{\mathcal{K}_0} \mathfrak{I} =: c.$$

Using (11) we infer that $\mathfrak{I}$ is bounded on $\mathcal{K}_0$. Consider $\{u_k\} \subset \mathcal{K}_0$ such that $\mathfrak{I}(u_k) \to c$. By the compactness of $\mathcal{K}_0$ we may assume, passing eventually to a subsequence, that there exists $u \in \mathcal{K}_0$ such that $u_k \to u$ in C. We obtain that $\mathcal{L}(u_k) \to \mathcal{L}(u)$ and $\psi(u) \leq \liminf_{k\to\infty} \psi(u_k)$. Consequently, one has that $\mathfrak{I}(u) \leq c$ and u is a minimum of $\mathfrak{I}$ on C. Then, the proof is accomplished by virtue of Proposition 1.1 in Ref. 23. □

4. Positive radial solutions for problem (1)

First, we need to consider Cauchy problems of type

$$\begin{aligned} (r^{N-1}\phi(u'(r)))' + r^{N-1}[\lambda\varphi(r)p(u(r))] = 0, \qquad (r \in [0, R]) \\ u(0) = \xi, \quad u'(0) = 0, \end{aligned} \tag{14}$$

where $R > 0$ is fixed, λ and $\xi > 0$, $\varphi : [0, R] \to \mathbb{R}$ is continuous, $p : \mathbb{R} \to \mathbb{R}$ is a locally Lipschitz function and ϕ satisfies (H_ϕ) and, in addition, it is of class C^1 and $\phi' > 0$.

Proposition 4.1. [Ref. 5] *Under the above assumptions on the data, the Cauchy problem* (14) *has a unique solution* $u(\lambda, \xi, \cdot)$ *and the mapping* $(\lambda, \xi) \mapsto u(\lambda, \xi, \cdot)$ *is continuous from* $(0, \infty) \times (0, \infty)$ *to* C^1.

Recall, setting $r = |x|$ and $v(x) = u(r)$, the Dirichlet problem (1) reduces to the mixed boundary value problem (8). As the results in the previous sections apply with

$$\phi(s) = \frac{s}{\sqrt{1 - s^2}} \qquad (s \in (-1, 1)),$$

note that u is a positive solution of (8) if and only if u is a nontrivial solution of

$$\left(r^{N-1}\frac{u'}{\sqrt{1 - u'^2}}\right)' + r^{N-1}\lambda\varphi(r)|u|^q = 0, \qquad u'(0) = 0 = u(R); \tag{15}$$

in this case, u is strictly decreasing.

Let $\varphi_M := \max_{[0,R]} |\varphi|$. The main result of this section is the following.

Theorem 4.1. *There exists $\Lambda > 2N/\varphi_M R^{q+1}$ such that problem (1) has zero, at least one or at least two positive solution according to $\lambda \in (0, \Lambda)$, $\lambda = \Lambda$ or $\lambda > \Lambda$. Moreover, Λ is strictly decreasing with respect to R.*

Proof. We define

$$\begin{aligned} S_j &= \{\lambda > 0 : (8) \text{ has at least } j \text{ positive solutions}\} \\ &= \{\lambda > 0 : (15) \text{ has at least } j \text{ non-trivial solutions}\} \qquad (j = 1, 2). \end{aligned}$$

Let $\lambda > 0$ and u be a positive solution of (8). Integrating (8) on $[0, r]$ it follows that

$$-r^{N-1} \frac{u'(r)}{\sqrt{1 - u'^2(r)}} = \lambda \int_0^r t^{N-1} \varphi(t) u^q(t) dt \qquad \text{for all } r \in [0, R].$$

Using that u is strictly positive on $[0, R)$ and strictly decreasing on $[0, R]$, one has

$$-r^{N-1} u'(r) \leq -r^{N-1} \frac{u'(r)}{\sqrt{1 - u'^2(r)}} \leq \lambda u^q(0) \frac{\varphi_M r^N}{N}, \quad \forall\, r \in [0, R].$$

and then, integrating over $[0, R]$ we deduce that

$$u(0) \leq \lambda u^q(0) \varphi_M R^2 / 2N.$$

This, together with $u(0) < R$ and $q \geq 1$ imply that

$$\lambda > 2N/\varphi_M R^{q+1}. \tag{16}$$

Now, the energy functional $\mathfrak{I} : C \to (-\infty, +\infty]$ associated to (15) is

$$\mathfrak{I}(u) = \frac{R^N}{N} - \int_0^R r^{N-1} \sqrt{1 - u'^2} dr - \frac{\lambda}{q+1} \int_0^R r^{N-1} \varphi(r) |u|^{q+1} dr \ (u \in \mathcal{K}_0)$$

and $\mathfrak{I} \equiv +\infty$ on $C \setminus \mathcal{K}_0$. If

$$(q+1) R^N < \lambda N \int_0^R r^{N-1} \varphi(r) (R - r)^{q+1} dr, \tag{17}$$

then the minimum of $\mathfrak{I}$ is negative and on account of Theorem 3.1 one has that (8) has at least one positive solution, which is a radial positive solution of (1). It is clear that the above condition (17) is satisfied if λ is sufficiently large. In particular, $S_1 \neq \emptyset$ and we can define

$$\Lambda = \Lambda(R) := \inf S_1.$$

It is not hard to prove that $\Lambda > 2N/\varphi_M R^{q+1}$ (see (16)) and $\Lambda \in S_1$. Now, let $\lambda_0 > \Lambda$. One has that (15) with $\lambda = \lambda_0$ has a lower solution u_1 (= a positive solution of (8) with $\lambda = \Lambda$) and an upper solution u_2 (which can be constructed). Moreover $u_1 < u_2$ on $[0, R]$. So, by Proposition 2.1 (i) we have $\lambda_0 \in S_1$ and

$$S_1 = [\Lambda, \infty).$$

Next, if u_0 is a solution of (15) with $\lambda = \lambda_0$ such that $u_1 \leq u_0 \leq u_2$ (i.e. $u_0 \in \Omega_{u_1,u_2}$), then there exists $\varepsilon > 0$ such that $\overline{B}(u_0, \varepsilon) \subset \Omega_{u_1,u_2}$.

Let u_0 be such that $u_0 \in \Omega_{u_1,u_2}$. If (15) has a second positive solution contained in Ω_{u_1,u_2} the proof is finished. If not, using Proposition 2.1 (ii) we deduce that

$$d_{LS}[I - \mathcal{N}_0, B(u_0, \rho), 0] = 1 \qquad \text{for all } 0 < \rho \leq \varepsilon,$$

where $\mathcal{N}_0$ is the fixed point operator associated to (15) with $\lambda = \lambda_0$. On the other hand, from Lemma 2.1 one has

$$d_{LS}[I - \mathcal{N}_0, B_\rho, 0] = 1 \qquad \text{for all } \rho \geq a(R + 1),$$

and from Lemma 2.2, we have

$$d_{LS}[I - \mathcal{N}_0, B_\rho, 0] = 1 \qquad \text{for all } \rho \text{ sufficiently small.}$$

Now, consider $\rho_1, \rho_2 > 0$ sufficiently small and $\rho_3 \geq a(R + 1)$ such that $\overline{B}(u_0, \rho_1) \cap \overline{B}_{\rho_2} = \emptyset$ and $\overline{B}(u_0, \rho_1) \cup \overline{B}_{\rho_2} \subset B_{\rho_3}$. Then, from the additivity-excision property of the Leray-Schauder degree theory it follows that

$$d_{LS}[I - \mathcal{N}_0, B_{\rho_3} \setminus [\overline{B}(u_0, \rho_1) \cup \overline{B}_{\rho_2}], 0] = -1,$$

which, together with the existence property of the Leray-Schauder degree, imply that $\mathcal{N}_0$ has a fixed point $\widetilde{u}_0 \in B_{\rho_3} \setminus [\overline{B}(u_0, \rho_1) \cup \overline{B}_{\rho_2}]$. We infer that (8) has a second positive solution and hence, (1) has a second radial positive solution.

Finally, let u_0 be a solution of (15) with $\lambda = \lambda_0 := \Lambda(R_0)$ and $R = R_0$. Let us fix $R > R_0$. Then, from Proposition 4.1 with $p(s) = |s|^q$, one has that $u_0 = u(\lambda_0, \xi_0, \cdot)|_{[0,R_0]}$, where $\xi_0 = u_0(0)$. Using that $u(\lambda_0, \xi_0, \cdot)$ is strictly decreasing on $[0, R]$ and $u(\lambda_0, \xi_0, R_0) = 0$, it follows that $u(\lambda_0, \xi_0, R) < 0$. Using again Proposition 4.1, we infer that there exists $\varepsilon > 0$ such that $u(\lambda, \xi_0, R) < 0$ for all $\lambda \in [\lambda_0 - \varepsilon, \lambda_0 + \varepsilon]$. In particular, $u(\lambda, \xi_0, \cdot)$ is a lower solution of (15). On the other hand (15) has an upper solution β such that $u(\lambda, \xi_0, \cdot) \leq \beta$ on $[0, R]$. Then, applying Proposition 2.1 we deduce that (15) has at least one nonzero solution which is also a strictly positive solution of (8). Consequently, $\Lambda(R_0) > \lambda(R)$ and the proof is complete. □

5. A "universal" existence result for problem (2)

A key ingredient in our approach for problem (2) is the following result essentially due to R. Bartnik and L. Simon[1] (see Lemma 2.2 in Ref. 11).

Theorem 5.1. *If $h \in L^\infty(\Omega)$, then the Dirichlet problem*

$$\mathcal{M}(u) = h(x) \ \ in\ \Omega, \ \ u|_{\partial\Omega} = 0, \tag{18}$$

has a unique solution $u_h \in W^{2,p}(\Omega)$ for all $1 \le p < \infty$.

We define

$$K_0 = \{u \in W^{1,\infty}(\Omega) : \|\nabla u\|_\infty \le 1,\ u|_{\partial\Omega} = 0\}, \tag{19}$$

which is a convex and compact subset of $C(\overline{\Omega})$. Also, notice that, on account of $K_0 \subset W_0^{1,p}(\Omega) \subset C(\overline{\Omega})$ for any $p > N$, from Theorem 7.10 in Ref. 16 it holds

$$\|u\|_\infty \le c_1(\Omega, p)\|\nabla u\|_p \quad \text{for all } u \in W_0^{1,p}(\Omega),$$

which implies

$$\|u\|_\infty \le c(\Omega) \quad \text{for all } u \in K_0, \tag{20}$$

with $c(\Omega)$ a positive constant.

Lemma 5.1. [Ref. 2] *If $h \in L^\infty(\Omega)$, then the solution u_h of problem* (18) *is the unique solution in K_0 of the variational inequality*

$$\int_\Omega [\sqrt{1-|\nabla u|^2} - \sqrt{1-|\nabla v|^2} + h(v-u)] \ge 0 \ \ for\ all\ \ v \in K_0. \tag{21}$$

Next, we introduce the functional $\Psi : C(\overline{\Omega}) \to (-\infty, +\infty]$ given by

$$\Psi(u) = \begin{cases} \displaystyle\int_\Omega [1 - \sqrt{1-|\nabla u|^2}] & \text{for} \quad u \in K_0 \\ +\infty & \text{for} \quad u \in C(\overline{\Omega}) \setminus K_0. \end{cases} \tag{22}$$

This is convex and lower semicontinuous. Viewing that the Carathéodory function f satisfies condition (L^∞), we can consider the Nemytskii operator

$$N_f : C(\overline{\Omega}) \to L^1(\Omega), \quad N_f(u) = f(\cdot, u(\cdot)),$$

which is continuous, takes bounded sets into bounded sets and

$$N_f(C(\overline{\Omega})) \subset L^\infty(\Omega). \tag{23}$$

Let $\mathcal{F} : C(\overline{\Omega}) \to \mathbb{R}$ be defined by

$$\mathcal{F}(u) = \int_\Omega F(x, u),$$

where

$$F(x,s) = \int_0^s f(x,\xi)d\xi \quad (x \in \Omega,\ s \in \mathbb{R}).$$

A standard reasoning shows that $\mathcal{F}$ is of class C^1 on $C(\overline{\Omega})$ and its derivative is given by

$$\langle \mathcal{F}'(u), v\rangle = \int_\Omega N_f(u)v \quad \text{for all } u, v \in C(\overline{\Omega}).$$

Then, the energy functional $I := \Psi + \mathcal{F}$ has the structure required by the critical point theory of Szulkin[23]. Accordingly, $u \in K_0 = D(\Psi)$ is a *critical point* of I if

$$\Psi(v) - \Psi(u) + \langle \mathcal{F}'(u), v - u\rangle \geq 0, \quad \forall\, v \in K_0.$$

Also, I is said *to satisfy the Palais-Smale condition* (in short, (PS)-*condition*) if any sequence $\{u_n\} \subset K_0$ for which $I(u_n) \to c \in \mathbb{R}$ and

$$\Psi(v) - \Psi(u_n) + \langle \mathcal{F}'(u_n), v - u_n\rangle \geq -\varepsilon_n \|v - u_n\|_\infty \text{ for all } v \in K_0,$$

where $\varepsilon_n \to 0_+$, possesses a convergence subsequence. Notice that, the compactness of $K_0 \subset C(\overline{\Omega})$ implies that I satisfies (PS)-condition.

We have the following "universal" existence result for problem (2).

Theorem 5.2. *A function $u \in C(\overline{\Omega})$ is a critical point of I if and only if it is a solution of problem* (2). *Moreover, I is bounded from below and attains its infimum at some $u_0 \in K_0$, which solves problem* (2).

Proof. Let u be a critical point of I. Then it satisfies the variational inequality (21) with $h := N_f(u) \in L^\infty(\Omega)$ (see (23)). From Lemma 5.1 we infer that u is a solution of problem (18), with the above h. But, this means that u solves (2). The reversed implication follows in a similar way.

Now, let $\alpha_{c(\Omega)} \in L^\infty(\Omega)$ be such that

$$|f(x,s)| \leq \alpha_{c(\Omega)}(x) \quad \text{for a.e. } x \in \Omega,\ \forall\, s \in \mathbb{R} \text{ with } |s| \leq c(\Omega),$$

with $c(\Omega)$ the constant entering in (20). Then, for an arbitrarily chosen $u \in K_0$, one has

$$|f(x,\tau u(x))| \leq \alpha_{c(\Omega)}(x) \quad \text{for a.e. } x \in \Omega,\ \forall\, \tau \in \mathbb{R} \text{ with } |\tau| \leq 1.$$

Thus, we estimate

$$\begin{aligned} |\mathcal{F}(u)| &\leq \int_\Omega |F(x,u)| = \int_\Omega \left| \int_0^1 \frac{d}{d\tau} F(x,\tau u(x))d\tau \right| dx \\ &\leq \int_\Omega \int_0^1 |f(x,\tau u(x))|\, |u(x)|\, d\tau\, dx \leq c(\Omega)\|\alpha_{c(\Omega)}\|_\infty |\Omega|. \end{aligned}$$

We deduce that I is bounded from below on $C(\overline{\Omega})$. As I satisfies the (PS)-condition, the proof follows from Theorem 1.7 in Ref. 23. □

6. Multiplicity in the one-parameter Dirichlet problem (3)

Here we deal with problem (3) under hypothesis (H). According to the previous section, the energy functional $I_\lambda : C(\overline{\Omega}) \to (-\infty, +\infty]$ associated to (3) is

$$I_\lambda(u) = \begin{cases} \Psi(u) + \dfrac{1}{q+1}\displaystyle\int_\Omega \mu(x)|u|^{q+1} - \lambda \int_\Omega G(x,u) & \text{if } u \in K_0 \\ +\infty & \text{otherwise.} \end{cases}$$

Note that I_λ satisfies the (PS)-condition and $I_\lambda(0) = 0$, for all $\lambda > 0$.

Theorem 6.1. *If*

$$\exists\ R_1 > 0 :\ g(x,s) < 0 \quad \textit{for a.e. } x \in \Omega,\ \forall\, s \in (-R_1, 0), \tag{24}$$

then there exists $\Lambda_1 > 0$ such that for all $\lambda > \Lambda_1$, one has

$$\inf_{C(\overline{\Omega})} I_\lambda < 0 \tag{25}$$

and (3) has a nontrivial solution which is a minimizer of the functional I_λ.

Proof. Let $x_0 \in \Omega$ and $r \in (0, R_1)$ be such that $\overline{B}_r(x_0) \subset \Omega$. We define

$$\eta_0(x) := -\min\left\{r, \|\nabla\eta\|_\infty^{-1}\right\}\eta(x) \quad (x \in \Omega),$$

where

$$\eta(x) = \begin{cases} \exp\left(\frac{r^2}{|x-x_0|^2 - r^2}\right) & \text{for } \ x \in B_r(x_0) \\ 0 & \text{for } \ x \in \Omega \setminus B_r(x_0). \end{cases}$$

One has that $\eta_0 \in K_0$ and $-R_1 < \eta_0 \leq 0$, which on account of (24), give

$$\int_\Omega G(x,\eta_0) = -\int_{B_r(x_0)} \int_{\eta_0(x)}^0 g(x,\xi)d\xi dx > 0.$$

Taking

$$\Lambda_1 > \max\left\{0, \frac{\Psi(\eta_0) + \frac{1}{q+1}\int_\Omega \mu(x)|\eta_0|^{q+1}}{\int_\Omega G(x,\eta_0)}\right\},$$

one gets that $I_\lambda(\eta_0) < 0$, for all $\lambda > \Lambda_1$. Then, the result follows in virtue of Theorem 5.2. □

Analogously, we can prove the following

Theorem 6.2. *If*

$$\exists\ R_2 > 0 : \ g(x,s) > 0 \quad \textit{for a.e. } x \in \Omega,\ \forall\, s \in (0, R_2), \tag{26}$$

then there is some $\Lambda_2 > 0$ *such that* (25) *holds true, for all* $\lambda > \Lambda_2$, *and problem* (3) *has a nontrivial solution which is a minimizer of* I_λ.

We set $g_-(\cdot,s) := g(\cdot,s^-)$ and $g_+(\cdot,s) := g(\cdot,s^+)$, for all $s \in \mathbb{R}$ (recall that $s^- = \min\{0,s\}$, resp. $s^+ = \max\{0,s\}$). It is clear that g_- and g_+ are Carathéodory functions and satisfy the growth condition (L^∞).

Lemma 6.1. *If* u *is a solution of*

$$\mathcal{M}(u) = \mu(x)|u|^{q-1}u - \lambda g_-(x,u) \ \textit{ in } \Omega, \quad u|_{\partial\Omega} = 0, \tag{27}$$

then $u \leq 0$ *in* Ω *and it solves problem* (3).

Proof. Let $v \in W^{1,\infty}(\Omega)$ with $v|_{\partial\Omega} = 0$, be arbitrarily chosen. Then, multiplying (27) by v, integrating over Ω and using the integration by parts formula, one gets

$$-\int_\Omega \frac{\nabla u \cdot \nabla v}{\sqrt{1-|\nabla u|^2}} = \int_\Omega \mu(x)|u|^{q-1}uv - \lambda \int_\Omega g_-(x,u)v. \tag{28}$$

Next, we set $\mathcal{O} := \{x \in \Omega : u(x) > 0\}$. From Theorem A.1 in Ref. 17 we know that $u^+ \in W^{1,\infty}(\Omega)$ and $\nabla u^+ = \nabla u$ in $\mathcal{O}$, $\nabla u^+ = 0_{\mathbb{R}^N}$ in $\Omega \setminus \mathcal{O}$. So, taking $v = u^+$ in (28) and using hypothesis (H), it follows

$$-\int_{\mathcal{O}} \frac{|\nabla u^+|^2}{\sqrt{1-|\nabla u^+|^2}} = \int_{\mathcal{O}} \mu(x)|u^+|^{q+1} \geq 0. \tag{29}$$

If we denote $\mathcal{O}_\nabla := \{x \in \mathcal{O} \ : \ |\nabla u^+(x)| > 0\}$ and assume that *meas* $\mathcal{O}_\nabla > 0$, then from (29) we obtain the contradiction

$$0 > -\int_{\mathcal{O}_\nabla} \frac{|\nabla u^+|^2}{\sqrt{1-|\nabla u^+|^2}} \geq 0.$$

So, *meas* $\mathcal{O}_\nabla = 0$ and, since $\nabla u^+ = 0_{\mathbb{R}^N}$ in $\Omega \setminus \mathcal{O}$, we get that $\nabla u^+ = 0_{\mathbb{R}^N}$ a.e. on Ω. As Ω is a domain, we infer $u^+ =$ const., hence $u^+ \equiv 0$ in Ω. But, this implies $u \leq 0$ in Ω; it is also clear that it solves problem (3). $\square$

Using a similar argument one obtains

Lemma 6.2. *If* u *is a solution of problem*

$$\mathcal{M}(u) = \mu(x)|u|^{q-1}u - \lambda g_+(x,u) \ \textit{ in } \Omega, \quad u|_{\partial\Omega} = 0, \tag{30}$$

then $u \geq 0$ *in* Ω *and it solves* (3).

Corollary 6.1. *Assume that* (H) *and* (24) *are fulfilled. Then, there exists* $\Lambda_1 > 0$ *such that for all* $\lambda > \Lambda_1$, *problem* (3) *has at least one nontrivial non-positive solution.*

Proof. It follows from Theorem 6.1 with g_- instead of g and Lemma 6.1. □

Also, from Theorem 6.2 with g_+ instead of g and Lemma 6.2, we get

Corollary 6.2. *Assume that* (H) *and* (26) *hold true. Then, there exists* $\Lambda_2 > 0$ *such that for all* $\lambda > \Lambda_2$, *problem* (3) *has at least one nontrivial non-negative solution.*

Example 6.1. For any positive $a, b \in \mathbb{R}$, there exists $\Lambda > 0$ such that the Dirichlet problem

$$\mathcal{M}(u) + \lambda(au^+ + bu^-) = 0 \text{ in } \Omega, \quad u|_{\partial\Omega} = 0,$$

has at least two nontrivial solutions whenever $\lambda > \Lambda$.

The main result concerning problem (3) is the following

Theorem 6.3. *Assume that* (H), (24) *and* (26) *are satisfied. If either*

$$\lim_{s \to 0} \frac{G(x,s)}{|s|^2} = 0 \textit{ uniformly with } x \in \Omega, \tag{31}$$

or

$$\mu_m := \textit{essinf}_{\,\Omega}\, \mu > 0 \textit{ and } \lim_{s \to 0} \frac{G(x,s)}{|s|^{q+1}} = 0 \textit{ uniformly with } x \in \Omega, \tag{32}$$

then there exists $\Lambda > 0$ *such that for all* $\lambda > \Lambda$, *problem* (3) *has at least four nontrivial solutions, two of them are non-negative and the other two are non-positive.*

Proof. Let $I_\lambda^- : C(\overline{\Omega}) \to (-\infty, +\infty]$ be the action functional associated to (27); this is

$$I_\lambda^-(u) = \begin{cases} \Psi(u) + \dfrac{1}{q+1}\displaystyle\int_\Omega \mu(x)|u|^{q+1} - \lambda \int_\Omega G_-(x,u) & \text{if } u \in K_0 \\ +\infty & \text{for } u \in C(\overline{\Omega}) \setminus K_0. \end{cases}$$

Using Theorem 6.1, we can find $\Lambda_1 > 0$ such that, for all $\lambda > \Lambda_1$, the functional I_λ^- has a nontrivial minimizer $e_{\lambda,1} \in K_0$ with

$$I_\lambda^-(e_{\lambda,1}) < 0. \tag{33}$$

Also, the functional $I_\lambda^+ : C(\overline{\Omega}) \to (-\infty, +\infty]$ defined by

$$I_\lambda^+(u) = \begin{cases} \Psi(u) + \dfrac{1}{q+1}\displaystyle\int_\Omega \mu(x)|u|^{q+1} - \lambda \int_\Omega G_+(x,u) & \text{if } u \in K_0 \\ +\infty & \text{for } u \in C(\overline{\Omega}) \setminus K_0 \end{cases}$$

being associated to problem (30), from Theorem 6.2, we can find $\Lambda_2 > 0$ so that, for all $\lambda > \Lambda_2$, there exists a minimizer $e_{\lambda,2}$ of I_λ^+ with $I_\lambda^+(e_{\lambda,2}) < 0$.

Let $\lambda > \Lambda := \max\{\Lambda_1, \Lambda_2\}$ be fixed. We shall produce a second nontrivial critical point $u_{\lambda,1}$ of I_λ^- by the Mountain Pass Theorem (Theorem 3.2 in Ref. 23). The same argument works to prove the existence of a second nontrivial critical point $u_{\lambda,2}$ for the functional I_λ^+.

As $I_\lambda^-(0) = 0$, I_λ^- satisfies (PS)-condition and taking into account (33), it suffices to prove that there exist $\alpha > 0$ and $\rho \in (0, \|e_{\lambda,1}\|_\infty)$, with

$$I_\lambda^-(u) \geq \alpha \ \text{ for all } u \in K_0, \ \text{ with } \|u\|_\infty = \rho. \tag{34}$$

First, we assume that (31) is satisfied. From Poincaré inequality and since

$$1 - \sqrt{1 - s^2} \geq \frac{s^2}{2} \ \text{ for all } \ s \in [0,1],$$

one has that

$$\Psi(u) = \int_\Omega [1 - \sqrt{1 - |\nabla u|^2}] \geq c_1 \int_\Omega |u|^2 \qquad (u \in K_0), \tag{35}$$

with c_1 a positive constant. Using (31), we can find $\delta > 0$, such that

$$G_-(x,s) \leq \frac{c_1}{2\lambda}|s|^2 \quad \text{for a.e. } \ x \in \Omega, \ \forall\, s \in \mathbb{R} \text{ with } |s| \leq \delta. \tag{36}$$

Then, using (35) and (36), for all $u \in K_0$ with $\|u\|_\infty \leq \delta$, we obtain

$$I_\lambda^-(u) \geq \frac{c_1}{2} \int_\Omega |u|^2.$$

Let $\rho \in (0, \min\{\delta, \|e_{\lambda,1}\|_\infty\})$. We *claim* that

$$\inf_{\{u \in K_0 :\ \|u\|_\infty = \rho\}} \int_\Omega |u|^2 > 0, \tag{37}$$

which will imply (34) with appropriate α and ρ. To prove this, suppose by contradiction that there exists a sequence $\{u_n\} \subset K_0$ satisfying $\|u_n\|_\infty = \rho$ for all $n \in \mathbb{N}$, with

$$\int_\Omega |u_n|^2 \to 0 \quad \text{as } n \to \infty.$$

Since $\{u_n\}$ is bounded in $W^{1,\infty}(\Omega)$, passing to a subsequence if necessary, we may assume that $\{u_n\}$ is convergent in $C(\overline{\Omega})$ to some $u \in C(\overline{\Omega})$. This yields that $\|u\|_\infty = \rho$ and

$$\int_\Omega |u_n|^2 \to \int_\Omega |u|^2 \quad \text{as } n \to \infty.$$

It follows that $u = 0$, which is a contradiction with $\|u\|_\infty = \rho > 0$. So, (37) holds true, as claimed.

Next, if (32) holds true, then there exists $\delta > 0$, such that

$$G_-(x, s) \leq \frac{\mu_m}{2\lambda(q+1)}|s|^{q+1} \quad \text{for a.e. } x \in \Omega, \ \forall s \in \mathbb{R} \text{ with } |s| \leq \delta.$$

Then, for $u \in C(\overline{\Omega})$ with $\|u\|_\infty \leq \delta$, one has

$$I_\lambda^-(u) \geq \frac{1}{q+1}\int_\Omega \mu(x)|u|^{q+1} - \lambda\int_\Omega G_-(x,u) \geq \frac{\mu_m}{2(q+1)}\int_\Omega |u|^{q+1}$$

and the proof follows now similarly as in the case when (31) is satisfied.

Therefore, under each of the assumptions (31) or (32), by the Mountain Pass Theorem there exists a nontrivial critical point $u_{\lambda,1}$ of I_λ^- with $I_\lambda^-(u_{\lambda,1}) > 0$. Clearly, $u_{\lambda,1} \neq e_{\lambda,1}$ and from Theorem 5.2 and Lemma 6.1 one has that $e_{\lambda,1} \leq 0$ and $u_{\lambda,1} \leq 0$ are nontrivial solutions of problem (3).

Similarly, on account of Theorem 5.2 and Lemma 6.2, we obtain that $e_{\lambda,2}$ and $u_{\lambda,2}$ are nontrivial non-negative solutions of (3) and the proof is complete. □

Example 6.2. Let $\mu \in L^\infty(\Omega)$ be non-negative, $a, b, q \in \mathbb{R}$ be positive. If $c, d > 1$, then there exists $\Lambda > 0$ such that the Dirichlet problem

$$\mathcal{M}(u) = \mu(x)|u|^{q-1}u - \lambda\left[a(u^+)^c - b|u^-|^d\right] \text{ in } \Omega, \ \ u|_{\partial\Omega} = 0,$$

has at least four nontrivial solutions - two of them are non-negative and the other two are non-positive, for all $\lambda > \Lambda$. The same conclusion holds true for the problem

$$\mathcal{M}(u) = \mu(x)|u|^{q-1}u - \lambda\left[a(u^+)^r - b|u^-|^r\right] \text{ in } \Omega, \ \ u|_{\partial\Omega} = 0,$$

provided that $\mu_m > 0$ and $r > q \ (> 0)$.

7. The discontinuous Dirichlet problem (5)

To treat problem (5), we shall make use of the set K_0 given in (19) and $\Psi : C(\overline{\Omega}) \to (-\infty, +\infty]$ defined in (22). Next, viewing condition (6), we can define $\widehat{\mathcal{F}} : L^q(\Omega) \to \mathbb{R}$ by

$$\widehat{\mathcal{F}}(v) = \int_\Omega F(x, v), \quad \forall\, v \in L^q(\Omega)$$

and, on account of the embedding $C(\overline{\Omega}) \subset L^q(\Omega)$, we introduce the functional $I_F = \widehat{\mathcal{F}}|_{C(\overline{\Omega})}$. From Theorem 2.1 in Ref. 7, one has that $\widehat{\mathcal{F}}$ is locally Lipschitz in $L^q(\Omega)$ and

$$\partial\widehat{\mathcal{F}}(v) \subset \left[\underline{f}(\cdot, v(\cdot)), \overline{f}(\cdot, v(\cdot))\right], \quad \forall\, v \in L^q(\Omega)$$

(here and below, for a Banach space X and a locally Lipschitz function $\mathcal{G} : X \to \mathbb{R}$, the notation $\partial\mathcal{G}(v)$ stands for *the generalized Clarke gradient* of $\mathcal{G}$ at $v \in X$ – Ref. 9). By the continuity of the embedding $C(\overline{\Omega}) \subset L^q(\Omega)$ it is clear that I_F is locally Lipschitz on $C(\overline{\Omega})$. Also, note that since $C(\overline{\Omega})$ is dense in $L^q(\Omega)$, it holds (see Ref. 9, p. 47):

$$\partial\widehat{\mathcal{F}}(u) = \partial I_F(u), \quad \forall\, u \in C(\overline{\Omega}).$$

Lemma 7.1. [Ref. 6] *Let* $u \in K_0$. *If* $\ell \in \partial I_F(u)$, *then there exists* $\zeta_\ell \in L^\infty(\Omega)$ *such that* $\zeta_\ell(x) \in \left[\underline{f}(x, u(x)), \overline{f}(x, u(x))\right]$ *for a.e.* $x \in \Omega$ *and*

$$\langle \ell, w\rangle = \int_\Omega \zeta_\ell w$$

for all $w \in C(\overline{\Omega})$.

The action functional associated with problem (5) will be $\mathcal{I} := \Psi + I_F$. According to Kourogenis *et al.*[18], $u \in C(\overline{\Omega})$ is a critical point of $\mathcal{I}$ if

$$0 \in \overline{\partial}\Psi(u) + \partial I_F(u),$$

where $\overline{\partial}\Psi(u)$ stands for the subdifferential of Ψ at $u \in C(\overline{\Omega})$ in the sense of convex analysis. Also, $\mathcal{I}$ is said *to satisfy the* (PS)*-condition* if every sequence $\{u_n\} \subset C(\overline{\Omega})$ for which $\{\mathcal{I}(u_n)\}$ is bounded and

$$\Psi(v) - \Psi(u_n) + I_F^0(u_n; v - u_n) \geq -\varepsilon_n \|v - u_n\|_\infty, \quad \forall\, v \in C(\overline{\Omega}),$$

for a sequence $\{\varepsilon_n\} \subset \mathbb{R}_+$ with $\varepsilon_n \to 0$, possesses a convergent subsequence; $I_F^0(w, z)$ denotes the *generalized directional derivative* of I_F at $w \in C(\overline{\Omega})$ in the direction of $z \in C(\overline{\Omega})$ (see Ref. 9). Since K_0 is a compact subset of $C(\overline{\Omega})$ it follows that $\mathcal{I}$ satisfies the (PS)-condition.

The main result of this section is the following

Theorem 7.1. *Assume that* (6) *and* (7) *hold true. If a function* u *is a critical point of* $\mathcal{I}$, *then it is a solution of problem* (5). *Moreover,* $\mathcal{I}$ *is* *bounded from below and attains its infimum at some $u_0 \in K_0$, which is a critical point of $\mathcal{I}$ and hence, a solution of* (5).

Proof. Let u be a critical point of $\mathcal{I}$. Then $u \in K_0$ and there exist $e_u \in \overline{\partial}\Psi(u)$ and $\ell_u \in \partial I_F(u)$ such that

$$\langle e_u, w\rangle + \langle \ell_u, w\rangle = 0, \quad \forall\, w \in C(\overline{\Omega}).$$

Since $e_u \in \overline{\partial}\Psi(u)$ we infer

$$\Psi(w) - \Psi(u) + \langle \ell_u, w - u\rangle \geq 0, \quad \forall\, w \in C(\overline{\Omega}). \tag{38}$$

From Lemma 7.1 there exists $\zeta_u = \zeta(\ell_u) \in L^\infty(\Omega)$ such that

$$\zeta_u(x) \in \left[\underline{f}(x, u(x)), \overline{f}(x, u(x))\right], \quad \text{a.e. } x \in \Omega \tag{39}$$

and

$$\langle \ell_u, w\rangle = \int_\Omega \zeta_u w, \quad \forall\, w \in C(\overline{\Omega}). \tag{40}$$

By virtue of (40), inequality (38) becomes

$$\Psi(w) - \Psi(u) + \int_\Omega \zeta_u(w - u) \geq 0, \quad \forall\, w \in C(\overline{\Omega}). \tag{41}$$

On the other hand, for $h \in L^\infty(\Omega)$, from Lemma 5.1 we know that the unique solution u_h of (18) satisfies

$$\int_\Omega \left[\sqrt{1 - |\nabla u_h|^2} - \sqrt{1 - |\nabla w|^2} + h(w - u_h)\right] \geq 0, \quad \forall\, w \in K_0$$

and hence,

$$\Psi(w) - \Psi(u_h) + \int_\Omega h(w - u_h) \geq 0, \quad \forall\, w \in C(\overline{\Omega}).$$

From this and (41), we deduce that $u = u_h$, with $h = \zeta_u$. Thus, using (39), one gets that u solves problem (5).

Next, for arbitrary $u \in K_0$, by (6) and (20), the primitive F satisfies

$$|F(x, u(x))| \leq \mathcal{C}\left(c(\Omega) + c(\Omega)^q/q\right) =: C_2, \qquad \text{for a.e. } x \in \Omega$$

which yields

$$|I_F(u)| \leq \int_\Omega |F(x, u)| \leq C_2 \mathrm{vol}(\Omega), \quad \forall\, u \in K_0.$$

This implies that the functional $\mathcal{I}$ is bounded from below on $C(\overline{\Omega})$. Then, since $\mathcal{I}$ verifies the (PS)-condition, the proof is accomplished by virtue of Theorem 1 in Ref. 19. □

Acknowledgments

The support of Călin Şerban from the strategic grant POSDRU/159/1.5/S/ 137750 is gratefully acknowledged.

References

1. R. Bartnik and L. Simon, Spacelike hypersurfaces with prescribed boundary values and mean curvature, *Comm. Math. Phys.* **87** (1982-83), 131–152.
2. C. Bereanu, P. Jebelean and J. Mawhin, The Dirichlet problem with mean curvature operator in Minkowski space – a variational approach, *Adv. Nonlinear Stud.* **14** (2014), 315–326.
3. C. Bereanu, P. Jebelean and J. Mawhin, Corrigendum to "The Dirichlet problem with mean curvature operator in Minkowski space - a variational approach" [Adv. Nonlinear Stud. 14 (2014), no. 2, 315326], *Adv. Nonlinear Stud.* **16 (1)** (2016), 173–174.
4. C. Bereanu, P. Jebelean and P.J. Torres, Positive radial solutions for Dirichlet problems with mean curvature operators in Minkowski space, *J. Funct. Anal.* **264** (2013), 270–287.
5. C. Bereanu, P. Jebelean and P.J. Torres, Multiple positive radial solutions for a Dirichlet problem involving the mean curvature operator in Minkowski space, *J. Funct. Anal.* **265** (2013), 644–659.
6. C. Bereanu, P. Jebelean and C. Şerban, The Dirichlet problem for discontinuous perturbations of the mean curvature operator in Minkowski space, *Electron. J. Qual. Theory Differ. Equ.* **35** (2015), 1–7.
7. K.-C. Chang, Variational methods for nondifferentiable functionals and their applications to partial differential equations, *J. Math. Anal. Appl.* **80** (1981), 102–129.
8. S.-Y. Cheng and S.-T. Yau, Maximal spacelike hypersurfaces in the Lorentz-Minkowski spaces, *Ann. of Math.* **104** (1976), 407–419.
9. F.H. Clarke, *Optimization and Nonsmooth Analysis*, John Wiley and Sons, New York, 1983.
10. C. Corsato, F. Obersnel, P. Omari and S. Rivetti, On the lower and upper solution method for the prescribed mean curvature equation in Minkowski space, *Discrete Contin. Dyn. Syst.* **Supplement 2013** (2013), 159–169.
11. C. Corsato, F. Obersnel, P. Omari and S. Rivetti, Positive solutions of the Dirichlet problem for the prescribed mean curvature equation in Minkowski space, *J. Math. Anal. Appl.* **405** (2013), 227–239.

12. I. Coelho, C. Corsato and S. Rivetti, Positive radial solutions of the Dirichlet problem for the Minkowski-curvature equation in a ball, *Topol. Methods Nonlinear Anal.* **44** (2014), 23–40.
13. A.F. Filippov, Right hand side discontinuous differential equations, *Trans. Am. Math. Soc.* **42** (1964), 199–227.
14. F.J. Flaherty, The boundary value problem for maximal hypersurfaces, *Proc. Natl. Acad. Sci. USA* **76** (1979), 4765–4767.
15. C. Gerhardt, H-surfaces in Lorentzian manifolds, *Comm. Math. Phys.* **89** (1983), 523–553.
16. D. Gilbarg and N.S. Trudinger, *Elliptic Partial Differential Equations of Second Order*, Springer-Verlag, New York, 2001.
17. D. Kinderlehrer and G. Stampacchia, *An Introduction to Variational Inequalities and Their Applications*, Academic Press, 1980.
18. N.C. Kourogenis, J. Papadrianos and N.S. Papageorgiu, Extensions of nonsmooth critical point theory and applications, *Atti Sem. Mat. Fis. Univ. Modena* **50** (2002), 381–414.
19. C. Lefter, Critical point theorems for lower semicontinuous functionals, *An. Ştiinţ. Univ. Al. I. Cuza Iaşi Mat. (N.S.)* **47** (2001), 189–198.
20. I. Massabo and C.A. Stuart, Elliptic eigenvalue problems with discontinuous nonlinearities, *J. Math. Anal. Appl.* **66** (1978), 261–281.
21. J. Rauch, Discontinuous semilinear differential equations and multiple valued maps, *Proc. Amer. Math. Soc.* **64** (1977), 277–282.
22. C.A. Stuart and J.F. Toland, A variational method for boundary value problems with discontinuous nonlinearities, *J. London Math. Soc.* **21** (1980), 319–328.
23. A. Szulkin, *Minimax principles for lower semicontinuous functions and applications to nonlinear boundary value problems*, Ann. Inst. H. Poincaré Anal. Non Linéaire **3** (1986), 77–109.

Free boundary fluid-elasticity interactions: Adjoint sensitivity analysis

Lorena Bociu* and Kristina Martin
Mathematics Department, NC State University,
Raleigh, NC 27695, USA
**E-mail: lvbociu@ncsu.edu*
www.ncsu.edu

We consider a PDE-constrained optimization problem governed by a free boundary fluid-elasticity interaction. The challenge of applying optimization tools to free boundary fluid-structure interactions is the proper derivation of the adjoint sensitivity information with correct balancing conditions on the common interface. As the interaction is a coupling of Eulerian (fluid state) and Lagrangian (motion of the solid) quantities, sensitivity analysis on the system falls into the framework of shape analysis. This paper builds upon the sensitivity theory developed in [5,6,8] and provides a description for the adjoint linearized model, necessary in the derivation of optimality conditions for controls and derivative-based optimization algorithms in free boundary fluid-structure interactions.

Keywords: fluid-structure interactions, optimal control, sensitivity analysis, adjoint system

1. Introduction

Free and moving boundary fluid-structure interactions (FSI) are ubiquitous in nature, with most known examples coming from industrial processes, aero-elasticity, and biomechanics (e.g., flutter analysis of airplanes, parachute FSI, robo-bees, blood flow in the cardiovascular system, heart valve dynamics, etc.). Due to their highly nonlinear nature (the equations are nonlinear, and the common boundary is one of the unknowns in the system and has to be determined as part of the solution), they continue to be one of the most challenging topics to date, even though many publications have appeared focusing on the mathematical solvability[9–11,13,15,16,25–27,29–33,39–41,44,48] and numerical approximation[18,22–24,28,37,42,43] for these coupled systems. However, in most of these applications, the ultimate goal is not only the mathematical modeling and numerical simulation of the complex system, but rather the optimization or optimal control of the considered process, as well as related sensitivity analysis (with respect to relevant physical parameters) investigations. Regarding control problems in fluid-structure interactions, most of the literature is focused on the assumption of small but rapid oscillations of the solid body, so that the common interface may be assumed static [12,34–36]. Recently, PDE constrained optimization problems governed by free boundary interactions (i.e., one works with a steady-state problem, and the position of the boundary is a spatial unknown) have been considered, with most research studies mainly addressed in the context of the numerical analysis of the finite element methods [2,45–47].

In the recent paper [4], we considered the problem of minimizing flow turbulence in a free and moving boundary, nonlinear fluid-structure interaction model, and established existence of a distributed optimal control in the framework of the known local well-posedness theory [16]. Sensitivity equations associated with the above mentioned control problem were obtained using shape optimization and tangential calculus techniques [5,6,8]. In comparison to earlier approaches, like transporting the equations to a fixed reference configuration, or using transpiration techniques [22,23], the shape optimization route is most suited to incorporating the geometry of the problem into the analysis. This refined description brings up new terms in the matching of the normal stresses and velocities on the interface (in particular, the matrix of the interface's curvatures), terms that are missing if one just considers the coupling of linear Stokes flow and linear elasticity.

Sensitivity analysis represents the first step towards an optimization problem. The challenge of applying optimization tools to moving boundary FSI is the proper derivation of the adjoint sensitivity information with correct adjoint balancing conditions on the common interface. The description and study of the appropriate adjoint linearized model is an open question in the field. As the coupled fluid-structure state is the solution of a system of PDEs that are coupled through continuity relations defined on the free interface, sensitivity of the fluid state, which is an Eulerian quantity, with respect to the motion of the solid, which is a Lagrangian quantity, falls into moving shape analysis framework [19–21,38].

The goal of this paper is to build upon the theory developed in [5,6,8] and perform adjoint sensitivity analysis, with direct motivation coming from the development of necessary optimality conditions that describe optimal controls. We note here that most of the work in this paper is independent of the choice of cost function associated with the free boundary interaction, and therefore the results have wide applicability in the field. This article will focus on free boundary steady-state fluid-elasticity interactions. The theory for the general case of a moving boundary dynamical coupled systems of fluid and elasticity will be considered in a subsequent paper.

1.1. *Notation*

- Throughout the paper we use the Einstein summation convention.
- $(Df(a))_{ij} = \partial_j f_i(a) \in \mathbb{M}^3$ is the gradient matrix at $a \in X$ of any vector field $f = (f_i) : X \subset \mathbb{R}^3 \to \mathbb{R}^3$.
- $\text{div } f(a) = \partial_i f_i \in \mathbb{R}$ is the divergence of $f : X \subset \mathbb{R}^3 \to \mathbb{R}^3$ at $a \in X$.
- We identify the set of all second-order tensors with the set $\mathbb{M}^3$ of all square matrices of order three.
- $\mathbb{S}^3 \subset \mathbb{M}^3$ is the set of all symmetric matrices of order three.
- $\text{Div } T(a) = \partial_j T_{ij} e_i \in \mathbb{R}^3$ is the divergence of any second-order tensor field $T = (T_{ij}) : X \subset \mathbb{R}^3 \to \mathbb{M}^3$ at $a \in X$.
- $A^* =$ transpose of A, for any $A \in \mathbb{M}^3$.
- $A..B = \text{Tr}(A^* B) \in \mathbb{R}$ is the matrix inner product in $\mathbb{M}^3$.
- $d_\Omega(x) = \begin{cases} \inf_{y\in\Omega} |y-x|, & \Omega \neq \emptyset \\ \infty, & \Omega = \emptyset \end{cases}$ is the distance from x to $\Omega \subset \mathbb{R}^n$.
- $b_\Omega(x) = d_\Omega(x) - d_{\Omega^c}(x)$ is the oriented distance function from x to Ω for any $\Omega \subset \mathbb{R}^n$.
- $H := \Delta b_\Omega = \text{Tr}(D^2 b_\Omega)$ is the additive curvature of $\Gamma = \partial\Omega$ ([17] pg. 45).
- $\text{div }_\Gamma v = -\text{div } (v) + \langle (Dv)n, n\rangle$ is the tangential divergence of a vector-valued function v, where n is the outward normal to Γ.
- $D_\Gamma v = -Dv + (Dv \cdot n) \otimes n$ on Γ is the tangential gradient matrix.

2. Free boundary fluid-elasticity interaction: PDE model

Let $\mathcal{D} \subset \mathbb{R}^3$ be a bounded open domain composed of two sub-domains $\overline{\mathcal{D}} = \overline{\tilde{\Omega}_e \cup \tilde{\Omega}_f}$ and having smooth (fixed) boundary $\partial\mathcal{D} = \Gamma_f$. The elastic body occupies open sub-domain $\tilde{\Omega}_e$ with boundary $\tilde{\Gamma}_c$. The structure is surrounded by fluid, which occupies open sub-domain $\tilde{\Omega}_f = \mathcal{D} \setminus \overline{\tilde{\Omega}_e}$, with smooth boundary $\tilde{\Gamma}_c \cup \Gamma_f$, assuming that the interface $\tilde{\Gamma}_c$ has positive distance to the boundary Γ_f of the domain.

We recall here that the elasticity domain $\tilde{\Omega}_e$ is described according to a map acting in a fixed reference domain (Lagrangian formulation). Then the evolution of the fluid domain $\tilde{\Omega}_f$ is induced by the structural deformation through the common interface $\tilde{\Gamma}_c$. Given a material reference configuration for the solid $\Omega_e \subset \mathcal{D}$ with boundary Γ_c, let $\Omega_f = \mathcal{D} \setminus \overline{\Omega_e}$ be the reference fluid configuration. Then the volume $\mathcal{D}$ is described by a smooth, injective, orientation-preserving map: $\varphi : \bar{\mathcal{D}} \longrightarrow \bar{\mathcal{D}},\ x \mapsto \varphi = \varphi(x)$. For any $x \in \Omega_e$, $\varphi(x)$ represents the position of the material point x. On Ω_f, $\varphi(x)$ is defined as an arbitrary extension of the restriction of φ to Γ_c, which preserves the boundary Γ_f, i.e., $\varphi = I_{\Gamma_f}$ on Γ_f. We define $J(\varphi) := \det(D\varphi) > 0$.

The fluid is Newtonian, viscous, homogeneous and incompressible. Its behavior is described by its velocity field w and pressure p. The viscosity of the fluid is $\nu > 0$, and the fluid strain tensor is given by $\varepsilon(w) = [Dw + (Dw)^*]/2$, where Dw is the gradient matrix of w, and $(Dw)^*$ represents the transpose of Dw. The fluid state satisfies the steady-state Navier-Stokes equations with no-slip condition on the outer boundary Γ_f.

The fluid is coupled with the St. Venant-Kirchhoff equations which model large displacement-small deformation elasticity. We write the equations of elasticity on the deformed configuration $\tilde{\Omega}_e$, using the Cauchy stress tensor $\mathcal{T} : \overline{\tilde{\Omega}_e} \to \mathbb{S}^3$, $\mathcal{T} = [J(\varphi)^{-1}\mathcal{P} \cdot (D\varphi)^*] \circ \varphi^{-1}$ given in terms of the Green-St. Venant nonlinear strain tensor $\sigma(\varphi) = \frac{1}{2}[(D\varphi)^* D\varphi - I]$ and the Piola transform of the Cauchy stress tensor $\mathcal{P} = D\varphi[\lambda \mathrm{Tr}[\sigma(\varphi)]I + 2\mu\sigma(\varphi)]$, where $\lambda, \mu > 0$ are the Lamé parameters of the material[14]. This provides the accurate matching of the normal stress tensors $\mathcal{T}n_e$ and $\sigma(p, w)n_e = (-pI + 2\nu\varepsilon(w))n_e$, where n_e is the unit outer normal vector along $\tilde{\Gamma}_c$ with respect to $\tilde{\Omega}_e$. In the sequel we also use the notation $n_{\partial\tilde{\Omega}_f}$ as the outer normal vector to the fluid domain where n_{Γ_f} is the outer normal along Γ_f and n_f is the outer normal to the fluid domain along $\tilde{\Gamma}_c$. That is, $n_{\partial\tilde{\Omega}_f}|_{\Gamma_f} := n_{\Gamma_f}$ and $n_{\partial\tilde{\Omega}_f}|_{\tilde{\Gamma}_c} := n_f$. Note that along $\tilde{\Gamma}_c$, $n_e = -n_f$. Equivalently, the equilibrium equations for the elasticity can be written on the reference configuration Ω_e in terms of the Piola transform of the Cauchy stress tensor. However, this translates into a more complicated matching on the boundary of the reference configuration, given by $\mathcal{P}n_e = J(\varphi)(\sigma(p, w) \circ \varphi)(D\varphi)^{-*}n_e$ on Γ_c.

In summary, the steady state PDE model is given by

$$\begin{cases} -\nu\,\Delta w + Dw \cdot w + \nabla p = v|_{\tilde{\Omega}_f} & \text{on } \tilde{\Omega}_f \\ \operatorname{div} w = 0 & \text{on } \tilde{\Omega}_f \\ w = 0 & \text{on } \tilde{\Gamma}_c := \varphi(\Gamma_c) \\ -\operatorname{Div} \mathcal{T} = v|_{\tilde{\Omega}_e} & \text{on } \tilde{\Omega}_e = \varphi(\Omega_e) \\ \mathcal{T}n_e = \sigma(p, w)n_e & \text{on } \tilde{\Gamma}_c \\ w = 0,\ \varphi = I_{\Gamma_f} & \text{on } \Gamma_f. \end{cases} \tag{1}$$

We assume that the distributed force v acts on the entire domain $\mathcal{D}$, and has regularity $v \in H^3(\mathcal{D})$.

Well-posedness analysis for steady-state interactions of this form appeared in[25] and, more recently in[48]. The authors of[48] also provide differentiability of solutions with respect to volume forces, provided a small data assumption holds, and complement the theoretical results with numerical computations. In this paper, we work in the framework of[48] and assume that the force is small enough and therefore the solution to (1) is Fréchet differentiable with respect to the force v.

3. Optimization problem

We consider the optimal control problem:

$$\min J(w, v) = 1/2\|w - w_d\|^2_{L^2(\tilde{\Omega}_f)} + 1/2\|v\|^2_{H^3(\mathcal{D})} \tag{2}$$

subject to (1), with control $v \in H^3(\mathcal{D})$ and such that $w_d \in L^2(\tilde{\Omega}_f)$ is a desired fluid velocity. Assuming that $(\bar{w}, \bar{v})$ is an optimal pair for (2), one is interested in deriving the first-order necessary optimality conditions associated with problem (2). Due to the nonlinearity of the state equations, the Lagrangian functional (which, formally, equals cost functional minus weak form of the system) associated with the coupled system is not convex-concave, and min-max theory does not apply, i.e., one can not reduce the cost function gradient to the derivative of the Lagrangian with respect to the control, at its saddle point. As a consequence, the optimality conditions must be derived from differentiability arguments on the cost functional with respect to the control, where the main challenge stems from the dependence of the cost integrals in J on the unknown domain $\tilde{\Omega}_f$, which also depends on the control v. Therefore, in order to compute the directional derivative of J with respect to v in the direction of v', one has to consider for small parameter $s \geq 0$, the perturbed functional $J(v + sv')$ and then calculate the derivative at $s = 0$ of the function $s \to J(v + sv')$.

With the following notation for the s-derivatives at $s = 0$,

$$\varphi' = \frac{\partial}{\partial s}\varphi_s\Big|_{s=0}, \quad U' = \varphi' \circ \varphi^{-1}, \quad w' = \frac{\partial}{\partial s}w_s\Big|_{s=0}, \quad \text{and} \quad p' = \frac{\partial}{\partial s}p_s\Big|_{s=0},$$

and using formula (4.7) on pg. 484 of[17] for differentiation of domain integrals, we can compute the directional derivative of J as

$$\partial J(v;v') = \lim_{s\to 0}\frac{J(v+sv')-J(v)}{s} = \frac{\partial}{\partial s}J(v+sv')\bigg|_{s=0}$$
$$= \frac{\partial}{\partial s}\left[\frac{1}{2}\int_{(\tilde{\Omega}_f)_s}|w_s-w_d|^2+\frac{1}{2}\|v+sv'\|^2_{H^3(\mathcal{D})}\right]\bigg|_{s=0}$$
$$= \int_{\tilde{\Omega}_f}(w-w_d)\cdot w' + \frac{1}{2}\int_{\tilde{\Gamma}_c}|w-w_d|^2 U'\cdot n_f + (v,v')_{H^3(\mathcal{D})} \tag{3}$$

From[5–8], we know that (U', w', p') satisfy the linear coupled system:

$$\begin{cases} -\nu\Delta w' + (Dw')w + (Dw)w' + \nabla p' = v'\big|_{\tilde{\Omega}_f} & \text{in } \tilde{\Omega}_f \\ \text{div } w' = 0 & \text{in } \tilde{\Omega}_f \\ w' + (Dw)U' = 0 & \text{on } \tilde{\Gamma}_c \\ -\text{Div } \bar{\mathcal{T}}'(U') = v'\big|_{\tilde{\Omega}_e} & \text{in } \tilde{\Omega}_e \\ \bar{\mathcal{T}}'(U')\cdot n_e = (-p'I + 2\nu\varepsilon(w'))\cdot n_e + \mathcal{B}(U') & \text{on } \tilde{\Gamma}_c \\ w' = 0,\ U' = 0 & \text{on } \Gamma_f \end{cases} \tag{4}$$

where $\Theta = D\varphi\circ\varphi^{-1}$, and $\overline{DU'}$ and the "linearized" stress tensor $\bar{\mathcal{T}}'(U')$ are defined via

$$\overline{DV} := \Theta^*(DV)\Theta, \tag{5}$$

$$\bar{\mathcal{T}}'(V) := (DV)\mathcal{T} + \frac{1}{\det\Theta}\Theta\cdot\{\lambda\text{tr}(\overline{DV})I + \mu[\overline{DV} + (\overline{DV})^*]\}\Theta^*,$$

and the boundary term $\mathcal{B}(U')$ has the following form:

$$\begin{aligned}\mathcal{B}(U') =& (\mathcal{T} + pI - 2\nu\varepsilon(w))\cdot[(D_{\tilde{\Gamma}_c}U')^* n_e + (D^2 b_\Omega)U'_{\tilde{\Gamma}_c}] + (D\mathcal{T})U'\cdot n_e \\ &+ \text{div }(U')\mathcal{T}\cdot n_e - \mathcal{T}\cdot(DU')^*\cdot n_e - \\ &- \langle U', n_e\rangle(-\text{Div }_{\tilde{\Gamma}_c}(\mathcal{T}) + [\partial_\nu pI - 2\nu\partial_\nu\varepsilon(w)]\cdot n_e).\end{aligned} \tag{6}$$

For sake of exposition, we introduce the following notation:

$$\begin{cases}\Sigma(V) = \lambda\text{tr}(\overline{DV})I + \mu[\overline{DV} + (\overline{DV})^*], \\ \tilde{\Sigma}(V) = \Theta\Sigma(V)\Theta^*,\end{cases} \tag{7}$$

$$\begin{cases} B_1 := \mathcal{T} + pI - 2\nu\varepsilon(w), \\ B_2 := -\text{Div }_{\tilde{\Gamma}_c}(\mathcal{T}) + [\partial_\nu pI - 2\nu\partial_\nu\varepsilon(w)]\cdot n_e.\end{cases} \tag{8}$$

Then the linearized stress tensor $\bar{\mathcal{T}}'$ can be rewritten as

$$\bar{\mathcal{T}}'(V) := \frac{1}{\det \Theta} \tilde{\Sigma}(V) + (DV)\mathcal{T}, \tag{9}$$

and the boundary term $\mathcal{B}(U')$ can we written more compactly as

$$\begin{aligned} \mathcal{B}(U') =& B_1 \cdot [(D_{\tilde{\Gamma}_c} U')^* n_e + (D^2 b_\Omega) U'_{\tilde{\Gamma}_c}] + (D\mathcal{T})U' \cdot n_e \\ &+ \text{div}\,(U')\mathcal{T} \cdot n_e - \mathcal{T} \cdot (DU')^* \cdot n_e - \langle U', n_e \rangle B_2. \end{aligned} \tag{10}$$

Remark 3.1. The new linear system (4) that characterizes the s-derivatives can be seen as a 'total linearization', as the coupled equations and the common, moving interface were "differentiated" together. As the force v becomes a function of parameter s (i.e., $v_s = v + sv'$), the geometry of the problem moves with the flow of a vector field that depends on the deformation of elasticity φ_s as well. More specifically, the perturbation $(\tilde{\Gamma}_c)_s$ of the boundary is built by the flow of the vector field $V(s,x) = \frac{\partial}{\partial s}\varphi_s \circ \varphi_s^{-1}$, i.e.,

$$(\Gamma_c)_s(t) = T_s(V)((\tilde{\Gamma}_c)_0), \text{where} \quad T_s(V) : \tilde{\Omega}_f \to (\tilde{\Omega}_f)_s, \;\; T_s(V) = \varphi_s \circ \varphi^{-1}.$$

We would like to point out that (φ', w', p') are 'shape' derivatives with respect to the speed V, which is a vector field that depends on φ_s and is not given a priori. This contrasts with the standard theory on shape derivatives, where the domain is perturbed by an a priori given vector field and then the speed method is applied. Nonetheless, one can think of the s-derivatives as 'pseudo-shape derivatives', in the sense that much of the theory of shape calculus remains applicable.

Remark 3.2. We note that the new linear coupled system for w', p' and U' is quite different from the usual coupling of linear models. In particular, extra terms are present in the normal stress tensors matching on the common interface, some of them involving boundary curvatures and boundary acceleration, proving that the connection on the interface between elasticity and fluid is tightly linked to the geometry of the boundary.

The sensitivity equations (4) provide the needed characterization for the s-derivatives (U', w', p') that appear in the formula (3) of the directional derivative $\partial J(v; v')$ of the cost functional. However, the expression in (3) does not represent the gradient of J, since it is not explicitly linear in v'. In fact, v' does not even appear in the chain rule computation, since it is hidden in the equations (4) for the s-derivatives w', p', and U'. The idea is to introduce a suitable adjoint problem that eliminates the s-derivatives and provide an explicit representation of the gradient of J at v, denoted by $J'(v;v')$.

Our **main result** provides the description for the adjoint of the linearized model (4), and the formula for the gradient $J'(v;v')$ of the cost functional J.

Theorem 3.1. *For the optimal control problem:*

$$\min J(w,v) = 1/2\|w-w_d\|^2_{L^2(\tilde{\Omega}_f)} + 1/2\|v\|^2_{H^3(\mathcal{D})},$$

subject to (1), the gradient of the cost functional is given by

$$J'(v;v') = (v',v)_{\mathcal{D}} + (v'|_{\tilde{\Omega}_f}, Q) + (v'|_{\tilde{\Omega}_e}, R),$$

where Q, P, and R solve the following adjoint problem:

$$\begin{cases} -\nu\Delta Q + (Dw)^* Q - (DQ)w + \nabla P = w - w_d & \tilde{\Omega}_f \\ div\ Q = 0 & \tilde{\Omega}_f \\ \langle Q, n_e\rangle = \langle R, n_e\rangle & \tilde{\Gamma}_c \\ -Div\ \bar{\mathcal{T}}'(R) = 0 & \tilde{\Omega}_e \\ [\bar{\mathcal{T}}'(R) - P(Dw)^*]n_e + div_{\tilde{\Gamma}_c}[B_1 R]n_e - (D\mathcal{T}^{\Delta}\cdot n_e)^* R & \\ -H\langle \mathcal{T} n_e, R\rangle n_e + \nabla_{\tilde{\Gamma}_c}\langle \mathcal{T} n_e, R\rangle & \\ -Div_{\tilde{\Gamma}_c}(n_e \otimes \mathcal{T} R) + \langle B_2, R\rangle n_e = \frac{1}{2}|w-w_d|^2 n_f & \tilde{\Gamma}_c \\ Q\cdot n_{\Gamma_f} = 0 & \Gamma_f \end{cases} \tag{11}$$

4. Proof of Theorem 3.1

We repeatedly make use of Lemma 5.2 in[8], so we restate it here for convenience.

Lemma 4.1. *Let Ω be an open domain with boundary $\partial\Omega = \Gamma$, let ξ and θ be two vector fields in $\mathbb{R}^3$, and let $\psi \in C^1(\Omega)$, with n the outer unit normal to Ω. Then the following identity is satisfied:*

$$\int_\Omega \langle \xi, (D\psi)\theta\rangle = -\int_\Omega \langle (D\xi)\theta + (div\ \theta)\xi, \psi\rangle + \int_\Gamma \langle (\theta\cdot n)\xi, \psi\rangle. \tag{12}$$

Moreover if we assume that $(D\psi)n = 0$ on Γ, then

$$\int_\Gamma \langle \xi, (D_\Gamma\psi)\theta\rangle = -\int_\Gamma \langle Div_\Gamma(\xi\otimes\theta), \psi\rangle. \tag{13}$$

Using test functions Q, P, and R, we obtain the weak formulation for the linear coupled system (4):

$$\int_{\tilde{\Omega}_f} \langle -\nu\Delta w' + (Dw')w + (Dw)w' + \nabla p', Q\rangle - \int_{\tilde{\Omega}_f} P\text{div } w'$$
$$- \int_{\tilde{\Omega}_e} \langle \text{Div } \bar{\mathcal{T}}'(U'), R\rangle = \int_{\tilde{\Omega}_f} \langle v'|_{\tilde{\Omega}_f}, Q\rangle + \int_{\tilde{\Omega}_e} \langle v'|_{\tilde{\Omega}_e}, R\rangle \qquad (14)$$

We integrate by parts in the first term of the LHS of (14), taking advantage of (12) and using the facts that $w' = 0$ on Γ_f and div $w' = 0$ on $\tilde{\Omega}_f$. We obtain

$$\int_{\tilde{\Omega}_f} \langle -\nu\Delta w' + (Dw')w + (Dw)w' + \nabla p', Q\rangle =$$
$$\int_{\tilde{\Omega}_f} \langle w', -\nu\Delta Q + (Dw)^* Q - (DQ)w\rangle - \int_{\partial\tilde{\Omega}_f} \langle (Dw')n_{\partial\tilde{\Omega}_f}, \nu Q\rangle$$
$$+ \int_{\tilde{\Gamma}_c} \langle w', \nu(DQ)n_f\rangle - \int_{\tilde{\Omega}_f} p'(\text{div } Q) + \int_{\partial\tilde{\Omega}_f} p'\langle n_{\partial\tilde{\Omega}_f}, Q\rangle. \qquad (15)$$

We integrate by parts in the second term of the LHS of (14) and use the facts that $w' = 0$ on Γ_f, and $w' + (Dw)U' = 0$ on $\tilde{\Gamma}_c$. We obtain

$$- \int_{\tilde{\Omega}_f} P\text{div } w' = - \int_{\partial\tilde{\Omega}_f} P\langle w', n_{\partial\tilde{\Omega}_f}\rangle + \int_{\tilde{\Omega}_f} \langle w', \nabla P\rangle$$
$$= - \int_{\tilde{\Gamma}_c} P\langle w', n_f\rangle + \int_{\tilde{\Omega}_f} \langle w', \nabla P\rangle$$
$$= \int_{\tilde{\Gamma}_c} P\langle (Dw)U', n_f\rangle + \int_{\tilde{\Omega}_f} \langle w', \nabla P\rangle. \qquad (16)$$

We integrate by parts in the third term of the LHS of (14). Also recalling the definition (9) of the linearized stress tensor $\bar{\mathcal{T}}'(U')$, we obtain

$$- \int_{\tilde{\Omega}_e} \langle \text{Div } \bar{\mathcal{T}}'(U'), R\rangle = \int_{\tilde{\Omega}_e} \bar{\mathcal{T}}'(U')..DR - \int_{\tilde{\Gamma}_c} \langle \bar{\mathcal{T}}'(U')n_e, R\rangle$$
$$= \int_{\tilde{\Omega}_e} [(DU')\mathcal{T}]..DR + \int_{\tilde{\Omega}_e} \frac{1}{\det\Theta}\tilde{\Sigma}(U')..DR - \int_{\tilde{\Gamma}_c} \langle \bar{\mathcal{T}}'(U')n_e, R\rangle \ (17)$$

In manipulating the terms on the RHS of (17) we repeatedly use the definition of matrix inner product and the invariance under cyclic permutations property of the trace. For the first term on the RHS of (17) we recall that

$\mathcal{T}$ is symmetric and use integration by parts to obtain

$$\int_{\tilde{\Omega}_e} [(DU')\mathcal{T}]..DR = \int_{\tilde{\Omega}_e} DU'..[(DR)\mathcal{T}]$$
$$= -\int_{\tilde{\Omega}_e} U', \langle \text{Div } [(DR)\mathcal{T}] \rangle + \int_{\tilde{\Gamma}_c} \langle U', [(DR)\mathcal{T}]n_e \rangle \tag{18}$$

For the second term on the RHS of (17) we first note that

$$\frac{1}{\det \Theta} \tilde{\Sigma}(U')..DR = \Sigma(U')..\left[\frac{1}{\det \Theta} \overline{DR}\right]$$

where $\overline{DR}$ is defined according to (5) and $\Sigma(U')$ is defined according to (7). Using properties of the trace we obtain

$$\int_{\tilde{\Omega}_e} [\lambda \text{tr}(\overline{DU'})I]..\left[\frac{1}{\det \Theta} \overline{DR}\right] = \int_{\tilde{\Omega}_e} DU'..\left[\frac{1}{\det \Theta} \lambda \Theta \{\text{tr}(\overline{DR})I\} \Theta^*\right]$$
$$= -\int_{\tilde{\Omega}_e} \left\langle U', \text{Div } \left[\frac{1}{\det \Theta} \lambda \Theta \{\text{tr}(\overline{DR})I\} \Theta^*\right]\right\rangle$$
$$+ \int_{\tilde{\Gamma}_c} \left\langle U', \left[\frac{1}{\det \Theta} \lambda \Theta \{\text{tr}(\overline{DR})I\} \Theta^*\right] n_e \right\rangle \tag{19}$$

Similarly, we have

$$\int_{\tilde{\Omega}_e} \mu[\overline{DU'} + (\overline{DU'})^*]..\left[\frac{1}{\det \Theta} \overline{DR}\right]$$
$$= -\int_{\tilde{\Omega}_e} \left\langle U', \text{Div } \left\{\frac{1}{\det \Theta} \mu \Theta [\overline{DR} + (\overline{DR})^*] \Theta^*\right\}\right\rangle$$
$$+ \int_{\tilde{\Gamma}_c} \left\langle U', \left\{\frac{1}{\det \Theta} \mu \Theta [\overline{DR} + (\overline{DR})^*] \Theta^*\right\} n_e \right\rangle \tag{20}$$

Combining (19) with (20) we obtain

$$\int_{\tilde{\Omega}_e} \frac{1}{\det \Theta} \tilde{\Sigma}(U')..DR = -\int_{\tilde{\Omega}_e} \left\langle U', \text{Div } \left[\frac{1}{\det \Theta} \tilde{\Sigma}(R)\right]\right\rangle$$
$$+ \int_{\tilde{\Gamma}_c} \left\langle U', \left[\frac{1}{\det \Theta} \tilde{\Sigma}(R)\right] n_e \right\rangle$$

Using (18) and (21) back into (17), we obtain the following integration by parts formula for elasticity:

$$-\int_{\tilde{\Omega}_e} \langle \text{Div } \bar{\mathcal{T}}'(U'), R \rangle = -\int_{\tilde{\Omega}_e} \langle U', \text{Div } \bar{\mathcal{T}}'(R) \rangle$$
$$+ \int_{\tilde{\Gamma}_c} \langle U', \bar{\mathcal{T}}'(R)n \rangle - \langle \bar{\mathcal{T}}'(U')n_e, R \rangle. \tag{21}$$

Taking (15), (16), and (21) together and substituting the boundary condition from (4) we obtain that (14) is equivalent to

$$
\begin{aligned}
&\int_{\tilde{\Omega}_f} \langle w', -\nu\Delta Q + (Dw)^* Q - (DQ)w + \nabla P\rangle - \int_{\partial\tilde{\Omega}_f} \langle (Dw')n_f, \nu Q\rangle \\
&+ \int_{\tilde{\Gamma}_c} \langle w', \nu(DQ)n_f\rangle - \int_{\tilde{\Omega}_f} p'(\text{div } Q) + \int_{\partial\tilde{\Omega}_f} p'\langle n_{\partial\tilde{\Omega}_f}, Q\rangle \\
&- \int_{\tilde{\Omega}_e} \big\langle U', \text{Div } \bar{\mathcal{T}}'(R)\big\rangle + \int_{\tilde{\Gamma}_c} \big\langle U', \bar{\mathcal{T}}'(R)n_e + P(Dw)^* n_f\big\rangle \\
&- \int_{\tilde{\Gamma}_c} \langle (-p'I + 2\nu\varepsilon(w'))n_e + \mathcal{B}(U'), R\rangle \\
&= \int_{\tilde{\Omega}_f} \langle v'|_{\tilde{\Omega}_f}, Q\rangle + \int_{\tilde{\Omega}_e} \langle v'|_{\tilde{\Omega}_e}, R\rangle
\end{aligned} \tag{22}
$$

The "bilinear form" from which we will recover the adjoint system for (Q, P, R) is taken directly from the weak formulation of the state equations presented above in (22) :

$$
\begin{aligned}
&a((\alpha, \beta), \gamma; (Q, P), R) := \int_{\tilde{\Omega}_f} \langle \alpha, -\nu\Delta Q + (Dw)^* Q - (DQ)w + \nabla P\rangle \\
&- \int_{\partial\tilde{\Omega}_f} \langle (D\alpha)n_f, \nu Q\rangle + \int_{\tilde{\Gamma}_c} \langle \alpha, \nu(DQ)n_f\rangle - \int_{\tilde{\Omega}_f} \beta(\text{div } Q) + \int_{\partial\tilde{\Omega}_f} \beta\langle n_{\partial\tilde{\Omega}_f}, Q\rangle \\
&- \int_{\tilde{\Omega}_e} \big\langle \gamma, \text{Div } \bar{\mathcal{T}}'(R)\big\rangle + \int_{\tilde{\Gamma}_c} \big\langle \gamma, \bar{\mathcal{T}}'(R)n_e + P(Dw)^* n_f\big\rangle \\
&- \int_{\tilde{\Gamma}_c} \langle (-\beta I + 2\nu\varepsilon(\alpha))n_e + \mathcal{B}(\gamma), R\rangle .
\end{aligned} \tag{23}
$$

Then for all α, β, γ, we set

$$
a((\alpha, \beta), \gamma; (Q, P), R) = \int_{\tilde{\Omega}_f} (w - w_d)\cdot\alpha + \frac{1}{2}\int_{\tilde{\Gamma}_c} |w - w_d|^2 \gamma \cdot n_f, \tag{24}
$$

from which we will recover the directional derivative of the cost as

$$
\partial J(v; v') = a(w', p', \varphi'; Q, P, R) + (v, v').
$$

We proceed to consider the following cases in order to recover the system for the adjoint variables (Q, P, R).

4.1. *Case 1:* $\alpha = \gamma = 0$

Consider $\alpha = \gamma = 0$. Then $\forall\beta$ we have

$$
0 = \int_{\partial\tilde{\Omega}_f} \beta\langle n_{\partial\tilde{\Omega}_f}, Q\rangle - \int_{\tilde{\Omega}_f} \beta(\text{div } Q) + \int_{\tilde{\Gamma}_c} \beta\langle n_e, R\rangle \tag{25}
$$

Take β to have compact support on $\partial\tilde{\Omega}_f$. Then (25) reduces to

$$\int_{\tilde{\Omega}_f} \beta(\text{div } Q) = 0,$$

from which we recover div $Q = 0$ in $\tilde{\Omega}_f$.

Next take β such that $\beta|_{\Gamma_f} = 0$. For all such β, we have $\int_{\tilde{\Gamma}_c} \beta(\langle n_f, Q\rangle + \langle n_e, R\rangle) = 0 \Rightarrow \int_{\tilde{\Gamma}_c} \beta\langle n_e, R - Q\rangle = 0$. This implies that

$$\langle Q, n_e\rangle = \langle R, n_e\rangle \text{ on } \tilde{\Gamma}_c. \tag{26}$$

Substituting (26) back in to (25), we recover

$$\langle Q, n_{\Gamma_f}\rangle = 0 \text{ on } \Gamma_f. \tag{27}$$

4.2. *Case 2:* $\beta = \gamma = 0$

Secondly, consider $\beta = \gamma = 0$. Then $\forall\alpha$, we have

$$\begin{aligned}\int_{\tilde{\Omega}_f} \langle w - w_d, \alpha\rangle = &\int_{\tilde{\Omega}_f} \langle \alpha, -\nu\Delta Q + (Dw)^*Q - (DQ)w + \nabla P\rangle \\ &- \int_{\partial\tilde{\Omega}_f} \langle (D\alpha)n_{\partial\tilde{\Omega}_f}, \nu Q\rangle + \int_{\tilde{\Gamma}_c} \langle \alpha, \nu(DQ)n_f\rangle - \langle 2\nu\varepsilon(\alpha)n_e, R\rangle\end{aligned} \tag{28}$$

In particular, if α has compact support on $\tilde{\Omega}_f$, then we obtain:

$$\int_{\tilde{\Omega}_f} \langle w - w_d, \alpha\rangle = \int_{\tilde{\Omega}_f} \langle -\nu\Delta Q + (Dw)^*Q - (DQ)w + \nabla P, \alpha\rangle,$$

from which we recover the following equation for Q and P:

$$-\nu\Delta Q + (Dw)^*Q - (DQ)w + \nabla P = w - w_d \text{ on } \tilde{\Omega}_f.$$

After substituting the recovered domain equation back into (8), we have

$$\begin{aligned}0 = &- \int_{\Gamma_f} \langle (D\alpha)n_{\Gamma_f}, \nu Q\rangle + \int_{\tilde{\Gamma}_c} \langle \alpha, \nu(DQ)n_f\rangle \\ &+ \int_{\tilde{\Gamma}_c} \langle (D\alpha)n_e, \nu Q\rangle - \langle 2\nu\varepsilon(\alpha)n_e, R\rangle\end{aligned} \tag{29}$$

Now consider α such that $D\alpha \cdot n_{\partial\tilde{\Omega}_f} = 0$. In this case we have

$$\begin{aligned}
0 &= \nu \int_{\tilde{\Gamma}_c} \langle \alpha, (DQ)n_f\rangle - \langle (D\alpha)^* n_e, R\rangle \\
&= \nu \int_{\tilde{\Gamma}_c} \langle \alpha, (DQ)n_e\rangle + \langle (D\alpha)^* n_e, R\rangle \\
&= \nu \int_{\tilde{\Gamma}_c} \langle \alpha, (DQ)n_e\rangle + \langle (D_{\tilde{\Gamma}_c}\alpha + (D\alpha)n_e \otimes n_e)^* \; n_e, R\rangle \\
&= \nu \int_{\tilde{\Gamma}_c} \langle \alpha, (DQ)n_e\rangle + \langle D_{\tilde{\Gamma}_c}\alpha, R\rangle \\
&= \nu \int_{\tilde{\Gamma}_c} \langle \alpha, (DQ)n_e - \text{Div}\,_{\tilde{\Gamma}_c}(n_e \otimes R)\rangle, \qquad (30)
\end{aligned}$$

where in the line above we used (13). We recover the following boundary condition

$$(DQ)n_e - \text{Div}\,_{\tilde{\Gamma}_c}(n_e \otimes R) = 0 \text{ on } \tilde{\Gamma}_c. \qquad (31)$$

4.3. *Case 3:* $\alpha = \beta = 0$

Lastly, consider $\alpha = \beta = 0$. Then from (24) we obtain

$$\begin{aligned}
\int_{\tilde{\Gamma}_c} \left\langle \gamma, \frac{1}{2}|w - w_d|^2 n_f \right\rangle &= -\int_{\tilde{\Omega}_e} \langle \gamma, \text{Div}\,\bar{\mathcal{T}}'(R)\rangle \\
&+ \int_{\tilde{\Gamma}_c} \langle \gamma, [\bar{\mathcal{T}}'(R) - P(Dw)^*]n_e\rangle - \langle \mathcal{B}(\gamma), R\rangle. \qquad (32)
\end{aligned}$$

As previously, we first consider the case where γ has compact support on $\mathcal{D}$. Then $\gamma_{\tilde{\Gamma}_c} = 0$ and moreover $\mathcal{B}(\gamma)|_{\tilde{\Gamma}_c} = 0$, as $\mathcal{B} : H^{1/2}(\tilde{\Gamma}_c) \to H^{-1/2}(\tilde{\Gamma}_c)$ is tangential (shown in[5] Proposition 6.2). Then we have for all γ

$$0 = \int_{\tilde{\Omega}_e} \langle \gamma, -\text{Div}\,\bar{\mathcal{T}}'(R)\rangle,$$

from which we recover

$$-\text{Div}\,\bar{\mathcal{T}}'(R) = 0 \text{ in } \tilde{\Omega}_e.$$

Substituting the recovered domain equation back in to (32) we have

$$\int_{\tilde{\Gamma}_c} \left\langle \gamma, \frac{1}{2}|w - w_d|^2 n_f \right\rangle = \int_{\tilde{\Gamma}_c} \langle \gamma, [\bar{\mathcal{T}}'(R) - P(Dw)^*]n_e\rangle - \langle \mathcal{B}(\gamma), R\rangle, \quad (33)$$

where $\mathcal{B}$ is defined in (10). In order to recover the corresponding adjoint boundary equation we work term by term. In particular, (33) is true when

$D\gamma \cdot n_e = 0$, so we work under this setting. We make use of a tangential chain rule from[17] ((5.23) pg. 497):

$$\nabla_\Gamma \langle \gamma, n \rangle = (D_\Gamma \gamma)^* n + (D^2 b)\gamma_\Gamma,$$

and also the tangential Green's formula from[17] ((5.27) pg. 498):

$$\int_\Gamma f \text{div }_\Gamma v + \nabla_\Gamma f \cdot v \, d\Gamma = \int_\Gamma H f v \cdot n \, d\Gamma. \tag{34}$$

We obtain

$$\begin{aligned}&\int_{\tilde{\Gamma}_c} \langle B_1 \cdot [(D_{\tilde{\Gamma}_c}\gamma)^* n_e + (D^2 b_{\tilde{\Omega}_e})\gamma_{\tilde{\Gamma}_c}], R\rangle \\ &= \int_{\tilde{\Gamma}_c} \langle B_1 \cdot \nabla_{\tilde{\Gamma}_c}\langle \gamma, n_e\rangle, R\rangle = -\int_{\tilde{\Gamma}_c} \langle \gamma, \text{div }_{\tilde{\Gamma}_c}[B_1 R] n_e\rangle,\end{aligned} \tag{35}$$

which relies on the facts that B_1 is symmetric and $B_1 n_e = 0$ on $\tilde{\Gamma}_c$, which implies that $\int_{\tilde{\Gamma}_c} H\langle \gamma, n_e\rangle\langle B_1 R, n_e\rangle = 0$.

We then consider the term $\int_{\tilde{\Gamma}_c} \langle \mathcal{T}(D\gamma)^* n_e, R\rangle$ and obtain

$$\int_{\tilde{\Gamma}_c} \langle \mathcal{T}(D\gamma)^* n_e, R\rangle = \int_{\tilde{\Gamma}_c} \langle \gamma, -\text{Div }_{\tilde{\Gamma}_c}(n_e \otimes \mathcal{T} R)\rangle, \tag{36}$$

using (13) and the fact that $\mathcal{T}$ is symmetric. Next, we have

$$\int_{\tilde{\Gamma}_c} \langle \langle \gamma, n_e\rangle B_2, R\rangle = \int_{\tilde{\Gamma}_c} \langle \gamma, \langle B_2, R\rangle n_e\rangle. \tag{37}$$

Consider the term $\int_{\tilde{\Gamma}_c} \langle \text{div }(\gamma)\mathcal{T} n_e, R\rangle$. Using the definition of the tangential divergence,we obtain

$$\begin{aligned}\int_{\tilde{\Gamma}_c} \langle \text{div }(\gamma)\mathcal{T} n_e, R\rangle &= \int_{\tilde{\Gamma}_c} \langle \text{div }_{\tilde{\Gamma}_c}(\gamma)\mathcal{T} n_e, R\rangle \\ &= \int_{\tilde{\Gamma}_c} \langle \gamma, H\langle \mathcal{T} n_e, R\rangle n_e - \nabla_{\tilde{\Gamma}_c}\langle \mathcal{T} n_e, R\rangle\rangle,\end{aligned} \tag{38}$$

where we have used the tangential Green's formula (34).

Finally, we consider the term $\int_{\tilde{\Gamma}_c} \langle \{(D\mathcal{T})\gamma\} \cdot n_e, R\rangle$. The term $D\mathcal{T}$ is a three entries tensor representing the gradient matrix of T, and was introduced in[8] as it arose in the derivation of the s-derivative of the Cauchy stress tensor $\mathcal{T}_s$. In particular, since

$$\forall i,j, (\mathcal{T}^s_{ij} \circ \varphi_s)' = \mathcal{T}'_{ij} \circ \varphi + \langle (\nabla \mathcal{T}_{ij}) \circ \varphi, \varphi' \rangle = \mathcal{T}'_{ij} \circ \varphi + [(\partial_k \mathcal{T}_{ij}) \circ \varphi]\varphi'_k,$$

$D\mathcal{T}$ was defined as $(D\mathcal{T}.\vec{e})_{ij} = (\partial_k \mathcal{T}_{ij})e_k$. Introducing the following notation $(D\mathcal{T}^\Delta \cdot \vec{f})_{ik} := \partial_k \mathcal{T}_{ij} f_j$, we can integrate by parts (in components) and write

$$\int_{\tilde{\Gamma}_c} \langle \{(D\mathcal{T})\gamma\} \cdot n_e, R\rangle = \int_{\tilde{\Gamma}_c} (\partial_k \mathcal{T}_{ij}\gamma_k)(n_e)_j R_i$$
$$= \int_{\tilde{\Gamma}_c} \gamma_k(\partial_k \mathcal{T}_{ij}(n_e)_j R_i) = \int_{\tilde{\Gamma}_c} \langle \gamma, (D\mathcal{T}^{\Delta} \cdot n_e)^* R\rangle. \tag{39}$$

Combining (35), (36), (37), (38), and (39) into (33) we obtain:

$$\int_{\tilde{\Gamma}_c} \left\langle \gamma, \frac{1}{2}|w - w_d|^2 n_f \right\rangle = \int_{\tilde{\Gamma}_c} \langle \gamma, [\bar{\mathcal{T}}'(R) - P(Dw)^*]n_e\rangle$$
$$- \int_{\tilde{\Gamma}_c} \langle \gamma, -\text{div}_{\tilde{\Gamma}_c}[B_1 R]n_e + (D\mathcal{T}^{\Delta} \cdot n_e)^* R + H\langle \mathcal{T} n_e, R\rangle n_e\rangle$$
$$- \int_{\tilde{\Gamma}_c} \langle \gamma, -\nabla_{\tilde{\Gamma}_c}\langle \mathcal{T} n_e, R\rangle + \text{Div}_{\tilde{\Gamma}_c}(n_e \otimes \mathcal{T} R) - \langle B_2, R\rangle n_e\rangle. \tag{40}$$

Combining (26) and (40) we recover the following interaction for R, Q, and P on the common interface $\tilde{\Gamma}_c$:

$$\frac{1}{2}|w - w_d|^2 n_f = [\bar{\mathcal{T}}'(R) - P(Dw)^*]n_e + \text{div}_{\tilde{\Gamma}_c}[B_1 R]n_e - (D\mathcal{T}^{\Delta} \cdot n_e)^* R$$
$$-H\langle \mathcal{T} n_e, R\rangle n_e + \nabla_{\tilde{\Gamma}_c}\langle \mathcal{T} n_e, R\rangle - \text{Div}_{\tilde{\Gamma}_c}(n_e \otimes \mathcal{T} R) + \langle B_2, R\rangle n_e. \tag{41}$$

5. Conclusion

In this paper, we derived the adjoint sensitivity system for a PDE-constrained optimization problem governed by a free boundary fluid-elasticity interaction. The state system is described by the steady-state interaction between a viscous, incompressible fluid, modeled by the Navier-Stokes equation, and the Lamé nonlinear elasticity equations. Since the interaction is a coupling of Eulerian and Lagrangian quantities, sensitivity analysis on the system falls into the framework of shape analysis. Both the sensitivity system, as well as the adjoint sensitivity system are more complicated than usual couplings of linear Stokes and elasticity. In particular, they involve the geometry of the free boundary, through its curvatures and acceleration. Well-posedness analysis for the sensitivity system is discussed in[7], and we anticipate that the technique used can be adjusted to investigate the well-posedness of the adjoint system (11). The work presented in this paper is paramount in the derivation of optimality conditions for controls and derivative-based optimization algorithms in free boundary fluid-elasticity interactions.

Acknowledgments

The authors would like to thank the anonymous referees for their insightful comments and corrections.

References

1. F. Abergel and R. Temam, *On Some Control Problems in Fluid Mechanics*, Theoret. Comput. Fluid Dynamics **1**, (1990), 303-325.
2. H. Antil, R. H. Nochetto, and P. Sodré, *Optimal Control of a Free Boundary Problem: Analysis with Second-Order Sufficient Conditions*, SIAM J. Control Optim. **52**, 5, (2014), 2771-2799.
3. G. Avalos and M. Dvorak, *A new maximality argument for a coupled fluid-structure interaction, with implications for a divergence-free finite element method*, Appl. Math., **35**, (2008), 259–280.
4. L. Bociu, L. Castle, K. Martin, and D. Tondykov, *Optimal Control in a Free Boundary Fluid-Elasticity Interaction*, AIMS Proceedings, (2015), 122-131.
5. L. Bociu, D. Toundykov, and J.-P. Zolésio, *Well-Posedness Analysis for a Linearization of a Fluid-Elasticity Interaction*, SIAM J. Math. Anal., **47**, 3, (2015), 1958-2000.
6. L. Bociu, J.-P. Zolésio, *Linearization of a coupled system of nonlinear elasticity and viscous fluid*, "Modern Aspects of the Theory of Partial Differential Equations", in the series "Operator Theory: Advances and Applications", **216**, (Springer, Basel, 2011), 93-120.
7. L. Bociu, J.-P. Zolésio, *Existence for the linearization of a steady state fluid - nonlinear elasticity interaction*, DCDA-S, (2011), 184-197.
8. L. Bociu, J.-P. Zolésio, *Sensitivity analysis for a free boundary fluid-elasticity interaction*, EECT **2**, (2012), 55-79.
9. M. Boulakia, *Existence of weak solutions for the three dimensional motion of an elastic structure in an incompressible fluid*, J. Math. Fluid Mech. **9**, 2, (2007), 262-294.
10. M. Boulakia, *Existence of weak solutions for the motion of an elastic structure in an incompressible viscous fluid*, C. R. Math. Acad. Sci. Paris, **336**, 12, (2003), 985-990.
11. M. Boulakia and S. Guerrero, *A regularity result for a solid-fluid system associated to the compressible Navier-Stokes equations*, Ann. Inst. H. Poincaré Anal. Non Linéaire, **26**, 3, (2009), 777-813.
12. F. Bucci and I. Lasiecka, *Optimal boundary control with critical penalization for a PDE model of fluid-solid interactions*, Calc. Var. Partial

Differential Equations, **37**, 1-2, (2010), 217-235.

13. S. Canic, A. Mikelic, G. Guidoboni, *Existence of a Unique Solution to a Nonlinear Moving-Boundary Problem of Mixed type Arising in Modeling Blood Flow*, IMA, **153**, (2011), 235-256.
14. P.G. Ciarlet, Mathematical Elasticity Volume I: Three-dimensional Elasticity, (Noth Holland, Amsterdam-New York-Oxford-Tokyo, 1988).
15. D. Coutand and S. Shkoller, *Motion of an Elastic Solid inside and Incompressible Viscous Fluid*, Arch. Rational Mech. Anal. **176**, 1, (2005), 303-352.
16. D. Coutand and S. Shkoller, *The interaction between quasi-linear elastodynamics and the Navier-Stokes Equations*, Arch. Rational Mech. Anal.. **179**, 3, (2006), 25-102.
17. M.C. Delfour and J.-P. Zolésio, Shapes and Geometries: Analysis, Differential Calculus and Optimization, Advances in Design and Control, (SIAM 2001).
18. J. Donea, S. Giuliani, and J.P. Halleux, *An arbitrary Lagrangian-Eulerian finite element method for transient dynamic fluid-structure interactions*, Comput. Methods Appl. Mech. Eng., **33**, (1982), 689-723.
19. R. Dziri and J.-P. Zolésio, *Dynamical shape control in non-cylindrical Navier-Stokes equations*, J. Convex Anal., **6**, 2, (1999), 293-318.
20. R. Dziri and J.-P. Zolésio, *Drag reduction for non-cylindrical Navier-Stokes flows*, Optim. Method Softw., **26**, 4-5, (2011), 575-600.
21. R. Dziri and J.-P. Zolésio, *Tube derivative of noncylindrical shape functionals and variational formulations. Free and moving boundaries*, Lect. Notes Pure Appl. Math., **252**, (Chapman and Hall/CRC, Boca Raton, FL, 2007), 411-442.
22. M.A. Fernandez and M. Moubachir, *An exact block-Newton algorithm for solving fluid-structure interaction problems*, C. R. Math Acad. Sci Paris, **336**, 8, (2003), 681-686.
23. M.A. Fernandez and P. Le Tallec, *Linear stability analysis in fluid-structure interaction with transpiration.II. Numerical analysis and applications*, Comput. Methods Appl. Mech. Eng., **192**, 43, (2003), 4837-4873.
24. L. Formaggia, A. Quarteroni, A. Veneziani Eds., Cardiovascular Mathematics. Modeling and simulation of the circulatory system. Vol I., MS & A, (Springer-Verlag Italia, Milano, 2009).
25. C. Grandmont, *Existence for a Three-Dimensional Steady State Fluid-Structure Interaction Problem*, J. Math. Fluid. Mech., **4** (2002), 76-94.
26. C. Grandmont and Y. Maday, *Existence for unsteady fluid-structure*

interaction problem, Math. Model. Numer. Anal., **34**, (2000), 609-636.
27. C. Grandmont and Y. Maday, *Fluid-structure interaction: A theoretical point of view*, Rev. Europeenne Elem. Finis, **9**, 6-7, (2001), 633-653.
28. T.J. Hughes, W.K. Liu, and T.K. Zimmermann, *Lagrangian-Eulerian finite element formulation for incompressible viscous flows*, Comput. Methods in Appl. Mech. Eng., **29**, (1981), 329-349.
29. M. Ignatova, Igor Kukavica, Irena Lasiecka, and Amjad Tuffaha *On well-posedness for a free boundary fluid- structure model*, J. Math. Phys., **53**, (2012), 115624-13.
30. M. Ignatova, Igor Kukavica, Irena Lasiecka, and Amjad Tuffaha *On well-posedness and small data global existence for an interface damped free boundary fluid-structure model*, Nonlinearity, **27**, (2014), 467-499.
31. T. Kim, S. Canic and G. Guidoboni, *Existence and Uniqueness of a Solution to a Three-Dimensional Axially Symmetric Biot Problem arising in Modeling Blood Flow*, CPAA, **9**, 4, (2010), 839-865.
32. I. Kukavica and A. Tuffaha, *Solutions to a fluid-structure interaction free boundary problem*, Discrete Contin. Dyn. Syst., **32**, (2012), 1355-1389.
33. I. Kukavica and A. Tuffaha, *Regularity of solutions to a free boundary problem of fluid structure interaction*, Indiana Univ. Math. J., **61**, (2012), 1817-1859.
34. I. Lasiecka, R. Triggiani, and J. Zhang, *The fluid-structure interaction model with both control and disturbance at the interface: a game theory problem via an abstract approach*, Appl. Anal., **90**, 6, (2011), 971-1009.
35. I. Lasiecka and A. Tuffaha, *A Bolza optimal synthesis problem for singular estimate control system*, Cont. Cybernetics, **38**, 4, (2009), 1429-1460.
36. I. Lasiecka and A. Tuffaha, *Boundary feedback control in fluid-structure interactions*, Proceedings of the 47th IEEE Conference on Decision and Control, (Cancun, Mexico, 2008), 203-208.
37. P. Le Tallec and J. Mouro, *Fluid structure interaction with large structural displacements*, Comput. Methods Applied Mech. Engineering, **190**, 24-25, (2001), 3039-3067.
38. M. Moubachir and J.-P. Zolésio, Moving Shape Analysis and Control: Applications to Fluid Structure Interactions, (Chapman and Hall/CRC Pure and Applied Mathematics, 2006).
39. B. Muha, S. Canic, *Existence of a solution to a fluid-multi-layered-structure interaction problem*, J Differ Equations,**256**, (2014), 658-706.
40. B. Muha, S. Canic, *Existence of a weak solution to a nonlinear fluid-*

structure interaction problem modeling the flow of an incompressible, viscous fluid in a cylinder with deformable walls, Arch. Rational Mech. Anal., **207**, 3, 2013, 919-968.
41. B. Muha, S. Canic, *A Nonlinear, 3D Fluid-Structure Interaction Problem Driven by the Time-Dependent Dynamic Pressure Data: A Constructive Existence Proof*, CIS, **13**, 3, (2013), 357-397.
42. T. Nomura and T.J. Hughes, *An arbitrary Lagrangian-Eulerian finite element method for interaction of fluid and rigid body*, Comput. Methods Applied Mech. Engineering, **95**, (1992), 115-138.
43. S. Piperno and C. Farhat, *Design of efficient partitioned procedures for the transient solution of aeroelastic problems*, Rev. Eur. Elem. Finis, **9**, 6-7, (2001), 655-680.
44. J.-P. Raymond and M. Vanninathan, *A fluid-structure model coupling the Navier-Stokes equations and the Lamé system*, J. Math. Pures Appl., **102**, (2014), 546-596.
45. T. Richter and T. Wick, *Optimal Control and Parameter Estimation for Stationary Fluid-Structure Interaction Problems*, SIAM J. Sci. Comput., **35**, 5, (2013), 1085-1104.
46. K. G. Van Der Zee, E.H. Van Brummelen, and R. De Borst, *Goal-Oriented Error Estimation and Adaptivity for Free-Boundary Problems: The Domain-Map Linearization Approach*, SIAM J. Sci. Comput., **32**, 2, (2010), 1064-1092.
47. K. G. Van Der Zee, E.H. Van Brummelen, and R. De Borst, *Goal-Oriented Error Estimation and Adaptivity for Free-Boundary Problems: The Shape-Linearization Approach*, SIAM J. Sci. Comput., **32**, 2, (2010), 1093-1118.
48. T. Wick and W. Wollner, *On the differentiability of fluid-structure interaction problems with respect to the problem data*, RICAM-Report, **16**, (2014).

Non-smooth regularization of a forward-backward parabolic equation

Elena Bonetti

Dipartimento di Matematica "F. Enriques", Università degli Studi di Milano
Via Saldini 50, 20133 Milano, Italy
E-mail: elena.bonetti@unimi.it

Pierluigi Colli

Dipartimento di Matematica "F. Casorati", Università di Pavia
Via Ferrata 5, 27100 Pavia, Italy
E-mail: pierluigi.colli@unipv.it

Giuseppe Tomassetti

Dipartimento di Ingegneria Civile, Università di Roma "Tor Vergata"
Via di Tor Vergata 110, 00133 Roma, Italy
E-mail: tomassetti@ing.uniroma2.it

In this note we review some results recently obtained for a model which describes diffusion of species using a suitable regularization of a "forward-backward" parabolic equation. Existence and uniqueness of solutions, as well as continuous dependence on data, are derived for a system of partial differential equations, which renders the evolution of physical quantities as concentration and chemical potential.

Keywords: Diffusion of species; Forward-backward parabolic equation; Non-smooth regularization

1. The model and the problem

This contribution is focused on a model applying to several physical situations and dealing with diffusion of different species in solids, described in terms of concentration. As a possible application, which was by the way our starting point, we can mention the hydrogen storage in metals (cf. Refs. 1,3,22).

In the following, we move from the classical approach in Thermodynamics leading to the well-known Cahn–Hilliard equation, and introduce the model making some comments on its thermodynamical consistency.

1.1. *Models for diffusion and phase separation*

A mathematical model describing species diffusion in solids may be obtained by combining three ingredients

- the *mass-balance law* for the *concentration* u:

$$\dot{u} + \operatorname{div} \mathbf{h} = 0; \tag{1}$$

- a *constitutive relation* involving the *flux of diffusant* $\mathbf{h}$ and the gradient of *chemical-potential* μ:

$$\mathbf{h} = -m\nabla\mu, \qquad m > 0 \ \textit{mobility constant}; \tag{2}$$

- a possibly nonlinear relation between μ and u:

$$\mu = \psi'(u) \tag{3}$$

  through the derivative ψ' of a *coarse-grain free energy* ψ.

In presence of *phase separation*, the function ψ is assumed to be non-convex and a typical choice is the standard *double-well potential*:

$$\psi(u) = k(u-1)^2u^2, \qquad k > 0. \tag{4}$$

With this choice the system (1)–(3) turns out *backward parabolic* in the region where $0 < u < 1$, an undesirable feature which calls for a suitable regularization. To this aim, the usual approach is the following *elliptic regularization* of (3):

$$\mu = \psi'(u) - a\Delta u, \qquad a > 0, \tag{5}$$

which results into the celebrated *Cahn–Hilliard system*[4,14]. Other choices have been considered and the literature on this subject is very rich. Among the others, Novick-Cohen and Pego[20] investigated the *viscous regularization*

$$\mu = \psi'(u) + \alpha\dot{u}, \qquad \alpha > 0, \tag{6}$$

whereas a combination of energetic and viscous regularization leads to the so-called *viscous Cahn–Hilliard equation*:

$$\mu = \psi'(u) - a\Delta u + \alpha\dot{u},$$

derived by Novick-Cohen[19] and analytically investigated in Refs. 12 and 13. Other recent contributions (see Refs. 5,7,9,16) deal with Cahn–Hilliard or viscous Cahn–Hilliard equations with singular potentials. Finally, Refs. 6,10,11,18,24,25 approach the optimal control problems for some Cahn–Hilliard systems, possibly including dynamic boundary conditions. We also point out Ref. 8, in which a wide class of evolution equations is obtained as asymptotic limit of a Cahn–Hilliard system.

1.2. *Cahn–Hilliard equation from a microforce balance*

Let us discuss the derivation of diffusion models proposed by Gurtin[17] and Podio-Guidugli[21]. The main ingredients for this derivation are: (i) a system of *microforces* (distinct from Newtonian forces) obeying their own balance laws and associated to the evolution of the observable fields of interest; (ii) the constitutive prescriptions relating microforces to the actual evolution of the system; (iii) a dissipation principle, which excludes thermodynamically inconsistent prescriptions.

Within this common framework, the balance statement is written for the vectorial *microstress* $\boldsymbol{\xi}$ and two scalar fields, π and γ denoting, respectively, the *internal* and the *external microforce*:

$$\operatorname{div}\boldsymbol{\xi} + \pi + \gamma = 0. \tag{7}$$

Note that (7) yields the elliptic regularization (5) when adopting the constitutive prescriptions

$$\boldsymbol{\xi} = a\nabla u, \qquad \pi = \mu - \psi(u), \qquad \gamma = 0. \tag{8}$$

Other constitutive prescriptions may be considered, provided that the thermodynamical consistency is satisfied. In this respect, two distinct options are proposed in Ref. 17 and in Ref. 21. In this note we follow the former, based on the dissipation inequality

$$\dot{\phi} \leq (\mu - \pi)\dot{u} + \boldsymbol{\xi}\cdot\nabla\dot{u} - \mathbf{h}\cdot\nabla\mu,$$

where ϕ denotes the *free-energy density*. Here, the free energy and the concentration fields are related through the prescription $\phi = \widehat{\phi}(u, \nabla u)$, thus assuming independence of the free energy of the time derivative of u. If we introduce the *equilibrium parts*

$$\boldsymbol{\xi}^{\mathrm{eq}} := \frac{\partial\widehat{\phi}}{\partial\nabla u}(u,\nabla u) \qquad \text{and} \qquad \pi^{\mathrm{eq}} := \mu - \frac{\partial\widehat{\phi}}{\partial u}(u,\nabla u), \tag{9}$$

of microstress and internal microforce, the test for consistency of a certain constitutive choice reduces to verify that, for whatever process, the *non-equilibrium parts*

$$\boldsymbol{\xi}^{\mathrm{ne}} := \boldsymbol{\xi} - \boldsymbol{\xi}^{\mathrm{eq}}, \qquad \pi^{\mathrm{ne}} := \pi - \pi^{\mathrm{eq}} \tag{10}$$

satisfy, together with the flux of diffusant $\mathbf{h}$, the *reduced dissipation inequality*:

$$0 \leq -\pi^{\rm ne}\dot{u} + \boldsymbol{\xi}^{\rm ne} \cdot \nabla\dot{u} - \mathbf{h} \cdot \nabla\mu. \tag{11}$$

In particular, the constitutive prescriptions (8) comply with (9)–(10) provided we take

$$\widehat{\phi}(u, \nabla u) = \frac{1}{2}a|\nabla u|^2 + \psi(u),$$

and choose, consistently with (11),

$$\boldsymbol{\xi}^{\rm ne} = \mathbf{0}, \qquad \pi^{\rm ne} = 0.$$

1.3. *Viscous and "non smooth" regularizations*

In order to deduce the viscous variant (6) we can adopt a similar approach: first, we exclude microscopic contact interactions by letting $\boldsymbol{\xi} = \mathbf{0}$, and we set γ to null, so that the microforce balance (7) reduces to

$$\pi = 0. \tag{12}$$

Second, we remove the dependence of free energy on concentration gradient by letting

$$\widehat{\phi}(u, \nabla u) = \psi(u). \tag{13}$$

Hence, the second of (9) specializes to

$$\pi^{\rm eq} = \mu - \psi'(u); \tag{14}$$

As to the "non-equilibrium part", the simplest constitutive choice is

$$\pi^{\rm ne} = -\alpha\dot{u}, \qquad \alpha > 0, \tag{15}$$

so that, in view of (10), we infer (6) from (12) and (14). We point out that (6) may be obtained with the help of a "dissipation functional" Φ: the so-called pseudo-potential of dissipation (see, e.g., Ref. 15) that is non-negative, convex and takes the minimum value zero for a null dissipation. In particular, the non-equilibrium part $\pi^{\rm ne}$ is viewed as

$$\pi^{\rm ne} = -\frac{\partial\Phi}{\partial\dot{u}} \tag{16}$$

when we prescribe $\Phi(\dot{u}) = (\alpha/2)|\dot{u}|^2$. Note in particular that the properties of Φ and (16) ensure that

$$\pi^{\rm ne}\dot{u} \leq 0, \tag{17}$$

and this inequality is necessary for the thermodynamical consistency. In fact, if (2) and (2) hold, then the reduced dissipation inequality (11) is fulfilled. Moreover, (15) can be generalized by a suitable choice of Φ: actually, in place of (15) we consider (cf. Ref. 23)

$$-(\pi^{\mathrm{ne}} + \alpha \dot{u}) \in \beta(\dot{u}), \tag{18}$$

where, as before, the coefficient α is positive and now β denotes a maximal monotone graph in $\mathbb{R} \times \mathbb{R}$ such that $0 \in \beta(0)$. In the case, β turns out to be the subdifferential $\partial\zeta$ of a non-negative convex function ζ, with $\zeta(0) = 0$, which can be included in the dissipation functional Φ. The reader can check that the choice (18) is consistent with the dissipation inequality (11).

1.4. *The system under investigation*

We aim to set the equations and conditions of the system.

First, we recall the balance equation (1) along with the constitutive prescription (2) for the flux of diffusant. Consequently, we obtain the partial differential equation

$$\dot{u} = m\Delta\mu\,, \tag{19}$$

to be satisfied in the cylindrical domain $\Omega \times (0, T)$, where Ω denotes a smooth bounded open set in $\mathbb{R}^3$ with smooth boundary Γ and $T > 0$ stands for some final time.

Then, we invoke the microforce balance, in its form (12), and infer that (see (10) and let us use the form of the inclusion)

$$\pi^{\mathrm{eq}} + \pi^{\mathrm{ne}} \ni 0.$$

In the light of prescriptions (14) and (18), we obtain

$$\mu \in \psi'(u) + \alpha\dot{u} + \beta(\dot{u}) \quad \text{pointwise in } \Omega \times (0, T). \tag{20}$$

Next, due to the second order operator in (19), we need to prescribe a boundary condition for μ. We assume a Dirichlet-type boundary condition, with a (time-dependent) boundary value of the chemical potential:

$$\mu(\cdot, t) = \mu_\flat(\cdot, t) \quad \text{on } \Gamma \times (0, T). \tag{21}$$

The reader can find a detailed discussion on the motivations to choose a Dirichlet boundary condition in Ref. 2, which is our basic reference for the present contribution.

We finally complete the system (19)–(20) with a prescription of the initial concentration: $u(\cdot, 0) = u_0$ in Ω.

1.5. *About nonlinearities of the system*

The requirements we set on the coarse-grain free energy ψ regard the C^2 smoothness on some open interval (possibly unbounded) of $\mathbb{R}$, the non-negativity, the convexity up to a concave (and at most quadratic) perturbation, and the singularity of the derivative ψ' at the extrema of the interval. A typical example is of course (4); another interesting case which works in a bounded interval is the logarithmic potential

$$\psi(u) = u \ln u + (1-u)\ln(1-u)) + k_1 u(1-u) + k_2\,, \quad u \in (-1,1), \quad (22)$$

where k_1 and k_2 denote positive constants large enough in order to break the convexity (k_1) and to make ψ non-negative (k_2). Note in particular that we ask ψ to be nonconvex: this is a feature that allows for phase separation.

The choice of the mapping β and that of the constant α in (20) affect the dissipative structure of the system. As to β, we assume that it is the subdifferential of a non-negative, convex, lower semicontinous potential ζ with minimum value $\zeta(0) = 0$. We do not set any restriction on the growth of ζ. Possible choices of β are

(1) the first simple choice is $\beta \equiv 0$ and leads to the PDE system studied by Novick Cohen and Pego[20];

(2) a second interesting choice, which allows for hysteresis (see Refs. 2,23), is $\beta(r) = \beta_0 \,\mathrm{sign}(r)$, where $\beta_0 > 0$ and the sign graph is defined by

$$\mathrm{sign}(r) = \begin{cases} \{+1\} & \text{if } r > 0, \\ [-1,+1] & \text{if } r = 0, \\ \{-1\} & \text{if } r < 0. \end{cases} \quad (23)$$

This β contains a vertical segment in correspondence of the value 0;

(3) the subdifferential of the indicator function $I_{[r_*,r^*]}$ of a closed bounded interval $[r_*, r^*]$, that is

$$\beta(r) = \partial I_{[r_*,r^*]}(r) = \begin{cases} \{0\} & \text{if } r \in (r_*, r^*), \\ [0,+\infty) & \text{if } r = r^*, \\ (-\infty, 0] & \text{if } r = r_*. \end{cases} \quad (24)$$

We remark that this choice of β forces the rate of change of concentration to live in the interval $[r_*, r^*]$. In our setting we require that $r_* \leq 0 \leq r^*$.

(4) the subdifferential of another indicator function (e.g. of $[0,+\infty)$):

$$\beta(r) = \partial I_{[0,+\infty)}(r) = \begin{cases} \{0\} & \text{if } r \in (0,+\infty), \\ (-\infty, 0] & \text{if } r = 0. \end{cases} \tag{25}$$

The presence of the related graph in (20) entails that the concentration at all points never decreases, that is, $\dot{u}$ always remains non-negative.

1.6. *Reduction to the homogeneous Dirichlet problem*

Recalling (21), it turns out that a time-dependent chemical potential $\mu_\flat(x,t)$ is given for all $x \in \Gamma$ at all times $t \in [0,T]$. Then, at each particular time we can harmonically extend $\mu_\flat$ to the interior of Ω (still denoting by $\mu_\flat$ the harmonic extension). Moreover, assuming that all energies densities per unit volume are renormalized to a reference value (and so are dimensionless), we introduce the characteristic time and lengthscale $T_0 = \alpha$ and $L_0 = \sqrt{\frac{m}{\alpha}}$, as well as the new variables and functions:

$$\widetilde{t} = \frac{t}{T_0}, \qquad \widetilde{x} = \frac{x}{L_0}, \qquad \widetilde{\mu} = \mu - \mu_\flat, \qquad \widetilde{\beta}(r) = \beta(r/T_0).$$

In terms of these new variables and, for the sake of convenience, dropping tildas, we obtain the following system of equations

$$\left.\begin{array}{l} \partial_t u = \Delta \mu, \\ \mu = \partial_t u + \xi + \mu_\flat + \psi'(u) \\ \xi \in \beta(\partial_t u) \end{array}\right\} \quad \text{in } \Omega \times (0,T) \tag{26}$$

with the boundary condition

$$\mu = 0 \quad \text{on } \Gamma \times (0,T) \tag{27}$$

and the initial condition

$$u(\cdot, 0) = u_0 \quad \text{in } \Omega. \tag{28}$$

In the next section, we state precisely the assumptions on the data of the problem and point out the main results (cf. Ref. 2): existence, uniqueness and continuous dependence of the solution with respect to the data of the problem.

2. The results

We consider the functional spaces

$$H := L^2(\Omega), \qquad V := H_0^1(\Omega), \qquad W := H^2(\Omega) \cap H_0^1(\Omega)$$

and endow them with their usual scalar products and norms. The norms $\|v\|_V$ and $\|\nabla v\|_H$ are equivalent for $v \in V$, thanks to the Poincaré inequality. In addition, let us point out that, after identifying H with its dual, the triplet (V, H, V') is a Hilbert triplet. We can use the notation $\langle \cdot, \cdot \rangle$ for the duality pairing between V' and V. We also set $Q := \Omega \times (0,T)$ and $\Sigma := \Gamma \times (0,T)$.

2.1. *Structural assumptions*

As for the data of the problem, we let the domain of ψ be the open interval $(a,b) \subseteq \mathbb{R}$, where a may be $-\infty$ and b may be $+\infty$, and require that

$$\psi \in C^2(a,b), \qquad \psi(r) \geq 0 \quad \text{for all } r \in (a,b), \tag{29}$$

$$\lim_{r \to a^+} \psi'(r) = -\infty, \qquad \lim_{r \to b^-} \psi'(r) = +\infty, \tag{30}$$

$$\psi''(r) \geq -K_1 \quad \text{for all } r \in (a,b), \tag{31}$$

for some positive constant K_1; we point out that in (30) a^+ becomes $-\infty$ if $a = -\infty$ and b^- reduces to $+\infty$ if $b = +\infty$. The initial concentration u_0 is assumed to satisfy $u_0 \in H$ and, in addition,

$$a_0 \leq u_0(x) \leq b_0 \quad \text{for some } a_0 > a, \ b_0 < b \text{ and for a.a. } x \in \Omega, \tag{32}$$

whence both u_0 and $\psi'(u_0)$ belong to $L^\infty(\Omega)$. For the harmonic extension $\mu_\flat$, we simply require that

$$\mu_\flat \in H^1(0,T;H) \cap L^\infty(Q). \tag{33}$$

Finally, as to the (possibly) multi-valued mapping β, we let $\beta = \partial\zeta$, with

$$\zeta : \mathbb{R} \to [0, +\infty] \text{ convex and lower-semicontinuous, } \zeta(0) = 0. \tag{34}$$

2.2. *Definition of solution*

A weak formulation of the problem is given in the set of the Hilbert triplet (V, H, V') as follows. We introduce the operator $A : V \to V'$, corresponding to the "weak realization" of the Laplace operator with homogenous Dirichlet conditions:

$$\langle A v_1, v_2 \rangle := \int_\Omega \nabla v_1 \cdot \nabla v_2 \, \mathrm{d}x, \quad v_1, v_2 \in V.$$

Definition 2.1. We say that a triplet (u, ξ, μ) is a weak solution to the problem (26)–(28) if

$$u \in C^1([0,T];H), \quad \xi \in C^0([0,T];H), \quad \mu \in C^0([0,T];V), \tag{35}$$

$$\psi'(u) \in H^1(0,T;H) \tag{36}$$

and the following equations are satisfied:

$$\partial_t u(t) + A\mu(t) = 0 \quad \text{in } V', \quad \text{for all } t \in [0,T], \tag{37}$$

$$\mu(t) = (\partial_t u + \xi + \mu_\flat + \psi'(u))(t) \quad \text{a.e. in } \Omega, \quad \text{for all } t \in [0,T], \tag{38}$$

$$\xi(t) \in \beta(\partial_t u(t)) \quad \text{a.e. in } \Omega, \quad \text{for all } t \in [0,T], \tag{39}$$

$$u(0) = u_0 \quad \text{a.e. in } \Omega. \tag{40}$$

Remark 2.1. If we denote by I the identity operator, it is not difficult to verify that the inverse graph $(I+\beta)^{-1}$ is a Lipschitz continuous function (with 1 as possible Lipschitz constant). Then, equation (38) and inclusion (39) imply that

$$\begin{aligned} \partial_t u(t) = (I+\beta)^{-1}\bigl(\mu(t) - \mu_\flat(t) - \psi'(u(t))\bigr) \\ \text{a.e. in } \Omega, \quad \text{for all } t \in [0,T]. \end{aligned} \tag{41}$$

Now, observe that (41) and (37) can be combined into

$$\begin{aligned} (I+\beta)^{-1}\bigl(\mu(t) - \mu_\flat(t) - \psi'(u(t))\bigr) + A\mu(t) = 0 \\ \text{in } V', \quad \text{for all } t \in [0,T], \end{aligned} \tag{42}$$

which is an elliptic equation, at any fixed time $t \in [0,T]$, in terms of the unknown $\mu(t)$; nonetheless, the other unknown $u(t)$ is present in the nonlinear contribution $\psi'(u(t))$. Thus, (42) gives a hint to solve the evolution problem, which is however not straightforward at all.

2.3. *Well-posedness*

Here is the statement of our main result, whose detailed proof is given in Ref. 2.

Theorem 2.1. *Under the assumptions* (29)–(34)*, there exists a unique solution to the problem* (26)–(28)*. Moreover, there holds*

$$u \in L^\infty(Q) \tag{43}$$

and the solution is strong, that is, $\mu \in C^0([0,T];W)$ *and equation* (37) *can be replaced by*

$$\partial_t u - \Delta\mu = 0 \quad \textit{a.e. in } Q. \tag{44}$$

Furthermore, if (u_1,ξ_1,μ_1), (u_2,ξ_2,μ_2) *denote two solutions corresponding to the respective data* $(u_{0i},\mu_{\flat i})$, $i=1,2$, *then the difference satisfies*

$$\begin{aligned} &\|u_1 - u_2\|_{C^1([0,T];H)} + \|\xi_1 - \xi_2\|_{C^0([0,T];H)} + \|\mu_1 - \mu_2\|_{C^0([0,T];W)} \\ &\le R\bigl(\|\mu_{\flat 1} - \mu_{\flat 2}\|_{C^0([0,T];H)} + \|u_{01} - u_{02}\|_H\bigr) \end{aligned} \tag{45}$$

for some constant R *depending only on the structural assumptions.*

References

1. E. Bonetti, P. Colli, and P. Laurençot, Global existence for a hydrogen storage model with full energy balance, *Nonlinear Anal.* **75** (2012) 3558-3573.
2. E. Bonetti, P. Colli and G. Tomassetti, A non-smooth regularization of a forward-backward parabolic equation, preprint arXiv:1508.03225 [math.AP] (2015), pp. 1-19.
3. E. Bonetti, M. Frémond, and C. Lexcellent, Hydrogen storage: modeling and analytical results, *Appl. Math. Optim.* **55** (2007) 31-59.
4. J. W. Cahn and J. E. Hilliard, Free energy of a nonuniform system I. Interfacial free energy, *J. Chem. Phys.* **2** (1958) 258-267.
5. L. Cherfils, A. Miranville and S. Zelik, The Cahn–Hilliard equation with logarithmic potentials, *Milan J. Math.* **79** (2011) 561-596.
6. P. Colli, M. H. Farshbaf-Shaker, G. Gilardi and J. Sprekels, Optimal boundary control of a viscous Cahn–Hilliard system with dynamic boundary condition and double obstacle potentials, *SIAM J. Control Optim.* **53** (2015) 2696-2721.
7. P. Colli and T. Fukao, Cahn–Hilliard equation with dynamic boundary conditions and mass constraint on the boundary, *J. Math. Anal. Appl.* **429** (2015) 1190-1213.
8. P. Colli and T. Fukao, Nonlinear diffusion equations as asymptotic limits of Cahn–Hilliard systems, *J. Differential Equations* **260** (2016) 6930-6959.
9. P. Colli, G. Gilardi and J. Sprekels, On the Cahn–Hilliard equation with dynamic boundary conditions and a dominating boundary potential, *J. Math. Anal. Appl.* **419** (2014) 972-994.
10. P. Colli, G. Gilardi and J. Sprekels, A boundary control problem for the pure Cahn–Hilliard equation with dynamic boundary conditions, *Adv. Nonlinear Anal.* **4** (2015) 311-325.
11. P. Colli, G. Gilardi and J. Sprekels, A boundary control problem for the viscous Cahn-Hilliard equation with dynamic boundary conditions, *Appl. Math. Optim.* **73** (2016) 195-225.
12. C. M. Elliott and H. Garcke, On the Cahn-Hilliard equation with degenerate mobility, *SIAM J. Math. Anal.* **27** (1996) 404-423.
13. C. M. Elliott and A. M. Stuart, Viscous Cahn–Hilliard equation. II. Analysis, *J. Differential Equations* **128** (1996) 387-414.
14. C. M. Elliott and S. Zheng, On the Cahn–Hilliard equation, *Arch. Rational Mech. Anal.* **96** (1986) 339-357.

15. M. Frémond, "Non-smooth Thermomechanics", Springer-Verlag, Berlin, 2002.
16. G. Gilardi, A. Miranville and G. Schimperna, On the Cahn–Hilliard equation with irregular potentials and dynamic boundary conditions, *Commun. Pure Appl. Anal.* **8** (2009) 881-912.
17. M. E. Gurtin, Generalized Ginzburg-Landau and Cahn-Hilliard equations based on a microforce balance, *Phys. D* **92** (1996) 178-192.
18. M. Hintermüller and D. Wegner, Distributed optimal control of the Cahn–Hilliard system including the case of a double-obstacle homogeneous free energy density, *SIAM J. Control Optim.* **50** (2012) 388-418.
19. A. Novick-Cohen, On the viscous Cahn-Hilliard equation, in "Material instabilities in continuum mechanics (Edinburgh, 1985–1986)", *Oxford Sci. Publ.*, Oxford Univ. Press, New York, 1988, pp. 329–342.
20. A. Novick-Cohen and R. L. Pego, Stable patterns in a viscous diffusion equation, *Trans. Amer. Math. Soc.* **324** (1991) 331-351.
21. P. Podio-Guidugli, Models of phase segregation and diffusion of atomic species on a lattice, *Ric. Mat.* **55** (2006) 105-118.
22. T. Roubíček and G. Tomassetti, Thermomechanics of hydrogen storage in metallic hydrides: modeling and analysis, *Discrete Contin. Dyn. Syst. Ser. B* **19** (2014) 2313-2333.
23. G. Tomassetti, Some remarks on a viscous regularization of the nonlinear diffusion equation, preprint arXiv:1505.07993 [math.AP] (2015), pp. 1-22.
24. Q.-F. Wang and S.-I. Nakagiri, Optimal control of distributed parameter system given by Cahn–Hilliard equation, *Nonlinear Funct. Anal. Appl.* **19** (2014) 19-33.
25. J. Zheng and Y. Wang, Optimal control problem for Cahn–Hilliard equations with state constraint, *J. Dyn. Control Syst.* **21** (2015) 257-272.

Approaching monotone inclusion problems via second order dynamical systems with linear and anisotropic damping

Radu Ioan Boţ* and Ernö Robert Csetnek**

University of Vienna
Faculty of Mathematics
Oskar-Morgenstern-Platz 1, A-1090 Vienna, Austria
*E-mail: *radu.bot@univie.ac.at,*
***ernoe.robert.csetnek@univie.ac.at*

We investigate second order dynamical systems of the from $\ddot{x}(t) + \Gamma(\dot{x}(t)) + \lambda(t)B(x(t)) = 0$, where $\Gamma : \mathcal{H} \to \mathcal{H}$ is an elliptic bounded self-adjoint linear operator defined on a real Hilbert space $\mathcal{H}$, $B : \mathcal{H} \to \mathcal{H}$ is a cocoercive operator and $\lambda : [0, +\infty) \to (0, +\infty)$ is a relaxation function depending on time. We prove via Lyapunov analysis that the generated trajectories weakly converge to a zero of the operator B. This opens the gate towards approaching through a second order dynamical system the problem of determining zeros of the sum of a maximally monotone operator and a cocoercive one, which captures as particular case the minimization of the sum of a nonsmooth convex function with a smooth convex one. Finally, when B is the gradient of a smooth convex function, we prove a rate of $\mathcal{O}(1/t)$ for the convergence of the function values along the ergodic trajectory to its minimum value.

Keywords: Dynamical systems; Lyapunov analysis; Monotone inclusions; Convex optimization problems; Continuous forward-backward method.

1. Introduction and preliminaries

Let $\mathcal{H}$ a real Hilbert space endowed with inner product $\langle \cdot, \cdot \rangle$ and corresponding norm $\| \cdot \| = \sqrt{\langle \cdot, \cdot \rangle}$. The problem of approaching the minimization of a potential function $f : \mathcal{H} \to \mathbb{R}$, supposed to be convex and differentiable, has been considered by several authors see[3,7,8,11]). These investigations addressed either the convergence of the generated trajectories to a critical point of f or the convergence of the function along the trajectories to its global minimum value. We recall in this context the heavy ball with friction dynamical system

$$\ddot{x} + \gamma \dot{x} + \nabla f(x) = 0, \tag{1}$$

which is a nonlinear oscillator with constant damping parameter $\gamma > 0$. When $\mathcal{H} = \mathbb{R}^2$, this system describes the motion of a heavy ball that keeps rolling over the graph of the function f under its own inertia until friction stops it at a critical point of f (see[11]). The time discretization of (1) leads to so-called inertial-type algorithms, which are numerical schemes sharing the feature that the current iterate of the generated sequence is defined by making use of the previous two iterates (see, for instance,[3–5,17,19,21]).

The minimization of f over a nonempty, convex and closed set $C \subseteq \mathcal{H}$ has been approached in the same spirit in[7,8], by considering the gradient-projection second order dynamical system

$$\ddot{x} + \gamma\dot{x} + x - P_C(x - \eta\nabla f(x)) = 0, \tag{2}$$

where $P_C : \mathcal{H} \to C$ denotes the projection onto the set C and $\eta > 0$. These investigations have been further expanded in[8] to more general systems of the form

$$\ddot{x} + \gamma\dot{x} + x - Tx = 0, \tag{3}$$

where $T : \mathcal{H} \to \mathcal{H}$ is a nonexpansive operator. When $\gamma^2 > 2$ the trajectory of (3) has been shown to converge weakly to an element in the fixed points set of T, provided it is nonempty.

The dynamical system which we investigate in this paper reads

$$\begin{cases} \ddot{x}(t) + \Gamma(\dot{x}(t)) + \lambda(t)B(x(t)) = 0 \\ x(0) = u_0, \dot{x}(0) = v_0, \end{cases} \tag{4}$$

where $u_0, v_0 \in \mathcal{H}$, $\Gamma : \mathcal{H} \to \mathcal{H}$ is a bounded self-adjoint linear operator, which is γ-elliptic, that is, $\langle \Gamma u, u\rangle \geq \gamma\|u\|^2$ for all $u \in \mathcal{H}$ (with $\gamma > 0$) and $B : \mathcal{H} \to \mathcal{H}$ is a β-*cocoercive* operator for $\beta > 0$, that is $\beta\|Bx - By\|^2 \leq \langle x - y, Bx - By\rangle$ for all $x, y \in \mathcal{H}$. This obviously implies that B is $1/\beta$-Lipschitz continuous, that is $\|Bx - By\| \leq \frac{1}{\beta}\|Bx - By\|$ for all $x, y \in \mathcal{H}$. The elliptic operator Γ induces an anisotropic damping, a similar construction being already used in[3] in the context of minimizing a convex and smooth function.

The assumption which we make on the function λ is

(λ1) $\lambda : [0, +\infty) \to (0, +\infty)$ is locally absolutely continuous, monotonically increasing on $[0, +\infty)$ with $\lim_{t\to+\infty} \lambda(t) < \beta\gamma^2$.

According to the above assumption, $\dot{\lambda}(t)$ exists for almost every $t \geq 0$ and $\dot{\lambda}$ is Lebesgue integrable on each interval $[0, b]$ for $0 < b < +\infty$. Since λ is monotonically increasing, as λ is assumed to take only positive values, (λ1) yields the existence of a lower bound $\underline{\lambda}$ and a positive real number θ

such that for every $t \in [0, +\infty)$ one has

$$0 < \underline{\lambda} \leq \lambda(t) \leq \frac{\beta\gamma^2}{1+\theta}. \tag{5}$$

Thus, the existence and uniqueness of global solutions in $C^2([0, +\infty); \mathcal{H})$ for (4) follows from the global version of the Picard-Lindelöf Theorem.

We begin our investigations by showing that under mild assumptions on the relaxation function λ the trajectory $x(t)$ converges weakly as $t \to +\infty$ to a zero of the operator B, provided it has a nonempty set of zeros. Further, we approach the problem of finding a zero of the sum of a maximally monotone operator and a cocoercive one via a second order dynamical system formulated by making use of the resolvent of the set-valued operator, see (22). Dynamical systems of implicit type have been already considered in the literature in[1,2,9,12,14–16]. We further specialize these investigations to the minimization of the sum of a nonsmooth convex function with a smooth convex function one. This allows us to recover and improve results given in[7,8] in the context of studying the dynamical system (2). Whenever B is the gradient of a smooth convex function, we show that the function values of the latter converge along the ergodic trajectories generated by (4) to its minimum value with a rate of convergence of $\mathcal{O}(1/t)$.

2. Convergence of the trajectories

In this section we address the convergence properties of the trajectories generated by the dynamical system (4) and use to this end as essential ingredient the continuous version of the Opial Lemma that we state below (see, for example,[2] Lemma 5.3,[1] Lemma 1.10).

Lemma 2.1. *Let $S \subseteq \mathcal{H}$ be a nonempty set and $x : [0, +\infty) \to \mathcal{H}$ a given map. Assume that*

(i) for every $x^* \in S$, $\lim_{t\to+\infty} \|x(t) - x^*\|$ exists;

(ii) every weak sequential cluster point of the map x belongs to S.

Then there exists $x_\infty \in S$ such that $x(t)$ converges weakly to x_∞ as $t \to +\infty$.

We come now to the main result of this section.

Theorem 2.1. *Let $B : \mathcal{H} \to \mathcal{H}$ be a β-cocoercive operator for $\beta > 0$ such that $\operatorname{zer} B := \{u \in \mathcal{H} : Bu = 0\} \neq \emptyset$, $\Gamma : \mathcal{H} \to \mathcal{H}$ be a γ-elliptic operator, $\lambda : [0, +\infty) \to (0, +\infty)$ be a function fulfilling $(\lambda 1)$ and $u_0, v_0 \in \mathcal{H}$. Let*

$x \in C^2([0,+\infty);\mathcal{H})$ be the unique global solution of (4). Then the following statements are true:

(i) the trajectory x is bounded and $\dot{x}, \ddot{x}, Bx \in L^2([0,+\infty);\mathcal{H})$;

(ii) $\lim_{t\to+\infty} \dot{x}(t) = \lim_{t\to+\infty} \ddot{x}(t) = \lim_{t\to+\infty} B(x(t) = 0$;

(iii) $x(t)$ converges weakly to an element in $\operatorname{zer} B$ as $t \to +\infty$.

Proof. (i) Take an arbitrary $x^* \in \operatorname{zer} B$ and consider for every $t \in [0,+\infty)$ the function $h(t) = \frac{1}{2}\|x(t)-x^*\|^2$. We have $\dot{h}(t) = \langle x(t)-x^*, \dot{x}(t)\rangle$ and $\ddot{h}(t) = \|\dot{x}(t)\|^2 + \langle x(t)-x^*, \ddot{x}(t)\rangle$ for every $t \in [0,+\infty)$. Taking into account (4), we get for every $t \in [0,+\infty)$

$$\ddot{h}(t) + \gamma\dot{h}(t) + \\ \lambda(t)\langle x(t)-x^*, B(x(t))\rangle + \langle x(t)-x^*, \Gamma(\dot{x}(t)) - \gamma\dot{x}(t)\rangle = \|\dot{x}(t)\|^2. \tag{6}$$

Now we introduce the function $p : [0,+\infty) \to \mathbb{R}$,

$$p(t) = \frac{1}{2}\langle(\Gamma - \gamma\,\mathrm{Id})(x(t)-x^*), x(t)-x^*\rangle, \tag{7}$$

where Id denotes the identity on $\mathcal{H}$. Since $\langle(\Gamma-\gamma\,\mathrm{Id})\,u,u\rangle \geq 0$ for all $u \in \mathcal{H}$, it holds

$$p(t) \geq 0 \text{ for all } t \geq 0. \tag{8}$$

Moreover, $\dot{p}(t) = \langle(\Gamma-\gamma\,\mathrm{Id})(\dot{x}(t)), x(t)-x^*\rangle$, which combined with (6), the cocoercivity of B and the fact that $Bx^* = 0$ yields for every $t \in [0,+\infty)$

$$\ddot{h}(t) + \gamma\dot{h}(t) + \beta\lambda(t)\|B(x(t))\|^2 + \dot{p}(t) \leq \|\dot{x}(t)\|^2.$$

Taking into account (4) one obtains for every $t \in [0,+\infty)$

$$\ddot{h}(t) + \gamma\dot{h}(t) + \frac{\beta}{\lambda(t)}\|\ddot{x}(t) + \Gamma(\dot{x}(t))\|^2 + \dot{p}(t) \leq \|\dot{x}(t)\|^2,$$

hence

$$\ddot{h}(t) + \gamma\dot{h}(t) \\ + \frac{\beta}{\lambda(t)}\|\ddot{x}(t)\|^2 + \frac{2\beta}{\lambda(t)}\langle\ddot{x}(t), \Gamma(\dot{x}(t))\rangle + \frac{\beta}{\lambda(t)}\|\Gamma(\dot{x}(t))\|^2 + \dot{p}(t) \leq \|\dot{x}(t)\|^2. \tag{9}$$

We have

$$\gamma\|u\| \leq \|\Gamma u\| \text{ for all } u \in \mathcal{H}, \tag{10}$$

which combined with (9) yields for every $t \in [0,+\infty)$

$$\ddot{h}(t) + \gamma\dot{h}(t) + \dot{p}(t) \\ + \frac{\beta}{\lambda(t)}\frac{d}{dt}\big(\langle\dot{x}(t), \Gamma(\dot{x}(t))\rangle\big) + \left(\frac{\beta\gamma^2}{\lambda(t)} - 1\right)||\dot{x}(t)||^2 + \frac{\beta}{\lambda(t)}||\ddot{x}(t)||^2 \leq 0.$$

By taking into account that for almost every $t \in [0, +\infty)$

$$\frac{1}{\lambda(t)}\frac{d}{dt}\left(\langle \dot{x}(t), \Gamma(\dot{x}(t))\rangle \right) \geq \frac{d}{dt}\left(\frac{1}{\lambda(t)}\langle \dot{x}(t), \Gamma(\dot{x}(t))\rangle\right) + \gamma\frac{\dot{\lambda}(t)}{\lambda^2(t)}\|\dot{x}(t)\|^2, \tag{11}$$

we obtain for almost every $t \in [0, +\infty)$

$$\beta\frac{d}{dt}\left(\frac{1}{\lambda(t)}\langle \dot{x}(t), \Gamma(\dot{x}(t))\rangle\right) + \left(\frac{\beta\gamma^2}{\lambda(t)} + \beta\gamma\frac{\dot{\lambda}(t)}{\lambda^2(t)} - 1\right)||\dot{x}(t)||^2$$
$$+ \ddot{h}(t) + \gamma\dot{h}(t) + \dot{p}(t) + \frac{\beta}{\lambda(t)}||\ddot{x}(t)||^2 \leq 0. \tag{12}$$

By using now assumption $(\lambda 1)$ and (5) we obtain that the following inequality holds for almost every $t \in [0, +\infty)$

$$\ddot{h}(t)+\gamma\dot{h}(t)+\dot{p}(t)+\beta\frac{d}{dt}\left(\frac{1}{\lambda(t)}\langle \dot{x}(t), \Gamma(\dot{x}(t))\rangle\right)+\theta||\dot{x}(t)||^2+\frac{1+\theta}{\gamma^2}||\ddot{x}(t)||^2 \leq 0. \tag{13}$$

This implies that the function $t \mapsto \dot{h}(t) + \gamma h(t) + p(t) + \frac{\beta}{\lambda(t)}\langle \dot{x}(t), \Gamma(\dot{x}(t))\rangle$ is monotonically decreasing. Hence there exists a real number M such that for every $t \in [0, +\infty)$

$$\dot{h}(t) + \gamma h(t) + p(t) + \frac{\beta}{\lambda(t)}\langle \dot{x}(t), \Gamma(\dot{x}(t))\rangle \leq M, \tag{14}$$

which yields, together with (7) and $(\lambda 1)$, that for every $t \in [0, +\infty)$ $\dot{h}(t) + \gamma h(t) \leq M$. This implies that

$$h \text{ is bounded} \tag{15}$$

and, consequently,

$$\text{the trajectory } x \text{ is bounded.} \tag{16}$$

On the other hand, from (14), by taking into account (7), $(\lambda 1)$ and (5), it follows that for every $t \in [0, +\infty)$

$$\dot{h}(t) + \frac{1+\theta}{\gamma}\|\dot{x}(t)\|^2 \leq M,$$

hence $\langle x(t) - x^*, \dot{x}(t)\rangle + \frac{1+\theta}{\gamma}\|\dot{x}(t)\|^2 \leq M$. This inequality, in combination with (16), yields

$$\dot{x} \text{ is bounded,} \tag{17}$$

which further implies that

$$\dot{h} \text{ is bounded.} \tag{18}$$

Integrating the inequality (13) we obtain that there exists a real number $N \in \mathbb{R}$ such that for every $t \in [0, +\infty)$

$$\begin{aligned}\dot{h}(t) + \gamma h(t) + p(t) + \frac{\beta}{\lambda(t)} \langle \dot{x}(t), \Gamma(\dot{x}(t))\rangle \\ + \theta \int_0^t ||\dot{x}(s)||^2 ds + \frac{1+\theta}{\gamma^2} \int_0^t ||\ddot{x}(s)||^2 ds \leq N.\end{aligned}$$

From here, via (18) and (7), we conclude that $\dot{x}(\cdot), \ddot{x}(\cdot) \in L^2([0, +\infty); \mathcal{H})$. Finally, from (4) and ($\lambda 1$) we deduce $Bx \in L^2([0, +\infty); \mathcal{H})$ and the proof of (i) is complete.

(ii) For every $t \in [0, +\infty)$ it holds

$$\frac{d}{dt}\left(\frac{1}{2}\|\dot{x}(t)\|^2\right) = \langle \dot{x}(t), \ddot{x}(t)\rangle \leq \frac{1}{2}\|\dot{x}(t)\|^2 + \frac{1}{2}\|\ddot{x}(t)\|^2$$

and[2] Lemma 5.2 together with (i) lead to $\lim_{t\to+\infty} \dot{x}(t) = 0$.

Further, for every $t \in [0, +\infty)$ we have

$$\frac{d}{dt}\left(\frac{1}{2}\|B(x(t))\|^2\right) = \left\langle B(x(t)), \frac{d}{dt}(Bx(t))\right\rangle \leq \frac{1}{2}\|B(x(t))\|^2 + \frac{1}{2\beta^2}\|\dot{x}(t)\|^2.$$

By using again[2] Lemma 5.2 and (i) we get $\lim_{t\to+\infty} B(x(t)) = 0$, while the fact that $\lim_{t\to+\infty} \ddot{x}(t) = 0$ follows from (4), and ($\lambda 1$).

(iii) As seen in the proof of part (i), the function $t \mapsto \dot{h}(t) + \gamma h(t) + p(t) + \frac{\beta}{\lambda(t)} \langle \dot{x}(t), \Gamma(\dot{x}(t))\rangle$ is monotonically decreasing, thus from (i), (ii), (7), (Γ) and ($\lambda 1$) we deduce that $\lim_{t\to+\infty}(\gamma h(t) + p(t))$ exists and it is a real number.

In the following we consider the scalar product defined by $\langle\langle x, y\rangle\rangle = \frac{1}{\gamma}\langle \Gamma x, y\rangle$ and the corresponding induced norm $|||x|||^2 = \frac{1}{\gamma}\langle \Gamma x, x\rangle$. Taking into account the definition of p, we have that $\lim_{t\to+\infty} \frac{1}{2}|||x(t) - x^*|||^2$ exists and it is a real number.

Let $\overline{x}$ be a weak sequential cluster point of x, that is, there exists a sequence $t_n \to +\infty$ (as $n \to +\infty$) such that $(x(t_n))_{n\in\mathbb{N}}$ converges weakly to $\overline{x}$. Since B is a maximally monotone operator (see for instance[13] Example 20.28), its graph is sequentially closed with respect to the weak-strong topology of the product space $\mathcal{H} \times \mathcal{H}$. By using also that $\lim_{n\to+\infty} B(x(t_n)) = 0$, we conclude that $B\overline{x} = 0$, hence $\overline{x} \in \operatorname{zer} B$.

The conclusion follows by applying the Opial Lemma in the Hilbert space $(\mathcal{H}, (\langle\langle\cdot,\cdot\rangle\rangle))$, by noticing that a sequence $(x_n)_{n\geq 0}$ converges weakly to $\overline{x} \in \mathcal{H}$ in $(\mathcal{H}, (\langle\langle\cdot,\cdot\rangle\rangle))$ if and only if $(x_n)_{n\geq 0}$ converges weakly to $\overline{x}$ in $(\mathcal{H}, (\langle\cdot,\cdot\rangle))$. $\square$

A standard instance of a cocoercive operator defined on a real Hilbert spaces is the one that can be represented as $B = \mathrm{Id} - T$, where $T : \mathcal{H} \to \mathcal{H}$ is a *nonexpansive operator*, that is, a 1-Lipschitz continuous operator. It is easy to see that in this case B is $1/2$-cocoercive. For this particular choice of the operator B, the dynamical system (4) becomes

$$\begin{cases} \ddot{x}(t) + \Gamma(\dot{x}(t)) + \lambda(t)\big(x(t) - T(x(t))\big) = 0 \\ x(0) = u_0, \dot{x}(0) = v_0, \end{cases} \tag{19}$$

while assumption ($\lambda1$) reads

($\lambda2$) $\lambda : [0, +\infty) \to (0, +\infty)$ is locally absolutely continuous, monotonically increasing on $[0, +\infty)$ with $\lim_{t\to+\infty} \lambda(t) < \frac{\gamma^2}{2}$.

Theorem 2.1 gives rise to the following result.

Corollary 2.1. *Let $T : \mathcal{H} \to \mathcal{H}$ be a nonexpansive operator such that $\operatorname{Fix} T = \{u \in \mathcal{H} : Tu = u\} \neq \emptyset$, $\Gamma : \mathcal{H} \to \mathcal{H}$ be a γ-elliptic operator, $\lambda : [0, +\infty) \to (0, +\infty)$ be a function fulfilling ($\lambda2$) and $u_0, v_0 \in \mathcal{H}$. Let $x \in C^2([0, +\infty); \mathcal{H})$ be the unique global solution of (19). Then the following statements are true:*

(i) the trajectory x is bounded and $\dot{x}, \ddot{x}, (\mathrm{Id} - T)x \in L^2([0, +\infty); \mathcal{H})$;

(ii) $\lim_{t\to+\infty} \dot{x}(t) = \lim_{t\to+\infty} \ddot{x}(t) = \lim_{t\to+\infty}(\mathrm{Id} - T)(x(t)) = 0$;

(iii) $x(t)$ converges weakly to a point in $\operatorname{Fix} T$ as $t \to +\infty$.

Remark 2.1. Taking $\Gamma = \gamma \,\mathrm{Id}$ for $\gamma > 0$ and $\lambda(t) = 1$ for all $t \in [0, +\infty)$, the dynamical system (19) turns out to be

$$\begin{cases} \ddot{x}(t) + \gamma\dot{x}(t) + x(t) - T(x(t)) = 0 \\ x(0) = u_0, \dot{x}(0) = v_0. \end{cases} \tag{20}$$

The convergence of the trajectories generated by (20) has been studied in[8] Theorem 3.2 under the condition $\gamma^2 > 2$. In this case ($\lambda2$) is obviously fulfilled. However, different to[8], we allow in Corollary 2.1 an anisotropic damping through the use of the elliptic operator Γ and also a variable relaxation function λ depending on time.

Finally, we discuss a consequence of the above corollary applied to second order dynamical systems governed by averaged operators. The operator $R : \mathcal{H} \to \mathcal{H}$ is said to be *α-averaged* for $\alpha \in (0, 1)$, if there exists a nonexpansive operator $T : \mathcal{H} \to \mathcal{H}$ such that $R = (1 - \alpha)\,\mathrm{Id} + \alpha T$. An important representative of this class are the *firmly nonexpansive* operators which are obtained for $\alpha = \frac{1}{2}$.

We consider the dynamical system

$$\begin{cases} \ddot{x}(t) + \Gamma(\dot{x}(t)) + \lambda(t)\big(x(t) - R(x(t))\big) = 0 \\ x(0) = u_0, \dot{x}(0) = v_0 \end{cases} \tag{21}$$

and make the assumption

($\lambda 3$) $\lambda : [0, +\infty) \to (0, +\infty)$ is locally absolutely continuous, monotonically increasing on $[0, +\infty)$ with $\lim_{t\to+\infty} \lambda(t) < \frac{\gamma^2}{2\alpha}$.

Corollary 2.2. *Let $R : \mathcal{H} \to \mathcal{H}$ be an α-averaged operator for $\alpha \in (0,1)$ such that $\operatorname{Fix} R \neq \emptyset$, $\Gamma : \mathcal{H} \to \mathcal{H}$ be a γ-elliptic operator, $\lambda : [0,+\infty) \to (0,+\infty)$ be a function fulfilling ($\lambda 3$) and $u_0, v_0 \in \mathcal{H}$. Let $x \in C^2([0,+\infty);\mathcal{H})$ be the unique global solution of (21). Then the following statements are true:*

(i) the trajectory x is bounded and $\dot{x}, \ddot{x}, (\mathrm{Id} - R)x \in L^2([0,+\infty);\mathcal{H})$;
(ii) $\lim_{t\to+\infty} \dot{x}(t) = \lim_{t\to+\infty} \ddot{x}(t) = \lim_{t\to+\infty} (\mathrm{Id} - R)(x(t)) = 0$;
(iii) $x(t)$ converges weakly to a point in $\operatorname{Fix} R$ as $t \to +\infty$.

Proof. Since R is α-averaged, there exists a nonexpansive operator $T : \mathcal{H} \to \mathcal{H}$ such that $R = (1-\alpha)\,\mathrm{Id} + \alpha T$. Corollary 2.1 leads to the conclusion, by taking into account also $\operatorname{Fix} R = \operatorname{Fix} T$. □

3. Forward-backward second order dynamical systems

In this section we address the structured monotone inclusion problem

$$\text{solve } 0 \in A(x) + B(x),$$

where $A : \mathcal{H} \rightrightarrows \mathcal{H}$ is a maximally monotone operator and $B : \mathcal{H} \to \mathcal{H}$ is a β-cocoercive operator for $\beta > 0$ via a second-order forward-backward dynamical system with anisotropic damping and variable relaxation parameter.

For readers convenience we recall at the beginning some standard notions and results in monotone operator theory which will be used in the following (see also[13]). For an arbitrary set-valued operator $A : \mathcal{H} \rightrightarrows \mathcal{H}$ we denote by $\operatorname{Gr} A = \{(x,u) \in \mathcal{H} \times \mathcal{H} : u \in Ax\}$ its graph. We use also the notation $\operatorname{zer} A = \{x \in \mathcal{H} : 0 \in Ax\}$ for the set of zeros of A. We say that A is monotone, if $\langle x - y, u - v\rangle \geq 0$ for all $(x,u),(y,v) \in \operatorname{Gr} A$. A monotone operator A is said to be maximally monotone, if there exists no proper monotone extension of the graph of A on $\mathcal{H}\times\mathcal{H}$. The resolvent of A, $J_A : \mathcal{H} \rightrightarrows \mathcal{H}$, is defined by $J_A = (\mathrm{Id} + A)^{-1}$. If A is maximally monotone, then $J_A : \mathcal{H} \to \mathcal{H}$ is single-valued and maximally monotone (see[13] Proposition 23.7 and Corollary 23.10). For an arbitrary $\gamma > 0$ we have (see[13] Proposition 23.2) $p \in J_{\gamma A}x$ if and only if $(p, \gamma^{-1}(x - p)) \in \operatorname{Gr} A$.

The operator A is said to be uniformly monotone if there exists an increasing function $\phi_A : [0, +\infty) \to [0, +\infty]$ that vanishes only at 0 and fulfills $\langle x-y, u-v\rangle \geq \phi_A(\|x - y\|)$ for every $(x, u) \in \operatorname{Gr} A$ and $(y, v) \in \operatorname{Gr} A$. A popular class of operators having this property is the one of the strongly monotone operators. We say that A is γ-strongly monotone for $\gamma > 0$, if $\langle x - y, u - v\rangle \geq \gamma\|x - y\|^2$ for all $(x, u), (y, v) \in \operatorname{Gr} A$.

For $\eta \in (0, 2\beta)$, we approach the monotone inclusion problem to solve via the dynamical system

$$\begin{cases} \ddot{x}(t) + \Gamma(\dot{x}(t)) + \lambda(t)\Big[x(t) - J_{\eta A}\Big(x(t) - \eta B(x(t))\Big)\Big] = 0 \\ x(0) = u_0, \dot{x}(0) = v_0. \end{cases} \tag{22}$$

and make the following assumption for the relaxation function

(λ4) $\lambda : [0, +\infty) \to (0, +\infty)$ is locally absolutely continuous, monotonically increasing on $[0, +\infty)$ with $\lim_{t\to+\infty} \lambda(t) < \frac{(4\beta-\eta)\gamma^2}{4\beta}$.

Theorem 3.1. *Let $A : \mathcal{H} \rightrightarrows \mathcal{H}$ be a maximally monotone operator and $B : \mathcal{H} \to \mathcal{H}$ be β-cocoercive operator for $\beta > 0$ such that* $\operatorname{zer}(A + B) \neq \emptyset$. *Let $\eta \in (0, 2\beta)$, $\Gamma : \mathcal{H} \to \mathcal{H}$ be a γ-elliptic operator, $\lambda : [0, +\infty) \to (0, +\infty)$ be a function fulfilling (λ4), $u_0, v_0 \in \mathcal{H}$ and $x \in C^2([0, +\infty); \mathcal{H})$ be the unique global solution of* (22). *Then the following statements are true:*

(i) the trajectory x is bounded and $\dot{x}, \ddot{x}, \big(\operatorname{Id} - J_{\eta A} \circ (\operatorname{Id} - \eta B)\big)x \in L^2([0, +\infty); \mathcal{H})$;

(ii) $\lim_{t\to+\infty} \dot{x}(t) = \lim_{t\to+\infty} \ddot{x}(t) = 0$;

(iii) $x(t)$ converges weakly to a point in $\operatorname{zer}(A + B)$ *as $t \to +\infty$;*

(iv) if $x^* \in$ $\operatorname{zer}(A + B)$, *then $B(x(\cdot)) - Bx^* \in L^2([0, +\infty); \mathcal{H})$,* $\lim_{t\to+\infty} B(x(t)) = Bx^*$ *and B is constant on* $\operatorname{zer}(A + B)$;

(v) if A or B is uniformly monotone, then $x(t)$ converges strongly to the unique point in $\operatorname{zer}(A + B)$ *as $t \to +\infty$.*

Proof. (i)-(iii) It is immediate that the dynamical system (22) can be written in the form

$$\begin{cases} \ddot{x}(t) + \Gamma(\dot{x}(t)) + \lambda(t)\big(x(t) - R(x(t))\big) = 0 \\ x(0) = u_0, \dot{x}(0) = v_0, \end{cases} \tag{23}$$

where $R = J_{\eta A} \circ (\operatorname{Id} - \eta B)$. According to[13] Corollary 23.8 and Remark 4.24(iii), $J_{\eta A}$ is 1/2-cocoercive. Moreover, by[13] Proposition 4.33, $\operatorname{Id} - \eta B$

is $\eta/(2\beta)$-averaged. Combining this with[20] Theorem 3(b), we derive that R is $\frac{2\beta}{4\beta-\eta}$-averaged. The statements (i)-(iii) follow now from Corollary 2.2 by noticing that $\operatorname{Fix} R = \operatorname{zer}(A+B)$ (see[13] Proposition 25.1(iv)).

(iv) The fact that B is constant on $\operatorname{zer}(A+B)$ follows from the cocoercivity of B and the monotonicity of A.

Take an arbitrary $x^* \in \operatorname{zer}(A+B)$. From the definition of the resolvent we have for every $t \in [0,+\infty)$

$$-B(x(t)) - \frac{1}{\eta\lambda(t)}\ddot{x}(t) - \frac{1}{\eta\lambda(t)}\Gamma(\dot{x}(t)) \in A\left(\frac{1}{\lambda(t)}\ddot{x}(t) + \frac{1}{\lambda(t)}\Gamma(\dot{x}(t)) + x(t)\right), \tag{24}$$

which combined with $-Bx^* \in Ax^*$ and the monotonicity of A leads to

$$\begin{aligned} 0 \le & \left\langle \frac{1}{\lambda(t)}\ddot{x}(t) + \frac{1}{\lambda(t)}\Gamma(\dot{x}(t)) + x(t) - x^*, -B(x(t)) + Bx^* \right\rangle \\ & + \left\langle \frac{1}{\lambda(t)}\ddot{x}(t) + \frac{1}{\lambda(t)}\Gamma(\dot{x}(t)) + x(t) - x^*, -\frac{1}{\eta\lambda(t)}\ddot{x}(t) - \frac{1}{\eta\lambda(t)}\Gamma(\dot{x}(t)) \right\rangle. \end{aligned} \tag{25}$$

After using the cocoercivity of B we obtain for every $t \in [0,+\infty)$

$$\begin{aligned} \beta\|B(x(t)) - Bx^*\|^2 \le & \frac{1}{2\beta}\left\|\frac{1}{\lambda(t)}\ddot{x}(t) + \frac{1}{\lambda(t)}\Gamma(\dot{x}(t))\right\|^2 + \frac{\beta}{2}\|B(x(t)) - Bx^*\|^2 \\ & + \left\langle x(t) - x^*, -\frac{1}{\eta\lambda(t)}\ddot{x}(t) - \frac{1}{\eta\lambda(t)}\Gamma(\dot{x}(t)) \right\rangle. \end{aligned}$$

For evaluating the last term of the above inequality we consider the functions $h : [0,+\infty) \to \mathbb{R}, h(t) = \frac{1}{2}\|x(t) - x^*\|^2$ and $p : [0,+\infty) \to \mathbb{R}$, $p(t) = \frac{1}{2}\langle (\Gamma - \gamma \operatorname{Id})(x(t) - x^*), x(t) - x^*\rangle$, already used in the proof of Theorem 2.1. For every $t \in [0,+\infty)$ we have $\langle x(t) - x^*, \ddot{x}(t)\rangle = \ddot{h}(t) - \|\dot{x}(t)\|^2$ and $\dot{p}(t) = \langle x(t) - x^*, \Gamma(\dot{x}(t))\rangle - \gamma\langle x(t) - x^*, \dot{x}(t)\rangle = \langle x(t) - x^*, \Gamma(\dot{x}(t))\rangle - \gamma\dot{h}(t)$, hence

$$\begin{aligned} & \left\langle x(t) - x^*, -\frac{1}{\eta\lambda(t)}\ddot{x}(t) - \frac{1}{\eta\lambda(t)}\Gamma(\dot{x}(t)) \right\rangle \\ & = -\frac{1}{\eta\lambda(t)}\left(\ddot{h}(t) + \gamma\dot{h}(t) + \dot{p}(t) - \|\dot{x}(t)\|^2\right). \end{aligned} \tag{26}$$

From here, for every $t \in [0,+\infty)$ it holds

$$\begin{aligned} \frac{\beta}{2}\|B(x(t)) - Bx^*\|^2 \le & \frac{1}{2\beta}\left\|\frac{1}{\lambda(t)}\ddot{x}(t) + \frac{1}{\lambda(t)}\Gamma(\dot{x}(t))\right\|^2 \\ & - \frac{1}{\eta\lambda(t)}\left(\ddot{h}(t) + \gamma\dot{h}(t) + \dot{p}(t) - \|\dot{x}(t)\|^2\right). \end{aligned} \tag{27}$$

By taking into account ($\lambda 4$) we obtain a lower bound $\underline{\lambda}$ such that for every $t \in [0, +\infty)$ one has $0 < \underline{\lambda} \leq \lambda(t)$. By multiplying (27) with $\lambda(t)$ we obtain for every $t \in [0, +\infty)$ that

$$\frac{\beta\underline{\lambda}}{2}\|B(x(t)) - Bx^*\|^2 + \frac{1}{\eta}\left(\ddot{h}(t) + \gamma\dot{h}(t) + \dot{p}(t)\right)$$
$$\leq \frac{1}{2\beta\underline{\lambda}}\left\|\ddot{x}(t) + \Gamma(\dot{x}(t))\right\|^2 + \frac{1}{\eta}\|\dot{x}(t)\|^2.$$

After integration we obtain that for every $T \in [0, +\infty)$

$$\frac{\beta\underline{\lambda}}{2}\int_0^T \|B(x(t)) - Bx^*\|^2 dt$$
$$+ \frac{1}{\eta}\left(\dot{h}(T) - \dot{h}(0) + \gamma h(T) - \gamma h(0) + p(T) - p(0)\right)$$
$$\leq \int_0^T \left(\frac{1}{2\beta\underline{\lambda}}\left\|\ddot{x}(t) + \Gamma(\dot{x}(t))\right\|^2 + \frac{1}{\eta}\|\dot{x}(t)\|^2\right) dt.$$

As $\dot{x}, \ddot{x} \in L^2([0, +\infty); \mathcal{H})$, $h(T) \geq 0, p(T) \geq 0$ for every $T \in [0, +\infty)$ and $\lim_{T\to+\infty} \dot{h}(T) = 0$, it follows that $B(x(\cdot)) - Bx^* \in L^2([0, +\infty); \mathcal{H})$.

Further, we have

$$\frac{d}{dt}\left(\frac{1}{2}\|B(x(t)) - Bx^*\|^2\right) \leq \frac{1}{2}\|B(x(t)) - Bx^*\|^2 + \frac{1}{2\beta^2}\|\dot{x}(t)\|^2$$

hence, in light of[2] Lemma 5.2, it follows that $\lim_{t\to+\infty} B(x(t)) = Bx^*$.

(v) Let x^* be the unique element of $\operatorname{zer}(A + B)$. For the beginning we suppose that A is uniformly monotone with corresponding function $\phi_A : [0, +\infty) \to [0, +\infty]$, which is increasing and vanishes only at 0.

By similar arguments as in the proof of statement (iv), for almost every $t \in [0, +\infty)$ we have

$$\phi_A\left(\left\|\frac{1}{\lambda(t)}\ddot{x}(t) + \frac{1}{\lambda(t)}\Gamma(\dot{x}(t)) + x(t) - x^*\right\|\right)$$
$$\leq \left\langle \frac{1}{\lambda(t)}\ddot{x}(t) + \frac{1}{\lambda(t)}\Gamma(\dot{x}(t)) + x(t) - x^*, -B(x(t)) + Bx^*\right\rangle$$
$$+ \left\langle \frac{1}{\lambda(t)}\ddot{x}(t) + \frac{1}{\lambda(t)}\Gamma(\dot{x}(t)) + x(t) - x^*, -\frac{1}{\eta\lambda(t)}\ddot{x}(t) - \frac{1}{\eta\lambda(t)}\Gamma(\dot{x}(t))\right\rangle,$$

which combined with the inequality $\langle x(t)-x^*, B(x(t))-Bx^*\rangle \geq 0$ yields

$$\begin{aligned}
&\phi_A\left(\left\|\frac{1}{\lambda(t)}\ddot{x}(t)+\frac{1}{\lambda(t)}\Gamma(\dot{x}(t))+x(t)-x^*\right\|\right)\\
&\leq \left\langle \frac{1}{\lambda(t)}\ddot{x}(t)+\frac{1}{\lambda(t)}\Gamma(\dot{x}(t)), -B(x(t))+Bx^*\right\rangle\\
&+\left\langle x(t)-x^*, -\frac{1}{\eta\lambda(t)}\ddot{x}(t)-\frac{1}{\eta\lambda(t)}\Gamma(\dot{x}(t))\right\rangle.
\end{aligned}$$

As λ is bounded by positive constants, by using (i)-(iv) it follows that the right-hand side of the last inequality converges to 0 as $t\to+\infty$. Hence $\lim_{t\to+\infty}\phi_A\left(\left\|\frac{1}{\lambda(t)}\ddot{x}(t)+\frac{1}{\lambda(t)}\Gamma(\dot{x}(t))+x(t)-x^*\right\|\right)=0$ and the properties of the function ϕ_A allow to conclude that $\frac{1}{\lambda(t)}\ddot{x}(t)+\frac{1}{\lambda(t)}\Gamma(\dot{x}(t))+x(t)-x^*$ converges strongly to 0 as $t\to+\infty$. By using again the boundedness of λ and (ii) we obtain that $x(t)$ converges strongly to x^* as $t\to+\infty$.

Finally, suppose that B is uniformly monotone with corresponding function $\phi_B:[0,+\infty)\to[0,+\infty]$, which is increasing and vanishes only at 0. The conclusion follows by letting t in the inequality $\langle x(t)-x^*, B(x(t))-Bx^*\rangle \geq \phi_B(\|x(t)-x^*\|)\ \forall t\in[0,+\infty)$ converge to $+\infty$ and by using that x is bounded and $\lim_{t\to+\infty}(B(x(t)-Bx^*)=0$. □

Remark 3.1. The statements in Theorem 3.1 remain valid also for $\eta:=2\beta$. Under this assumption, the cocoercivity of B implies that $\operatorname{Id}-\eta B$ is nonexpansive, hence $R=J_{\eta A}\circ(\operatorname{Id}-\eta B)$ used in the proof is nonexpansive, too, and so the statements in (i)-(iii) follow from Corollary 2.1. The statements (iv) and (v) can be proved in the similar way for $\eta=2\beta$, too.

We close this section by approaching from the perspective of second order dynamical systems optimization problems of the form

$$\min_{x\in\mathcal{H}} f(x)+g(x),$$

where $f:\mathcal{H}\to\mathbb{R}\cup\{+\infty\}$ is a proper, convex and lower semicontinuous function and $g:\mathcal{H}\to\mathbb{R}$ is a convex and (Fréchet) differentiable function with $1/\beta$-Lipschitz continuous gradient for $\beta>0$.

For a proper, convex and lower semicontinuous function $f:\mathcal{H}\to\mathbb{R}\cup\{+\infty\}$, its (convex) subdifferential at $x\in\mathcal{H}$ is defined as $\partial f(x)=\{u\in\mathcal{H}: f(y)\geq f(x)+\langle u, y-x\rangle\ \forall y\in\mathcal{H}\}$. When seen as a set-valued mapping, it is a maximally monotone operator (see[22]) and its resolvent is given by $J_{\eta\partial f}=\operatorname{prox}_{\eta f}$ (see[13]), where $\operatorname{prox}_{\eta f}:\mathcal{H}\to\mathcal{H}$,

$$\operatorname{prox}_{\eta f}(x)=\operatorname*{argmin}_{y\in\mathcal{H}}\left\{f(y)+\frac{1}{2\eta}\|y-x\|^2\right\}, \tag{28}$$

denotes the proximal point operator of f and $\eta > 0$. According to[13] Definition 10.5, f is said to be uniformly convex with modulus function $\phi : [0, +\infty) \to [0, +\infty]$, if ϕ is increasing, vanishes only at 0 and fulfills $f(\alpha x + (1-\alpha)y) + \alpha(1-\alpha)\phi(\|x-y\|) \leq \alpha f(x) + (1-\alpha)f(y)$ for all $\alpha \in (0,1)$ and $x, y \in \operatorname{dom} f := \{x \in \mathcal{H} : f(x) < +\infty\}$. If this inequality holds for $\phi = (\nu/2)|\cdot|^2$ for $\nu > 0$, then f is said to be ν-strongly convex.

The second order dynamical system we consider in order to approach the minimizers of $f + g$ reads

$$\begin{cases} \ddot{x}(t) + \Gamma(\dot{x}(t)) + \lambda(t)\left[x(t) - \operatorname{prox}_{\eta f}\Big(x(t) - \eta\nabla g(x(t))\Big)\right] = 0 \\ x(0) = u_0, \dot{x}(0) = v_0. \end{cases} \tag{29}$$

Corollary 3.1. *Let $f : \mathcal{H} \to \mathbb{R} \cup \{+\infty\}$ by a proper, convex and lower semicontinuous function and $g : \mathcal{H} \to \mathbb{R}$ be a convex and (Fréchet) differentiable function with $1/\beta$-Lipschitz continuous gradient for $\beta > 0$ such that* $\operatorname{argmin}_{x\in\mathcal{H}}\{f(x) + g(x)\} \neq \emptyset$. *Let $\eta \in (0, 2\beta]$, $\Gamma : \mathcal{H} \to \mathcal{H}$ be a γ-elliptic operator, $\lambda : [0, +\infty) \to (0, +\infty)$ be a function fulfilling ($\lambda 4$), $u_0, v_0 \in \mathcal{H}$ and $x \in C^2([0, +\infty); \mathcal{H})$ be the unique global solution of (29). Then the following statements are true:*

(i) the trajectory x is bounded and $\dot{x}, \ddot{x}, \big(\operatorname{Id} - \operatorname{prox}_{\eta f} \circ (\operatorname{Id} - \eta\nabla g)\big)x \in L^2([0, +\infty); \mathcal{H})$;

(ii) $\lim_{t\to+\infty} \dot{x}(t) = \lim_{t\to+\infty} \ddot{x}(t) = 0$;

(iii) $x(t)$ converges weakly to a minimizer of $f + g$ as $t \to +\infty$;

(iv) if x^* is a minimizer of $f + g$, then $\nabla g(x(\cdot)) - \nabla g(x^*) \in L^2([0, +\infty); \mathcal{H})$, $\lim_{t\to+\infty} \nabla g(x(t)) = \nabla g(x^*)$ and ∇g is constant on $\operatorname{argmin}_{x\in\mathcal{H}}\{f(x) + g(x)\}$;

(v) if f or g is uniformly convex, then $x(t)$ converges strongly to the unique minimizer of $f + g$ as $t \to +\infty$.

Proof. The statements are direct consequences of the corresponding ones from Theorem 3.1 (see also Remark 3.1), by choosing $A := \partial f$ and $B := \nabla g$, by taking into account that $\operatorname{zer}(\partial f + \nabla g) = \operatorname{argmin}_{x\in\mathcal{H}}\{f(x) + g(x)\}$ and by making use of the Baillon-Haddad Theorem, which says that ∇g is $1/\beta$-Lipschitz if and only if ∇g is β-cocoercive (see[13] Corollary 18.16). For statement (v) we also use the fact that if f is uniformly convex with modulus ϕ, then ∂f is uniformly monotone with modulus 2ϕ (see[13] Example 22.3(iii)). □

Remark 3.2. Let us consider again the setting in Remark 2.1, namely, when $\Gamma = \gamma \operatorname{Id}$ for $\gamma > 0$ and $\lambda(t) = 1$ for every $t \in [0, +\infty)$. For C a nonempty, convex, closed subset of $\mathcal{H}$, let $f = \delta_C$ be the indicator function of C, which is defined as being equal to 0 for $x \in C$ and to $+\infty$, else. The dynamical system (29) attached in this setting to the minimization of g over C becomes

$$\begin{cases} \ddot{x}(t) + \gamma\dot{x}(t) + x(t) - P_C\big(x(t) - \eta\nabla g(x(t))\big) = 0 \\ x(0) = u_0, \dot{x}(0) = v_0, \end{cases} \tag{30}$$

where P_C denotes the projection onto the set C.

In[8] Theorem 3.1 it has been shown that whenever $\gamma^2 > 2$ and $0 < \eta \leq 2\beta$ the trajectories of (30) asymptotically converge to o a minimizer of g over C, provided this exists. In this setting, (λ4) trivially holds. Moreover, in order to verify (λ4) when $\lambda(t) = 1$ for every $t \in [0, +\infty)$, one needs to equivalently assume that $\gamma^2 > \frac{4\beta}{4\beta-\eta}$. As $\frac{4\beta}{4\beta-\eta} \leq 2$, this provides a slight improvement over[8] Theorem 3.1 in what concerns the choice of γ. We refer the reader also to[7] for an analysis of the convergence rates of trajectories of the dynamical system (30) when g is endowed with supplementary properties.

For the two main convergence statements provided in this section it was essential to choose the step size η in the interval $(0, 2\beta]$ (see Theorem 3.1, Remark 3.1 and Corollary 3.1). This allowed us to guarantee for the generated trajectories the existence of the limit $\lim_{t\to+\infty} \|x(t)-x^*\|^2$, where x^* denotes a solution of the problem under investigation. However, when dealing with convex optimization problems, one can go also beyond this classical restriction concerning the choice of the step size (see also[1] Section 4.2). This is proved in the following corollary, which is valid under the assumption

(λ5) $\lambda : [0, +\infty) \to (0, +\infty)$ is locally absolutely continuous, monotonically increasing on $[0, +\infty)$ and there exist $a, \theta, \theta' > 0$ such that for $t \geq 0$

$$\frac{1}{\beta}\left(\theta' + \frac{a}{2}\|\Gamma - \gamma \operatorname{Id}\|\right) \leq \lambda(t) \leq \frac{\gamma^2}{\eta\theta + \frac{\eta}{\beta} + \frac{\eta}{2a}\|\Gamma - \gamma \operatorname{Id}\| + 1}.$$

Corollary 3.2. *Let $f : \mathcal{H} \to \mathbb{R} \cup \{+\infty\}$ by a proper, convex and lower semicontinuous function and $g : \mathcal{H} \to \mathbb{R}$ be a convex and (Fréchet) differentiable function with $1/\beta$-Lipschitz continuous gradient for $\beta > 0$ such that* $\operatorname{argmin}_{x\in\mathcal{H}}\{f(x) + g(x)\} \neq \emptyset$*. Let be $\eta > 0$, $\Gamma : \mathcal{H} \to \mathcal{H}$ be a γ-elliptic operator, $\lambda : [0, +\infty) \to (0, +\infty)$ be a function fulfilling (λ5), $u_0, v_0 \in \mathcal{H}$ and $x \in C^2([0, +\infty); \mathcal{H})$ be the unique global solution of (29). Then the following statements are true:*

(i) the trajectory x is bounded and $\dot{x}, \ddot{x}, \big(\operatorname{Id} - \operatorname{prox}_{\eta f} \circ (\operatorname{Id} - \eta \nabla g)\big)x \in L^2([0,+\infty);\mathcal{H})$;

(ii) $\lim_{t\to+\infty} \dot{x}(t) = \lim_{t\to+\infty} \ddot{x}(t) = 0$;

(iii) $x(t)$ converges weakly to a minimizer of $f+g$ as $t \to +\infty$;

(iv) if x^* is a minimizer of $f+g$, then $\nabla g(x(\cdot)) - \nabla g(x^*) \in L^2([0,+\infty);\mathcal{H})$, $\lim_{t\to+\infty} \nabla g(x(t)) = \nabla g(x^*)$ and ∇g is constant on $\operatorname{argmin}_{x\in\mathcal{H}}\{f(x)+g(x)\}$;

(v) if f or g is uniformly convex, then $x(t)$ converges strongly to the unique minimizer of $f+g$ as $t \to +\infty$.

Proof. Consider an arbitrary element $x^* \in \operatorname{argmin}_{x\in\mathcal{H}}\{f(x)+g(x)\} = \operatorname{zer}(\partial f + \nabla g)$. Consider the function $q : [0,+\infty) \to \mathbb{R}$, $q(t) = g(x(t)) - g(x^*) - \langle \nabla g(x^*), x(t) - x^* \rangle$. Using similar techniques as in Theorem 3.1 it follows that the function

$$t \mapsto \frac{d}{dt}\left(\frac{1}{\eta}h + q\right)(t) + \gamma\left(\frac{1}{\eta}h + q\right)(t) + \frac{1}{\eta}p(t) + \frac{1}{\eta}\left(\frac{1}{\lambda(t)}\langle \dot{x}(t), \Gamma(\dot{x}(t))\rangle\right) \tag{31}$$

is monotonically decreasing. Arguing as in the proof of Theorem 2.1, it follows that $\frac{1}{\eta}h + q$, h, q, $x, \dot{x}, \dot{h}, \dot{q}$ are bounded, $\dot{x}, \ddot{x}$ and $\big(\operatorname{Id} - \operatorname{prox}_{\eta f} \circ (\operatorname{Id} - \eta\nabla g)\big)x \in L^2([0,+\infty);\mathcal{H})$ and $\lim_{t\to+\infty} \dot{x}(t) = 0$. Since $\frac{d}{dt}\big(\operatorname{Id} - \operatorname{prox}_{\eta f} \circ (\operatorname{Id} - \eta\nabla g)\big)x \in L^2([0,+\infty);\mathcal{H})$, we derive from[2] Lemma 5.2 that $\lim_{t\to+\infty}\big(\operatorname{Id} - \operatorname{prox}_{\eta f} \circ (\operatorname{Id} - \eta\nabla g)\big)(x(t)) = 0$. As $\ddot{x}(t) = -\Gamma(\dot{x}(t)) - \lambda(t)\big(\operatorname{Id} - \operatorname{prox}_{\eta f} \circ (\operatorname{Id} - \eta\nabla g)\big)(x(t))$ for every $t \in [0,+\infty)$, we obtain that $\lim_{t\to+\infty} \ddot{x}(t) = 0$. Since $\nabla g(x(\cdot)) - \nabla g(x^*) \in L^2([0,+\infty);\mathcal{H})$, by applying again[2] Lemma 5.2, it yields $\lim_{t\to+\infty} \nabla g(x(t)) = \nabla g(x^*)$. In this way the statements (i), (ii) and (iv) are shown.

(iii) Since the function in (31) is monotonically decreasing, from (i), (ii) and (iv) it follows that the limit $\lim_{t\to+\infty}\left(\gamma\left(\frac{1}{\eta}h + q\right)(t) + \frac{1}{\eta}p(t)\right)$ exists and it is a real number. Consider again the renorming of the space already used in the proof of Theorem 2.1(iii).

As x^* has been chosen as an arbitrary minimizer of $f+g$ and taking into account the definition of p and the new norm, we conclude that for all $x^* \in \operatorname{argmin}_{x\in\mathcal{H}}\{f(x)+g(x)\}$ the limit $\lim_{t\to+\infty} E(t,x^*) \in \mathbb{R}$, exists, where

$$E(t,x^*) = \frac{1}{2\eta}|||x(t) - x^*|||^2 + g(x(t)) - g(x^*) - \langle \nabla g(x^*), x(t) - x^* \rangle .$$

In what follows we use a similar technique as in[14] (see, also,[1] Section 5.2). Since $x(\cdot)$ is bounded, it has at least one weak sequential cluster point.

We prove first that each weak sequential cluster point of $x(\cdot)$ is a minimizer of $f+g$. Let $x^* \in \operatorname{argmin}_{x\in\mathcal{H}}\{f(x)+g(x)\}$ and $t_n \to +\infty$ (as $n \to +\infty$) be such that $(x(t_n))_{n\in\mathbb{N}}$ converges weakly to $\overline{x}$. Since $(x(t_n), \nabla g(x(t_n))) \in \operatorname{Gr}(\nabla g)$, $\lim_{n\to+\infty} \nabla g(x(t_n)) = \nabla g(x^*)$ and $\operatorname{Gr}(\nabla g)$ is sequentially closed in the weak-strong topology, we obtain $\nabla g(\overline{x}) = \nabla g(x^*)$.

From (24) written for $t = t_n$, $A = \partial f$ and $B = \nabla g$, by letting n converge to $+\infty$ and by using that $\operatorname{Gr}(\partial f)$ is sequentially closed in the weak-strong topology, we obtain $-\nabla g(x^*) \in \partial f(\overline{x})$. This, combined with $\nabla g(\overline{x}) = \nabla g(x^*)$ delivers $-\nabla g(\overline{x}) \in \partial f(\overline{x})$, hence $\overline{x} \in \operatorname{zer}(\partial f + \nabla g) = \operatorname{argmin}_{x\in\mathcal{H}}\{f(x)+g(x)\}$.

Next we show that $x(\cdot)$ has at most one weak sequential cluster point, which will actually guarantee that it has exactly one weak sequential cluster point. This will imply the weak convergence of the trajectory to a minimizer of $f+g$.

Let x_1^*, x_2^* be two weak sequential cluster points of $x(\cdot)$. This means that there exist $t_n \to +\infty$ (as $n \to +\infty$) and $t_n' \to +\infty$ (as $n \to +\infty$) such that $(x(t_n))_{n\in\mathbb{N}}$ converges weakly to x_1^* (as $n \to +\infty$) and $(x(t_n'))_{n\in\mathbb{N}}$ converges weakly to x_2^* (as $n \to +\infty$). Since $x_1^*, x_2^* \in \operatorname{argmin}_{x\in\mathcal{H}}\{f(x)+g(x)\}$, we have $\lim_{t\to+\infty} E(t,x_1^*) \in \mathbb{R}$ and $\lim_{t\to+\infty} E(t,x_2^*) \in \mathbb{R}$, hence $\exists \lim_{t\to+\infty}(E(t,x_1^*) - E(t,x_2^*)) \in \mathbb{R}$. We obtain

$$\exists \lim_{t\to+\infty} \left(\frac{1}{\eta}\langle\langle x(t), x_2^* - x_1^*\rangle\rangle + \langle \nabla g(x_2^*) - \nabla g(x_1^*), x(t)\rangle \right) \in \mathbb{R},$$

which, when expressed by means of the sequences $(t_n)_{n\in\mathbb{N}}$ and $(t_n')_{n\in\mathbb{N}}$, leads to $\frac{1}{\eta}\langle\langle x_1^*, x_2^* - x_1^*\rangle\rangle + \langle \nabla g(x_2^*) - \nabla g(x_1^*), x_1^*\rangle = \frac{1}{\eta}\langle\langle x_2^*, x_2^* - x_1^*\rangle\rangle + \langle \nabla g(x_2^*) - \nabla g(x_1^*), x_2^*\rangle$. This is the same with

$$\frac{1}{\eta}|||x_1^* - x_2^*|||^2 + \langle \nabla g(x_2^*) - \nabla g(x_1^*), x_2^* - x_1^*\rangle = 0$$

and by the monotonicity of ∇g we conclude that $x_1^* = x_2^*$.

(v) This follows in analogy to the proof of the corresponding statement of Theorem 3.1(v) written for $A = \partial f$ and $B = \nabla g$. □

Remark 3.3. When $\Gamma = \gamma \operatorname{Id}$ for $\gamma > 0$, in order to verify the left-hand side of the second statement in assumption (λ5) one can take $\theta' := \beta \inf_{t\geq 0} \lambda(t)$. Thus, the bounds in ($\lambda$5) amounts in this case to the existence of $\theta > 0$ such that $\lambda(t) \leq \frac{\gamma^2}{\eta\theta + \frac{\eta}{\beta} + 1}$. Whenever $\lambda(t) = 1$ for every $t \in [0,+\infty)$, this is verified if and only if $\gamma^2 > \frac{\eta}{\beta} + 1$. In other words, ($\lambda$5) permits a more relaxed choice for the parameters γ, η and β, beyond the standard assumptions $0 < \eta \leq 2\beta$ and $\gamma^2 > 2$ considered in[8].

4. Convergence rate for the function values

In this section we furnish a rate for the convergence of a convex and (Fréchet) differentiable function with Lipschitz continuous gradient $g : \mathcal{H} \to \mathbb{R}$ along the ergodic trajectory generated by

$$\begin{cases} \ddot{x}(t) + \Gamma(\dot{x}(t)) + \lambda(t)\nabla g(x(t)) = 0 \\ x(0) = u_0, \dot{x}(0) = v_0 \end{cases} \tag{32}$$

to the minimum value of g. To this end we make the following assumption

(λ6) $\lambda : [0, +\infty) \to (0, +\infty)$ is locally absolutely continuous and there exists $\zeta > 0$ such that for almost every $t \in [0, +\infty)$ we have

$$0 < \zeta \leq \gamma\lambda(t) - \dot{\lambda}(t). \tag{33}$$

The following result is in the spirit of a convergence rate given for the objective function values on a sequence iteratively generated by an inertial-type algorithm recently obtained in[18] Theorem 1.

Theorem 4.1. *Let $g : \mathcal{H} \to \mathbb{R}$ be a convex and (Fréchet) differentiable function with $1/\beta$-Lipschitz continuous gradient for $\beta > 0$ such that $\operatorname{argmin}_{x\in\mathcal{H}} g(x) \neq \emptyset$. Let $\Gamma : \mathcal{H} \to \mathcal{H}$ be a γ-elliptic operator, $\lambda : [0, +\infty) \to (0, +\infty)$ be a function fulfilling (λ6), $u_0, v_0 \in \mathcal{H}$ and $x \in C^2([0, +\infty); \mathcal{H})$ be the unique global solution of (32).*

Then for every minimizer x^* of g and every $T > 0$ it holds

$$\begin{aligned} 0 \leq\, & g\left(\frac{1}{T}\int_0^T x(t)dt\right) - g(x^*) \\ \leq\, & \frac{1}{2\zeta T}\left[\|v_0 + \gamma(u_0 - x^*)\|^2 + \left(\gamma\|\Gamma - \gamma \operatorname{Id}\| + \frac{\lambda(0)}{\beta}\right)\|u_0 - x^*\|^2\right]. \end{aligned}$$

Proof. Let be $x^* \in \operatorname{argmin}_{x\in\mathcal{H}} g(x)$ and $T > 0$. Consider again the function $p : [0, +\infty) \to \mathbb{R}, p(t) = \frac{1}{2}\left\langle(\Gamma - \gamma \operatorname{Id})(x(t) - x^*), x(t) - x^*\right\rangle$ which was defined in (7). By using (32), the formula for the derivative of p, the positive semidefinitness of $\Gamma - \gamma \operatorname{Id}$, the convexity of g and (λ6) we get for almost every $t \in [0, +\infty)$

$$\begin{aligned} & \frac{d}{dt}\left(\frac{1}{2}\|\dot{x}(t) + \gamma(x(t) - x^*)\|^2 + \gamma p(t) + \lambda(t)g(x(t))\right) \\ \leq & -\zeta(g(x(t)) - g(x^*)) + \dot{\lambda}(t)g(x^*) \end{aligned}$$

After integration, by neglecting the nonnegative terms and by using that $g(x(T)) \geq g(x^*)$, it yields

$$\zeta \int_0^T (g(x(t)) - g(x^*))dt \leq \frac{1}{2}\|v_0 + \gamma(u_0 - x^*)\|^2 + \gamma p(0) + \lambda(0)(g(u_0) - g(x^*)).$$

The conclusion follows by using $p(0) = \frac{1}{2}\langle(\Gamma - \gamma \operatorname{Id})(u_0 - x^*), u_0 - x^*\rangle \leq \frac{1}{2}\|\Gamma - \gamma \operatorname{Id}\|\|u_0 - x^*\|^2$, $g(u_0) - g(x^*) \leq \frac{1}{2\beta}\|u_0 - x^*\|^2$, which is a consequence of the descent lemma and the inequality $g\left(\frac{1}{T}\int_0^T x(t)dt\right) - g(x^*) \leq \frac{1}{T}\int_0^T (g(x(t)) - g(x^*))dt$, which holds since g is convex. □

Remark 4.1. Under assumption (λ6) on the relaxation function λ, we obtain in the above theorem (only) the convergence of the function g along the ergodic trajectory to a global minimum value. If one is interested also in the (weak) convergence of the trajectory to a minimizer of g, this follows via Theorem 2.1 when λ is assumed to fulfill (λ1) (if x converges weakly to a minimizer of g, then from the Cesaro-Stolz Theorem one also obtains the weak convergence of the ergodic trajectory $T \mapsto \frac{1}{T}\int_0^T x(t)dt$ to the same minimizer). Take $a \geq 0$, $b > \frac{1}{\beta\gamma^2}$ and $0 \leq \rho \leq \gamma$. Then $\lambda(t) = \frac{1}{ae^{-\rho t}+b}$ is an example of a relaxation function which verifies assumption (λ1) and assumption (λ6) (with $0 < \zeta \leq \frac{\gamma b}{(a+b)^2}$).

Acknowledgments

The research of Ernö Robert Csetnek has been supported by FWF (Austrian Science Fund), Lise Meitner Programme, project M 1682-N25.

References

1. B. Abbas, H. Attouch, *Dynamical systems and forward-backward algorithms associated with the sum of a convex subdifferential and a monotone cocoercive operator*, Optimization 64(10), 2223–2252, 2015.
2. B. Abbas, H. Attouch, B.F. Svaiter, *Newton-like dynamics and forward-backward methods for structured monotone inclusions in Hilbert spaces*, Journal of Optimization Theory and its Applications 161(2), 331–360, 2014.
3. F. Alvarez, *On the minimizing property of a second order dissipative system in Hilbert spaces*, SIAM Journal on Control and Optimization 38(4), 1102–1119, 2000.

4. F. Alvarez, *Weak convergence of a relaxed and inertial hybrid projection-proximal point algorithm for maximal monotone operators in Hilbert space*, SIAM Journal on Optimization 14(3), 773–782, 2004.
5. F. Alvarez, H. Attouch, *An inertial proximal method for maximal monotone operators via discretization of a nonlinear oscillator with damping*, Set-Valued Analysis 9(1-2), 3–11, 2001.
6. F. Alvarez, H. Attouch, J. Bolte, P. Redont, *A second-order gradient-like dissipative dynamical system with Hessian-driven damping. Application to optimization and mechanics*, Journal de Mathématiques Pures et Appliquées (9) 81(8), 747–779, 2002.
7. A.S. Antipin, *Minimization of convex functions on convex sets by means of differential equations*, (Russian) Differentsial'nye Uravneniya 30(9), 1475–1486, 1994; translation in Differential Equations 30(9), 1365–1375, 1994.
8. H. Attouch, F. Alvarez, *The heavy ball with friction dynamical system for convex constrained minimization problems*, in: Optimization (Namur, 1998), 25–35, in: Lecture Notes in Economics and Mathematical Systems 481, Springer, Berlin, 2000.
9. H. Attouch, M. Marques Alves, B.F. Svaiter, *A dynamic approach to a proximal-Newton method for monotone inclusions in Hilbert spaces, with complexity $O(1/n^2)$*, Journal of Convex Analysis 23, 2016.
10. H. Attouch, M.-O. Czarnecki, *Asymptotic behavior of coupled dynamical systems with multiscale aspects*, Journal of Differential Equations 248(6), 1315-1344, 2010.
11. H. Attouch, X. Goudou, P. Redont, *The heavy ball with friction method. I. The continuous dynamical system: global exploration of the local minima of a real-valued function by asymptotic analysis of a dissipative dynamical system*, Communications in Contemporary Mathematics 2(1), 1–34, 2000.
12. H. Attouch, B.F. Svaiter, *A continuous dynamical Newton-like approach to solving monotone inclusions*, SIAM Journal on Control and Optimization 49(2), 574–598, 2011.
13. H.H. Bauschke, P.L. Combettes, *Convex Analysis and Monotone Operator Theory in Hilbert Spaces*, CMS Books in Mathematics, Springer, New York, 2011.
14. J. Bolte, *Continuous gradient projection method in Hilbert spaces*, Journal of Optimization Theory and its Applications 119(2), 235–259, 2003.
15. R.I. Boţ, E.R. Csetnek, *A dynamical system associated with the fixed points set of a nonexpansive operator*, Journal of Dynamics and Differ-

ential Equations, DOI: 10.1007/s10884-015-9438-x, 2015.
16. R.I. Boţ, E.R. Csetnek, *Approaching the solving of constrained variational inequalities via penalty term-based dynamical systems*, Journal of Mathematical Analysis and Applications 435(2), 1688-1700, 2016.
17. R.I. Boţ, E.R. Csetnek, C. Hendrich, *Inertial Douglas-Rachford splitting for monotone inclusion problems*, Applied Mathematics and Computation 256, 472–487, 2015.
18. E. Ghadimi, H.R. Feyzmahdavian, M. Johansson, *Global convergence of the Heavy-ball method for convex optimization*, arXiv:1412.7457.
19. A. Moudafi, M. Oliny, *Convergence of a splitting inertial proximal method for monotone operators*, Journal of Computational and Applied Mathematics 155, 447–454, 2003.
20. N. Ogura, I. Yamada, *Non-strictly convex minimization over the fixed point set of an asymptotically shrinking nonexpansive mapping*, Numerical Functional Analysis and Optimization 23(1-2), 113–137, 2002.
21. J.-C. Pesquet, N. Pustelnik, *A parallel inertial proximal optimization method*, Pacific Journal of Optimization 8(2), 273–305, 2012.
22. R.T. Rockafellar, *On the maximal monotonicity of subdifferential mappings*, Pacific Journal of Mathematics 33(1), 209–216, 1970.
23. R.T. Rockafellar, *Monotone operators and the proximal point algorithm*, SIAM Journal on Control and Optimization 14(5), 877–898, 1976.

On the solutions of a quadratic integral inclusion

Aurelian Cernea

Faculty of Mathematics and Informatics,
University of Bucharest,
Academiei 14, 010014 Bucharest, Romania
E-mail: acernea@fmi.unibuc.ro

The existence of solutions of a quadratic integral inclusion is obtained in the case when the inclusion is defined by a Carathéodory set-valued map with convex values.

Keywords: Quadratic integral inclusions; Carathéodory set-valued map; fixed point.

1. Introduction

In this note we study the following integral inclusion

$$x(t) \in (f(x))(t) \int_0^t k(t,s)F(s,x(s))ds, \quad t \in I := [0,T], \tag{1}$$

where $f : C(I,\mathbf{R}) \to C(I,\mathbf{R})$, $k : I \times I \to \mathbf{R}$, $F : I \times \mathbf{R} \to \mathcal{P}(\mathbf{R})$ is a set-valued map.

The present contribution is motivated by a recent paper of Anichini and Conti[1], where the existence of solutions of problem (1) is obtained in the case when the set-valued map is upper semicontinuous with convex values. The proof of the result in Ref. 1 is based on a fixed point theorem for the multivalued solution map with acyclic values.

The aim of this note is to improve the result in Ref. 1 by assuming that $F(.,.)$ is Carathéodory, i.e., $F(.,.)$ is upper semicontinuous only in the second variable instead the assumption that $F(.,.)$ is upper semicontinuous in both variable as in Ref. 1. The main tool in our approach is a Leray-Schauder nonlinear alternative for set-valued maps due to Dhage[3]. The price we pay is the assumption that the operator f is Lipschitz, instead of the hypothesis that f satisfies the Darbo property as in Ref. 1.

We note that the situation when the set-valued map is not convex valued was studied in our previous paper Ref. 2.

For motivation, examples and applications concerning problem (1) we refer to Ref. 1 and the references therein. We only mention that one of the first quadratic integral equation, proposed by M.S. Wertheim as a simplified model of certain equations arising in statistical mechanics, was studied in Pimbley[5].

The paper is organized as follows: in Section 2 we recall some preliminary facts that we need in the sequel and in Section 3 we prove our main result.

2. Preliminaries

In this section we sum up some basic facts that we are going to use later.

Let (X, d) be a metric space with the corresponding norm $|.|$ and let $I \subset \mathbf{R}$ be a compact interval. Denote by $\mathcal{L}(I)$ the σ-algebra of all Lebesgue measurable subsets of I, by $\mathcal{P}(X)$ the family of all nonempty subsets of X and by $\mathcal{B}(X)$ the family of all Borel subsets of X. If $A \subset I$ then $\chi_A : I \to \{0, 1\}$ denotes the characteristic function of A. For any subset $A \subset X$ we denote by $\overline{A}$ the closure of A.

As usual, we denote by $C(I, X)$ the Banach space of all continuous functions $x : I \to X$ endowed with the norm $|x|_C = \sup_{t\in I}|x(t)|$ and by $L^1(I, X)$ the Banach space of all (Bochner) integrable functions $x : I \to X$ endowed with the norm $|x|_1 = \int_I |x(t)|dt$.

Consider $T : X \to \mathcal{P}(X)$ a set-valued map. A point $x \in X$ is called a fixed point for T if $x \in T(x)$. A set-valued mapping $T : X \to \mathcal{P}(X)$ is called:

(i) bounded on bounded sets if $T(B) := \cup_{x\in B}T(x)$ is a bounded subset of X for all bounded sets B in X;

(ii) compact if $T(B)$ is relatively compact for any bounded sets B in X;

(iii) totally compact if $\overline{T(X)}$ is a compact subset of X;

(iv) upper semicontinuous if for any $x_0 \in X$, $T(x_0)$ is a nonempty closed subset of X and if for each open set D of X containing $T(x_0)$ there exists an open neighborhood V_0 of x_0 such that $T(V_0) \subset D$;

(v) completely continuous if it is upper semicontinuous and totally compact on X.

It is well known that a compact set-valued map T with nonempty compact values is upper semicontinuous if and only if T has a closed graph.

The main tool in the proof of our result is the next theorem proved in Ref. 3.

Theorem 2.1. *Let X be a Banach algebra, let $A : X \to X$ be a single* *valued map and $B : X \to \mathcal{P}(X)$ a set-valued map with compact convex values satisfying the following conditions:*

a) A is k-Lipschitz,
b) B is compact and upper semicontinuous,
c) $2Mk < 1$, where $M = \sup\{|v|; \quad v \in B(x)\}$.

Then, either

i) the operator inclusion $x \in AxBx$ has o solution,
or
ii) the set $\mathcal{C} = \{u \in X;\ \mu \in AuBu,\ \mu > 1\}$ is unbounded.

If X is a normed linear space, $F : I \times X \to \mathcal{P}(X)$ is a set-valued map with compact values and $x \in C(I, X)$ we define

$$S_F(x) := \{f \in L^1(I, X); \quad f(t) \in F(t, x(t)) \quad a.e.\ I\}.$$

A set-valued map $G : I \to \mathcal{P}(\mathbf{R})$ with nonempty compact convex values is said to be *measurable* if for any $x \in \mathbf{R}$ the function $t \to d(x, G(t))$ is measurable.

A set-valued map $F : I \times \mathbf{R} \to \mathcal{P}(\mathbf{R})$ is said to be *Carathéodory* if $t \to F(t, x)$ is measurable for any $x \in \mathbf{R}$ and $x \to F(t, x)$ is upper semicontinuous for almost all $t \in I$.

F is said to be L^1-*Carathéodory* if there exists $h \in L^1(I, \mathbf{R})$ such that $\sup\{|v| : v \in F(t, x)\} \le h(t)$ a.e. I, $\forall x \in \mathbf{R}$.

Theorem 2.2 (4). *Let X be a Banach space, let $F : I \times X \to \mathcal{P}(X)$ be a L^1-Carathéodory set-valued map with $S_F \neq \emptyset$ and let $\Gamma : L^1(I, X) \to C(I, X)$ be a linear continuous mapping.*

Then the set-valued map $\Gamma \circ S_F : C(I, X) \to \mathcal{P}(C(I, X))$ defined by

$$(\Gamma \circ S_F)(x) = \Gamma(S_F(x))$$

has compact convex values and has closed graph in $C(I, X) \times C(I, X)$.

Note that if $dimX < \infty$ and F is as in Theorem 2.2, then $S_F(x) \neq \emptyset$ for any $x \in C(I, X)$ (e.g., Ref. 4).

3. The main result

In order to prove our result we need the following assumptions.
Hypothesis. i) $F : I \times \mathbf{R} \to \mathcal{P}(\mathbf{R})$ has nonempty compact convex values and is Carathéodory.

ii) There exists $\varphi \in L^1(I, \mathbf{R})$ with $\varphi(t) > 0$ a.e. I such that

$$\sup\{|v|; \quad v \in F(t, x)\} \leq \varphi(t) \quad a.e.\ I, \quad \forall x \in \mathbf{R}.$$

iii) $f : C(I, \mathbf{R}) \to C(I, \mathbf{R})$ and there exists $L > 0$ such that

$$|f(x_1) - f(x_2)|_C \leq L|x_1 - x_2|_C \quad \forall x_1, x_2 \in C(I, \mathbf{R}).$$

iv) $k : I \times I \to \mathbf{R}$ is a continuous function such that there exists a bounded function $w : [0, \infty) \to [0, \infty)$ with $\lim_{t\to 0+} w(t) = 0$ and

$$\int_0^T |k(t_1, s) - k(t_2, s)|\varphi(s)ds \leq w(|t_1 - t_2|) \quad \forall t_1, t_2 \in I.$$

As an example for the operator f one may take $(f(x))(t) = \int_0^T k_1(t, s)x(s)ds$ with $k_1 : I \times I \to \mathbf{R}$ satisfying the condition in Hypothesis iv).

From Hypothesis iii) it follows that

$$|f(x)(t)| \leq |f(x)|_C \leq L|x|_C + |f(0)|_C \quad \forall\, x \in C(I, \mathbf{R}) \text{ and } t \in I.$$

Denote $m_0 := \max_{t,s\in I} |k(t, s)|$.

Theorem 3.1. *Assume that Hypothesis is satisfied and* $2m_0L|\varphi|_1 < 1$. *Then problem (1) has at least one solution.*

Proof. Let $X = C(I, \mathbf{R})$ and define, for $x \in X$,

$$N(x) = \{v \in X; \quad v(t) \in (f(x))(t) \int_0^t k(t, s)F(s, x(s))ds\},$$

$$(A(x))(t) = (f(x))(t),$$

$$B(x) = \{h \in X; \quad h(t) = \int_0^t k(t, s)v(s)ds \quad v \in S_F(x)\}.$$

Obviously, $N(x) = A(x)B(x)$ and A is L-Lipschitz. We intend to apply Theorem 2.1. We show that B satisfies the hypotheses of Theorem 2.1.

First, we show that $B(x) \subset C(I, \mathbf{R})$ is convex for any $x \in C(I, \mathbf{R})$. If $h_1, h_2 \in B(x)$ then there exist $v_1, v_2 \in S_F(x)$ such that for any $t \in I$ one has

$$h_i(t) = \int_0^t k(t, s)v_i(s)ds, \quad i = 1, 2.$$

Let $0 \leq \alpha \leq 1$. Then for any $t \in I$ we have

$$(\alpha h_1 + (1 - \alpha)h_2)(t) = \int_0^t k(t, s)[\alpha v_1(s) + (1 - \alpha)v_2(s)]ds.$$

The values of F are convex, thus $S_F(x)$ is a convex set and hence $\alpha h_1 + (1-\alpha)h_2 \in B(x)$.

Secondly, we show that B is bounded on bounded sets of $C(I,\mathbf{R})$. Let $S \subset C(I,\mathbf{R})$ be a bounded set. Then there exist $m > 0$ such that $|x|_C \le m \ \forall x \in S$. If $h \in B(x)$ there exists $v \in S_F(x)$ such that $h(t) = \int_0^t k(t,s)v(s)ds$. One may write for any $t \in I$

$$|h(t)| \le \int_0^t |k(t,s)|.|v(s)|ds \le m_0|\varphi|_1$$

and therefore

$$|h|_C \le m_0|\varphi|_1 \quad \forall v \in B(x),$$

i.e., $B(S)$ is bounded.

We show next that B maps bounded sets into equi-continuous sets. Let $S \subset C(I,\mathbf{R})$ be a bounded set and $h \in B(x)$ for some $x \in S$. There exists $v \in S_F(x)$ such that $h(t) = \int_0^t k(t,s)v(s)ds$. Then for any $t,\tau \in I$ we have

$$|h(t) - h(\tau)| = \left|\int_0^t k(t,s)v(s)ds - \int_0^\tau k(\tau,s)v(s)ds\right| \le$$

$$\left|\int_0^t k(t,s)v(s)ds - \int_0^t k(\tau,s)v(s)ds\right| + \left|\int_0^t k(\tau,s)v(s)ds - \int_0^\tau k(\tau,s)v(s)ds\right|$$

$$\le \int_0^t |k(t,s) - k(\tau,s)|.|v(s)|ds + \int_\tau^t |k(\tau,s)|.|v(s)|ds \le$$

$$\int_0^T |k(t,s) - k(\tau,s)|\varphi(s)ds + m_0\int_\tau^t \varphi(s)ds \le w(|t-\tau|) + m_0\int_\tau^t \varphi(s)ds.$$

It follows that $|h(t) - h(\tau)| \to 0$ as $t \to \tau$. Therefore, $B(S)$ is an equi-continuous set in $C(I,\mathbf{R})$. We apply now Arzelà-Ascoli's theorem we deduce that $B(S)$ is relatively compact, i.e., B is compact.

In the next step of the proof we prove that B has closed graph. Let $x_n \in C(I,\mathbf{R})$ be a sequence such that $x_n \to x^*$ and $h_n \in B(x_n) \ \forall n \in \mathbf{N}$ and such that $h_n \to h^*$. We prove that $h^* \in T(x^*)$. Since $h_n \in B(x_n)$, there exists $v_n \in S_F(x_n)$ such that $h_n(t) = \int_0^t k(t,s)v_n(s)ds$. Define $\Gamma : L^1(I,\mathbf{R}) \to C(I,\mathbf{R})$ by $(\Gamma(v))(t) := \int_0^t k(t,s)v(s)ds$. One has

$$\left|\int_0^t k(t,s)v_1(s)ds - \int_0^t k(t,s)v_2(s)ds\right| \le m_0|v_1 - v_2|_1.$$

We apply Theorem 2.2 to find that $\Gamma \circ S_F$ has closed graph and from the definition of Γ we get $h_n \in \Gamma \circ S_F(x_n)$. Since $x_n \to x^*$, $h_n \to h^*$ it follows the existence of $v^* \in S_F(x^*)$ such that $h^*(t) = \int_0^t k(t,s)v^*(s)ds$. Therefore, B is upper semicontinuous and compact.

We note that if $h \in B(x)$ there exists $v \in S_F(x)$ with $h(t) = \int_0^t k(t,s)v(s)ds$. We deduce that

$$|h(t)| = \left|\int_0^t k(t,s)v(s)ds\right| \leq \int_0^T |k(t,s)|.|v(s)|ds \leq m_0|\varphi|_1$$

and thus $|h|_C \leq m_0|\varphi|_1$ for any $h \in B(x)$.

Finally, in order to apply Theorem 2.1, it remains to prove that the corresponding statement ii) in Theorem 2.1 is not true. Assume that the set $\mathcal{C} = \{u \in X;\ \mu \in AuBu,\ \mu > 1\}$ is unbounded. Take $r > \frac{|f(0)|_C m_0|\varphi|_1}{1-Lm_0|\varphi|_1}$. Therefore, there exists $u \in \mathcal{C}$ with $|u|_C > r$. But, if $u \in \mathcal{C}$ then $u(t) = \frac{1}{\mu}Au(t)Bu(t)$ and thus

$$|u(t)| \leq |(f(u))(t)|.\left|\int_0^t k(t,s)v(s)ds\right| \leq (L|u|_C + |f(0)|_C)m_0|\varphi|_1,$$

i.e., $|u(t)| \leq \frac{|f(0)|_C m_0|\varphi|_1}{1-Lm_0|\varphi|_1}$ which contradicts the choice of r and the proof is complete. □

As an example we consider the quadratic inclusion

$$x(t) \in \int_0^1 x(s)ds \int_0^t \frac{t}{4(t+s)}F(s,x(s))ds, \quad t \in [0,1],$$

where $F : [0,1] \times \mathbf{R} \to \mathcal{P}(\mathbf{R})$ is defined by $F(t,x) = [\frac{|x|^3}{10(|x|^3+3)}, \frac{|sinx|}{9(|sinx|+1)} + \frac{8}{9}]$. In this case $T = 1$, $\varphi(t) \equiv 1$, $k(t,s) = \frac{t}{4(t+s)}$, $f(x) = \int_0^1 x(s)ds$, $m_0 = \frac{1}{4}$ and $L = 1$. Therefore, we can apply Theorem 3.1 in order to deduce the existence of solutions for this problem.

Acknowledgments

This research is supported by the CNCS grant PN-II-ID-PCE-2011-3-0198.

References

1. G. Anichini, G. Conti, *Existence of solutions for quadratic integral inclusions*, Libertas Math. N.S. 33, 57-67 (2013).
2. A. Cernea, *On the existence of solutions for a quadratic integral inclusion*, Libertas Math. N. S. 34, 47-52 (2014).
3. B.C. Dhage, *Existence results for neutral functional differential inclusions in Banach algebras*, Nonlin. Anal., 64, 1290-1306 (2006).

4. A. Lasota, Z. Opial, *An application of the Kakutani-Ky-Fan theorem in the theory of ordinary differential equations*, Bull. Acad. Polon. Sci. Math. Astronom. Physiques, 13, 781-786 (1965).
5. G.H. Pimbley Jr., *Positive solutions of a quadratic integral equation*, Arch. Rat. Mech. Anal., 24, 107-127 (1967).

On the bounded and stabilizing solution of a generalized Riccati differential equation with periodic coefficients arising in connection with a zero sum linear quadratic stochastic differential game

Vasile Dragan* and Toader Morozan**

Institute of Mathematics "Simion Stoilow" of the Romanian Academy, P.O.Box 1-764, RO-014700, Bucharest, Romania

*E-mail: *vasile.dragan@imar.ro, **toader.morozan@imar.ro*

In this work is considered a system of coupled matrix nonlinear differential equations arising in connection with the solution of a zero sum two players linear quadratic differential game for a system modeled by an Itô differential equation subject to random switching according with a standard homogeneous Markov process with a finite number of state. The system of differential equations under consideration contains as special cases the game theoretic Riccati differential equations arising in the solution of the H_∞ control problem from the deterministic and stochastic cases. A set of sufficient conditions which guarantee the existence of the bounded and stabilizing solution of this kind of Riccati differential equations is provided.

Keywords: generalized Riccati differential equations; stabilizing solution; periodic coefficients; zero sum differential games.

1. Problem formulation

Let us consider the following system of nonlinear differential equations:

$$-\dot{X}(t,i)=A_0^T(t,i)X(t,i)+X(t,i)A_0(t,i)+\sum_{k=1}^{r}A_k^T(t,i)X(t,i)A_k(t,i)$$
$$+\sum_{j=1}^{N}q_{ij}X(t,j)+M(t,i)-[X(t,i)B_0(t,i)+\sum_{k=1}^{r}A_k^T(t,i)X(t,i)B_k(t,i)][R(t,i) \quad (1)$$
$$+\sum_{k=1}^{r}B_k^T(t,i)X(t,i)B_k(t,i)]^{-1}[B_0^T(t,i)X(t,i)+\sum_{k=1}^{r}B_k^T(t,i)X(t,i)A_k(t,i)],$$

$1 \le i \le N$ with the unknown function $t \to \mathbf{X}(t) = (X(t,1),...,X(t,N))$, $X(t,i) = X^T(t,i)$. The superscript T stands for the transpose of a matrix or a vector.

Assumption 1.1. *In (1)* $t \to A_k(t,i) : \mathbb{R} \to \mathbb{R}^{n\times n}$, $t \to B_k(t,i) : \mathbb{R} \to \mathbb{R}^{n\times m}$, $0 \le k \le r$, $t \to M(t,i) : \mathbb{R} \to \mathcal{S}_n$, $t \to R(t,i) : \mathbb{R} \to \mathcal{S}_m$ *are continuous matrix valued functions which are periodic of period* $\theta > 0$.

Throughout the work, $\mathcal{S}_d \subset \mathbb{R}^{d\times d}$ denotes the linear space of symmetric matrices of size $d \times d$ and $\mathcal{S}_d^N = \mathcal{S}_d \times ... \times \mathcal{S}_d$. The scalars q_{ij} involved in (1) have the properties:

$$q_{ij} \ge 0,\ i \ne j; \quad \sum_{j=1}^{N} q_{ij} = 0, \quad 1 \le i \le N. \tag{2}$$

Since the system of differential equations (1) contains as special cases various types of matrix Riccati differential equations from both deterministic and stochastic framework, we shall refer to this system as **system of generalized Riccati differential equations** (SGRDE). The SGRDE of type (1) occur in connection with the linear quadratic control problems described by controlled systems with state space representation modeled by:

$$\begin{aligned} dx(t) &= [A_0(t,\eta_t)x(t) + B_0(t,\eta_t)u(t)]dt \\ &\quad + \sum_{k=1}^{r}[A_k(t,\eta_t)x(t) + B_k(t,\eta_t)u(t)]dw_k(t) \end{aligned} \tag{3}$$

$t > 0$, $x(0) = x_0 \in \mathbb{R}^n$ and the cost functional

$$J(x_0,u) = E\left[\int_0^\infty (x_u^T(t)M(t,\eta_t)x_u(t) + u^T(t)R(t,\eta_t)u(t))\right] dt \tag{4}$$

where $x_u(t)$, $t \ge 0$ is the solution of the initial value problem (3) determined by the input $u(t)$. In (3) $\{w(t)\}_{t\ge 0}$ ($w(t) = (w_1(t), ..., w_r(t))^T$) is the r-dimensional standard Wiener process defined on a given probability space $(\Omega, \mathcal{F}, \mathcal{P})$ and $\{\eta_t\}_{t\ge 0}$ is a standard right-continuous Markov process taking values in the finite set $\mathfrak{N} = \{1, 2, ..., N\}$ and having the transition semigroup $P(t) = e^{Qt}$, $t \ge 0$ where $Q \in \mathbb{R}^{N\times N}$ whose elements q_{ij} satisfy conditions (2). Additionally, $\{w(t)\}_{t\ge 0}$, $\{\eta_t\}_{t\ge 0}$ are assumed to be independent stochastic processes and $\pi_0(i) = \mathcal{P}\{\eta_0 = i\} > 0$, $1 \le i \le N$. The class of $\mathcal{U}_{adm}(x_0)$ of the admissible controls consists of all stochastic processes, $u = \{u(t)\}_{t\ge 0} \in L^2_{\eta w}([0,\infty); \mathbb{R}^m)$ with the property that $x_u(\cdot) \in L^2_{\eta w}([0,\infty); \mathbb{R}^n)$ and $\lim_{t\to\infty} E[|x_u(t)|^2 | \eta_0 = i] = 0$. For precise definition of the spaces, $L^2_{\eta w}(\mathcal{I}; \mathbb{R}^d)$ of stochastic processes adapted to the filtration generated by $\{(\eta_t, w(t))\}_{t\ge 0}$, we refer to Chapter 1 in the book[1].

Definition 1.1. A global solution $\tilde{\mathbf{X}} : \mathbb{R} \to \mathcal{S}_n^N$ of SGRDE (1) is named **stabilizing solution** if:

a) $\inf_{t\in\mathbb{R}} \left| det(R(t,i) + \sum_{k=1}^{r} B_k^T(t,i)\tilde{X}(t,i)B_k(t,i)) \right| > 0,\ i \in \mathfrak{N};$

b) the closed-loop system

$$dx(t) = (A_0(t,\eta_t) + B_0(t,\eta_t)\tilde{F}(t,\eta_t))x(t)dt + \\ + \sum_{k=1}^{r}(A_k(t,\eta_t) + B_k(t,\eta_t)\tilde{F}(t,\eta_t))x(t)dw_k(t) \tag{5}$$

is exponentially stable in mean square (ESMS), where

$$\tilde{F}(t,i) = -\left(R(t,i) + \sum_{k=1}^{r} B_k^T(t,i)\tilde{X}(t,i)B_k(t,i)\right)^{-1} (B_0^T(t,i)\tilde{X}(t,i) \\ + \sum_{k=1}^{r} B_k^T(t,i)\tilde{X}(t,i)A_k(t,i)),\quad 1 \le i \le N. \tag{6}$$

Invoking Theorem 5.6.5 from[1] in the case of SGRDE (1) we deduce:

Corollary 1.1. *Under the considered assumptions the SGRDE (1) has at most one bounded and stabilizing solution, additionally the bounded and stabilizing solution of (1), if it exists, is a periodic function of period θ.*

The problem of finding the necessary and sufficient conditions, or at least the sufficient conditions which guarantee the existence of the bounded and stabilizing solution $\tilde{\mathbf{X}}(t) = (\tilde{X}(t,1),...,\tilde{X}(t,N))$, $t \in \mathbb{R}$, of a SGRDE of type (1) is dependent upon the sign of the matrices

$$\mathbb{R}(t,\tilde{\mathbf{X}}(t),i) \triangleq R(t,i) + \sum_{k=1}^{r} B_k^T(t,i)\tilde{X}(t,i)B_k(t,i). \tag{7}$$

A set of necessary and sufficient conditions for the existence of the bounded and stabilizing solution $\tilde{\mathbf{X}}(\cdot)$ of (1) satisfying the sign condition

$$\mathbb{R}(t,\tilde{\mathbf{X}}(t),i) \ge \nu^2 I_m \tag{8}$$

$(t,i) \in \mathbb{R} \times \mathfrak{N}$ may be found in Theorem 5.6.6 from[1] for the finite dimensional case, while, for the infinite dimensional case, such conditions may be found in[3]. The problem of the existence of the bounded and stabilizing solution of SGRDE (1) satisfying the sign conditions of the form

$$\mathbb{R}(t,\tilde{\mathbf{X}}(t),i) \le -\nu^2 I_m \tag{9}$$

$(t,i) \in \mathbb{R}\times\mathfrak{N}$, was intensively studied in the literature in connection with the so called Bounded Real Lemma. For the readers convenience we refer to[4–6] in the case of systems affected by state multiplicative and/or disturbance multiplicative white noise perturbations and in[7] in the case of systems affected by Markov switching.

In the present work, our goal is to provide a set of sufficient conditions which guarantee the existence of the bounded and stabilizing solution $\tilde{\mathbf{X}}(\cdot)$ of SGRDE (1) satisfying the sign conditions:

$$sgn\mathbb{R}(t,\tilde{\mathbf{X}}(t),i) = sgn\ \ diag(-I_{m_1}, I_{m_2}) \tag{10}$$

that is $\mathbb{R}(t,\tilde{\mathbf{X}}(t),i)$ has m_1 negative eigenvalues and m_2 positive eigenvalues, $(t,i) \in \mathbb{R}\times\mathfrak{N}$, where $m_k \geq 1$, $k = 1,2$, are not depending upon (t,i) and $m_1+m_2 = m$. We shall extend to this more general framework the results proved in[8–11] in some special cases of (1). In the next developments the following assumption regarding the weights $M(t,i)$, $R(t,i)$ is done:

Assumption 1.2. *For each $(t,i) \in \mathbb{R}\times\mathfrak{N}$, $M(t,i) \geq 0$, $R(t,i)$ is invertible*

$$\begin{pmatrix} 0 & I_{m_2} \end{pmatrix} R(t,i) \begin{pmatrix} 0 & I_{m_2} \end{pmatrix}^T > 0 \tag{11}$$

2. Some preliminary results

We set $B_k(t,i) = \begin{pmatrix} B_{k1}(t,i) & B_{k2}(t,i) \end{pmatrix}$, $B_{kj}(t,i) \in \mathbb{R}^{n\times m_j}$, $0 \leq k \leq r$, $j = 1,2$, $R(t,i) = \begin{pmatrix} R_{11}(t,i) & R_{12}(t,i) \\ R_{12}^T(t,i) & R_{22}(t,i) \end{pmatrix}$, $R_{lj}(t,i) \in \mathbb{R}^{m_l\times m_j}$, $l,j = 1,2$. Based on these partitions, we introduce the subset $\mathcal{D}^{\mathfrak{R}}$ of continuous functions defined on some intervals from $\mathbb{R}$ taking values in the linear space $\mathcal{S}_n^N$, namely $\mathcal{D}^{\mathfrak{R}}$ contains all functions $\mathbf{X} : I_{\mathbf{X}} \subset \mathbb{R} \to \mathcal{S}_n^N$, $\mathbf{X}(t) = (X(t,1),...,X(t,N))$, $X(t,i)$ satisfying the conditions:

$$\mathbb{R}_{22}(t,i) = R_{22}(t,i) + \sum_{k=1}^{r} B_{k2}^T(t,i)X(t,i)B_{k2}(t,i) \geq \mu_2^2 I_{m_2} \tag{12}$$

$$\begin{aligned} &R_{11}(t,i) + \sum_{k=1}^{r} B_{k1}^T(t,i)X(t,i)B_{k1}(t,i) - (R_{12}(t,i) \\ &+\sum_{k=1}^{r} B_{k1}^T(t,i)X(t,i)B_{k2}(t,i))\mathbb{R}_{22}(t,i)^{-1}(R_{12}(t,i) \\ &\quad+\sum_{k=1}^{r} B_{k1}^T(t,i)X(t,i)B_{k2}(t,i))^T \leq -\mu_1^2 I_{m_1} \end{aligned} \tag{13}$$

for all $(t,i) \in I_{\mathbf{X}} \times \mathfrak{N}$, where μ_k, $k = 1,2$ do not depend upon (t,i), but may depend upon $\mathbf{X}(\cdot)$. Employing Lemma A1 from[12] we obtain:

Corollary 2.1. *Let $\tilde{\mathbf{X}}(\cdot) = (\tilde{X}(\cdot,1),...,\tilde{X}(\cdot,N))$ be a θ-periodic solution of the SGRDE (1). Under the Assumption* 1.2, *if $\tilde{X}(t,i) \geq 0$, $\forall (t,i) \in \mathbb{R} \times \mathfrak{N}$, then the following are equivalent:*

(i) the solution $\tilde{\mathbf{X}}(\cdot)$ satisfies the sign condition (10);

(ii) $\tilde{\mathbf{X}}(\cdot) \in \mathcal{D}^{\mathfrak{N}}$.

Taking $u(t) = \left(u_1^T(t)\ u_2^T(t)\right)^T$, $u_2(t) = K(t,\eta_t)x(t)$ in (3) and (4), we obtain:

$$\begin{aligned} dx(t) &= (A_{0\mathbf{K}}(t,\eta_t)x(t) + B_{01}(t,\eta_t)u_1(t))dt \\ &\quad + \sum_{k=1}^{r}(A_{k\mathbf{K}}(t,\eta_t)x(t) + B_{k1}(t,\eta_t)u_1(t))dw_k(t), \quad x(0) = x_0 \in \mathbb{R}^n \end{aligned} \tag{14}$$

$$J_{\mathbf{K}}(x_0,u_1) = E\left[\int_0^{\infty} \begin{pmatrix} x_{u_1}(t) \\ u_1(t) \end{pmatrix}^T \begin{pmatrix} M_{\mathbf{K}}(t,\eta_t) & L_{\mathbf{K}}(t,\eta_t) \\ L_{\mathbf{K}}^T(t,\eta_t) & R_{11}(t,\eta_t) \end{pmatrix} \begin{pmatrix} x_{u_1}(t) \\ u_1(t) \end{pmatrix} dt\right] \tag{15}$$

where $x_{u_1}(t)$ is the solution of (14) corresponding to the input $u_1(t)$ and

$$\begin{aligned} A_{k\mathbf{K}}(t,i) &= A_k(t,i) + B_{k2}(t,i)K(t,i), \quad 0 \leq k \leq r, \\ M_{\mathbf{K}}(t,i) &= M(t,i) + K^T(t,i)R_{22}(t,i)K(t,i) \\ L_{\mathbf{K}}(t,i) &= K^T(t,i)R_{12}^T(t,i). \end{aligned} \tag{16}$$

To the pair Σ formed by the system (14) and the quadratic functional (15) we associate the following SGRDE:

$$\begin{aligned} -\dot{Y}(t,i) &= A_{0\mathbf{K}}^T(t,i)Y(t,i) + Y(t,i)A_{0\mathbf{K}}(t,i) + \sum_{k=1}^{r} A_{k\mathbf{K}}^T(t,i)Y(t,i)A_{k\mathbf{K}}(t,i) \\ &\quad + \sum_{j=1}^{N} q_{ij}Y(t,j) - (Y(t,i)B_{01}(t,i) + \sum_{k=1}^{r} A_{k\mathbf{K}}^T(t,i)Y(t,i)B_{k1}(t,i) \\ &\quad + L_{\mathbf{K}}(t,i))(R_{11}(t,i) + \sum_{k=1}^{r} B_{k1}^T(t,i)Y(t,i)B_{k1}(t,i))^{-1}(B_{01}^T(t,i)Y(t,i) \\ &\quad + \sum_{k=1}^{r} B_{k1}^T(t,i)Y(t,i)A_{k\mathbf{K}}(t,i) + L_{\mathbf{K}}^T(t,i)) + M_{\mathbf{K}}(t,i). \end{aligned} \tag{17}$$

The concept of the stabilizing solution of SGRDE (17) may be defined in the same way as in the case of SGRDE (1).

Throughout this work $\mathcal{A}^{\Sigma}$ stands for the set of all continuous and θ-periodic functions $t \to \mathbf{K}(t) = (K(t,1), ..., K(t,N)) : \mathbb{R} \to \mathbb{R}^{m_2 \times n} \times ... \times \mathbb{R}^{m_2 \times n}$ having the properties:

(i) the system of stochastic linear differential equations

$$dx(t) = A_{0\mathbf{K}}(t,\eta_t)x(t)dt + \sum_{k=1}^{r} A_{k\mathbf{K}}(t,\eta_t)x(t)dw_k(t) \tag{18}$$

is ESMS;

(ii) the corresponding SGRDE (17) has a stabilizing and θ-periodic solution $\tilde{\mathbf{Y}}_{\mathbf{K}}(t) = (\tilde{Y}_{\mathbf{K}}(t,1), ..., \tilde{Y}_{\mathbf{K}}(t,N))$, $t \in \mathbb{R}$, satisfying the sign conditions

$$R_{11}(t,i) + \sum_{k=1}^{r} B_{k1}^T(t,i)\tilde{Y}_{\mathbf{K}}(t,i)B_{k1}(t,i) \leq -\mu^2 I_{m_1} \tag{19}$$

for all $(t,i) \in \mathbb{R} \times \mathfrak{N}$, μ being a constant which may depend upon $\mathbf{K}(\cdot)$.

Proposition 2.1. *Under the Assumptions 1.1 and 1.2 for each* $\mathbf{K}(\cdot) \in \mathcal{A}^{\Sigma}$ *there exist* $\rho = \rho(\mathbf{K}) > 0$ *such that* $J_{\mathbf{K}}(0,u_1) \leq -\rho E[\int_0^{\infty} |u_1(t)|^2 dt]$ *for all* $u_1 \in L^2_{\eta w}\{\mathbb{R}_+, \mathbb{R}^{m_1}\}$.

Proof. (hint) One shows that

$$\sup\{J_{\mathbf{K}}(0,u_1); u_1 \in L^2_{\eta w}\{\mathbb{R}_+, \mathbb{R}^{m_1}\}, \|u_1\| = 1\} < 0.$$

The conclusion is obtained using the equality $J_{\mathbf{K}}(0, \frac{u_1}{\|u_1\|}) = \frac{1}{\|u_1\|^2} J_{\mathbf{K}}(0,u_1)$ for all $0 \neq u_1 \in L^2_{\eta w}\{\mathbb{R}_+, \mathbb{R}^{m_1}\}$. We recall that $\|u_1\| = (E[\int_0^{\infty} |u_1(t)|^2 dt])^{1/2}$. □

For each $\mathbf{K}(\cdot) \in \mathcal{A}^{\Sigma}$ and each $\tau \in \mathbb{R}$, let $\mathbf{Y}^{\tau}_{\mathbf{K}}(t) = (Y^{\tau}_{\mathbf{K}}(t,1), ..., Y^{\tau}_{\mathbf{K}}(t,N))$ be the solution of (17) satisfying $Y^{\tau}_{\mathbf{K}}(\tau,i) = 0$, $1 \leq i \leq N$. Let $\mathfrak{I}_{\mathbf{K}}(\tau) \subset (-\infty; \tau]$ be the maximal interval where the solution $\mathbf{Y}^{\tau}_{\mathbf{K}}(\cdot)$ is well defined.

Using the comparison technique of solutions of (17) one may prove:

Proposition 2.2. *Under the Assumptions 1.1 and 1.2, the following hold:*

(i) $\mathfrak{I}_{\mathbf{K}}(\tau) = (-\infty, \tau]$ *for all* $\mathbf{K}(\cdot) \in \mathcal{A}^{\Sigma}$ *and all* $\tau \in \mathbb{R}$;

(ii) for each $\mathbf{K}(\cdot) \in \mathcal{A}^{\Sigma}$ *we have* $0 \leq Y^{\tau_1}_{\mathbf{K}}(t,i) \leq Y^{\tau_2}_{\mathbf{K}}(t,i) \leq \tilde{Y}_{\mathbf{K}}(t,i)$, *for all* $-\infty < t \leq \tau_1 < \tau_2$, $i \in \mathfrak{N}$, $\tilde{\mathbf{Y}}_{\mathbf{K}}(t)$ *being the stabilizing and* θ*-periodic solution of (17);*

$$(iii) \qquad R_{11}(t,i) + \sum_{k=1}^{r} B_{k1}^T(t,i)Y^{\tau}_{\mathbf{K}}(t,i)B_{k1}(t,i) \leq -\mu^2 I_{m_1} \tag{20}$$

for all $(t,i) \in (-\infty; \tau) \times \mathfrak{N}$, $\tau \in \mathbb{R}$.

Besides the cost functional (15) we consider also the quadratic functional on finite horizon time: $J_{\mathbf{K}}(t_0,\tau,x,i;u_1) =$

$$= E\left[\int_{t_0}^{\tau}\begin{pmatrix} x_{u_1}(t) \\ u_1(t)\end{pmatrix}^T \begin{pmatrix} M_{\mathbf{K}}(t,\eta_t) & L_{\mathbf{K}}(t,\eta_t) \\ L_{\mathbf{K}}^T(t,\eta_t) & R_{11}(t,\eta_t)\end{pmatrix}\begin{pmatrix} x_{u_1}(t) \\ u_1(t)\end{pmatrix} dt|\eta_{t_0}=i\right] \quad (21)$$

$0 \le t_0 < \tau$, $i \in \mathfrak{N}$, $x_0 \in \mathbb{R}^n$, $u_1 \in L^2_{\eta w}\{[t_0,\tau],\mathbb{R}^{m_1}\}$. Employing the Itô type formula and (20) one obtains:

Lemma 2.1. *Under the considered assumptions if* $\mathbf{K}(\cdot) \in \mathcal{A}^{\Sigma}$ *we have*

$$J_{\mathbf{K}}(t_0,\tau,x_0,i;u_1) \le x_0^T Y_{\mathbf{K}}^{\tau}(t_0,i)x_0 \quad (22)$$

for all $x_0 \in \mathbb{R}^n$, $i \in \mathfrak{N}$, $u_1 \in L^2_{\eta w}\{[t_0,\tau],\mathbb{R}^{m_1}\}$.

Proposition 2.3. *We set:*

$$\tilde{J}_{\mathbf{K}}(0,\tau,x_0;u_1) = \sum_{i=1}^{N}\pi_0(i)J_{\mathbf{K}}(0,\tau,x_0,i;u_1) \quad (23)$$

where $\pi_0(i) = \mathcal{P}\{\eta_0 = i\}$. *If* $\mathbf{K}(\cdot) \in \mathcal{A}^{\Sigma}$, *then there exist positive constants* ρ_0, ρ_1 *such that*

$$\tilde{J}_{\mathbf{K}}(0,\tau,x_0;u_1) \le \rho_0|x_0|^2 - \rho_1 E\left[\int_0^{\tau}|u_1(t)|^2dt\right] \quad (24)$$

for all $\tau > 0$ $u_1 \in L^2_{\eta w}\{[0,\tau],\mathbb{R}^{m_1}\}, x_0 \in \mathbb{R}^n$.

Proof. (hint) Let $\mathbf{Z}_{\tau}(t) = (Z_{\tau}(t,1),...,Z_{\tau}(t,N))$ be the solution of the linear differential equation

$$\dot{Z}(t,i) + A_{0\mathbf{K}}^T(t,i)Z(t,i) + Z(t,i)A_{0\mathbf{K}}(t,i) + \sum_{k=1}^{r}A_{k\mathbf{K}}^T(t,i)Z(t,i)A_{k\mathbf{K}}(t,i) + \sum_{j=1}^{N}q_{ij}Z(t,j) + M_{\mathbf{K}}(t,i) = 0 \quad (25)$$

satisfying the conditions $Z_{\tau}(\tau,i) = 0, i \in \mathfrak{N}$. Since (25) is a linear differential equation, it follows that $\mathbf{Z}_{\tau}(\cdot)$ is well defined for every $t \le \tau$. Furthermore, the exponential stability in mean square of the system (18) allows us to deduce that there exists $\gamma > 0$ not depending upon τ, t, i such that $0 \le Z_{\tau}(t,i) \le \gamma I_n$, $\tau > 0, t \le \tau, i \in \mathfrak{N}$. Applying the Itô type formula to the function $x^T Z_\tau(t,i)x$, involving (25), (23) and based on the splitting $x(t;x_0,u_1) = x(t;x_0,0) + x(t;0,u_1)$ one obtains

$$\tilde{J}_{\mathbf{K}}(0,\tau,x_0;u_1) = \sum_{i=1}^{N} \pi_0(i) x_0^T Z_\tau(0,i)x_0 + 2E\left[\int_0^\tau x^T(\tau,x_0,0)\mathbb{L}(t,\eta_t)u_1(t)dt\right]$$
$$+ \tilde{J}_{\mathbf{K}}(0,\tau,0;u_1), \quad \tau > 0, u_1 \in L^2_{\eta w}\{[0,\tau],\mathbb{R}^{m_1}\}, x_0 \in \mathbb{R}^n$$

where $\mathbb{L}(t,i) = \sum_{k=1}^{r} A_{k\mathbf{K}}^T(t,i)Z_\tau(t,i)B_{k1}(t,i) + L_{\mathbf{K}}(t,i)$ and $x(t;x_0,u_1)$ is the solution of the system (14) satisfying the initial condition $x(0;x_0,u_1) = x_0$. Further, by using Proposition 2.1 and reasoning as in the proof of Lemma 8.2.2 from[1] one obtains (24). □

3. The main result

For each $\tau > 0$, let $\mathbf{X}_\tau(t) = (X_\tau(t,1),...,X_\tau(t,N))$ be the solution of (1) satisfying the conditions $X_\tau(\tau,i) = 0$, $1 \le i \le N$. Let $\mathfrak{I}(\tau) \subset (-\infty;\tau]$ be the maximal interval of $t \le \tau$ for which $\mathbf{X}_\tau(\cdot)$ is well defined. If the Assumption 1.2 is fulfilled, $\mathfrak{I}(\tau)$ is not empty. Let $F_\tau(t,i)$ be the gains defined by

$$F_\tau(t,i) = -\left(R(t,i) + \sum_{k=1}^{r} B_k^T(t,i)X_\tau(t,i)B_k(t,i)\right)^{-1} (B_0^T(t,i)X_\tau(t,i)$$
$$+ \sum_{k=1}^{r} B_k^T(t,i)X_\tau(t,i)A_k(t,i)). \tag{26}$$

We set $F_{1\tau}(t,i) = \begin{pmatrix} I_{m_1} & 0 \end{pmatrix} F_\tau(t,i)$, $F_{2\tau}(t,i) = \begin{pmatrix} 0 & I_{m_2} \end{pmatrix} F_\tau(t,i)$, $t \in \mathfrak{I}(\tau)$, $i \in \mathfrak{N}$. If $t_0 \in \mathfrak{I}(\tau)$, we consider the quadratic cost functionals

$$J(t_0,\tau,x_0,i;u_1,u_2) = E\left[\int_{t_0}^{\tau} (x_u^T(t)M(t,\eta_t)x_u(t)\right.$$
$$\left. + \begin{pmatrix} u_1^T(t) & u_2^T(t) \end{pmatrix} R(t,\eta_t) \begin{pmatrix} u_1^T(t) & u_2^T(t) \end{pmatrix}^T)dt|\eta_{t_0} = i\right] \tag{27}$$

for all $x_0 \in \mathbb{R}^n$, $u_k \in L^2_{\eta w}\{[t_0,\tau],\mathbb{R}^{m_k}\}$, $k = 1,2$, $i \in \mathfrak{N}$.

Let us consider the controls,

$$\tilde{u}_{k\tau}(t) = F_{k\tau}(t,\eta_t)\tilde{x}_\tau(t) \tag{28}$$

$k = 1,2$, where $\tilde{x}_\tau(\cdot)$ is the solution of the initial value problem:

$$dx(t) = (A_0(t,\eta_t) + B_0(t,\eta_t)F_\tau(t,\eta_t))x(t)dt$$
$$+ \sum_{k=1}^{r} (A_k(t,\eta_t) + B_k(t,\eta_t)F_\tau(t,\eta_t))x(t)dw_k(t), t > t_0 \tag{29}$$
$$\tilde{x}_\tau(t_0) = x_0.$$

It is obvious that $\tilde{u}_{k\tau}(\cdot) \in L^2_{\eta w}\{[t_0,\tau],\mathbb{R}^{m_k}\}$, $k=1,2$. By using the idea of the proof in Proposition 5.10 from[11] one can prove the next result:

Lemma 3.1. *With the notations from above, the following hold:*

(i) $J(t_0,\tau,x_0,i;u_1,u_2)=x_0^T X_\tau(t_0,i)x_0+J(t_0,\tau,0,i;u_1-\tilde{u}_{1\tau},u_2-\tilde{u}_{2\tau})$ (30)

for all $\tau>0$, $t_0\in\mathfrak{I}(\tau)$, $u_k(\cdot)\in L^2_{\eta w}\{[t_0,\tau],\mathbb{R}^{m_k}\}$, $i\in\mathfrak{N}$, $x_0\in\mathbb{R}^n$.

(ii) $x_0^T X_\tau(t_0,i)x_0 \le J(t_0,\tau,x_0,i;\tilde{u}_{1\tau},u_2)$, *for all* $\tau>0, t_0\in\mathfrak{I}(\tau)$, $x_0\in\mathbb{R}^n$, $i\in\mathfrak{N}$, $u_2\in L^2_{\eta w}\{[t_0,\tau],\mathbb{R}^{m_2}\}$.

Proposition 3.1. *Assume: a) the Assumption 1.1 and Assumption 1.2 are fulfilled; b) the set $\mathcal{A}^\Sigma$ is not empty.*

Under these conditions we have $0\le X_\tau(t_0,i)\le \tilde{Y}_{\mathbf{K}}(t_0,i)$ for all $\tau>0$, $(t_0,i)\in\mathfrak{I}(\tau)\times\mathfrak{N}$, $\mathbf{K}\in\mathcal{A}^\Sigma$, $\tilde{\mathbf{Y}}_{\mathbf{K}}(t)=(\tilde{Y}_{\mathbf{K}}(t,1),...,\tilde{Y}_{\mathbf{K}}(t,N))$ being the stabilizing and θ-periodic solution of the corresponding SGRDE (17).

Proof. Let $\mathbf{K}(\cdot)\in\mathcal{A}^\Sigma$ be arbitrary but fixed and $u_{2\mathbf{K}}(t)=K(t,\eta_t)\xi(t)$, $\xi(t)$ being the solution of the initial value problem

$$d\xi(t)=[(A_0(t,\eta_t)+B_{02}(t,\eta_t)K(t,\eta_t))\xi(t)+B_{01}(t,\eta_t)\tilde{u}_{1\tau}(t)]dt$$

$$+\sum_{k=1}^{r}[(A_k(t,\eta_t)+B_{k2}(t,\eta_t)K(t,\eta_t))\xi(t)+B_{k1}(t,\eta_t)\tilde{u}_{1\tau}(t)]dw_k(t)$$

$t>t_0$, $\xi(t_0)=x_0$. It is obvious that $u_{2\mathbf{K}}\in L^2_{\eta w}\{[t_0,\tau],\mathbb{R}^{m_2}\}$ and the following equality holds:

$$J(t_0,\tau,x_0,i;\tilde{u}_{1\tau},u_{2\mathbf{K}})=J_{\mathbf{K}}(t_0,\tau,x_0,i;\tilde{u}_{1\tau}). \tag{31}$$

According with Proposition 2.2 we deduce that $\mathfrak{I}_{\mathbf{K}}(\tau)=(-\infty;\tau]$ and, hence, $\mathfrak{I}(\tau)\subset\mathfrak{I}_{\mathbf{K}}(\tau)$. Employing Lemma 3.1 (ii), Lemma 2.1 together with (31) we may obtain $x_0^T X_\tau(t_0,i)x_0\le x_0^T Y^\tau_{\mathbf{K}}(t_0,i)x_0$ for all $x_0\in\mathbb{R}^n$, $(t_0,i)\in\mathfrak{I}(\tau)\times\mathfrak{N}$. So, we obtain

$$X_\tau(t_0,i)\le Y^\tau_{\mathbf{K}}(t_0,i)\le\tilde{Y}_{\mathbf{K}}(t_0,i) \tag{32}$$

for all $\tau>0$, $(t_0,i)\in\mathfrak{I}(\tau)\times\mathfrak{N}$. Further, (20) and (32) yield

$$R_{11}(t,i)+\sum_{k=1}^{r}B_{k1}^T(t,i)X_\tau(t,i)B_{k1}(t,i)\le-\mu^2 I_{m_1} \tag{33}$$

for all $(t,i)\in\mathfrak{I}(\tau)\times\mathfrak{N}$, $\tau>0$. The differential equation (1) satisfied by $\mathbf{X}_\tau(\cdot)$ may be written in the form of a backward Lyapunov differential equation. From the properties of the solutions of a backward differential Lyapunov equation together with (33), we deduce that $X_\tau(t,i)\ge 0$ for all $\tau>0$, $(t,i)\in\mathfrak{I}(\tau)\times\mathfrak{N}$. □

Corollary 3.1. *Under the assumptions in Proposition 3.1, we have* $[0,\tau] \subset \mathfrak{I}(\tau)$*, for all* $\tau > 0$.

By using Proposition 3.1 one can easily prove the following result:

Proposition 3.2. *Under the assumptions from Proposition 3.1 we have* $X_{\tau_1}(t,i) \leq X_{\tau_2}(t,i)$*, for all* $0 \leq t \leq \tau_1 < \tau_2$, $i \in \mathfrak{N}$.

Before stating the next result, let us introduce the following auxiliary system:

$$\begin{aligned} dx(t) &= A_0(t,\eta_t)x(t)dt + \sum_{k=1}^{r} A_k(t,\eta_t)x(t)dw_k(t) \\ y(t) &= C(t,\eta_t)x(t) \end{aligned} \tag{34}$$

where $C(t,i)$ is obtained from factorization $M(t,i) = C^T(t,i)C(t,i)$. We recall that the system (34) is stochastic detectable if there exist the continuous and θ-periodic functions $t \to L(t,i) : \mathbb{R} \to \mathbb{R}^{n \times n_y}$ such that the closed-loop system

$$dx(t) = (A_0(t,\eta_t) + L(t,\eta_t)C(t,\eta_t))x(t)dt + \sum_{k=1}^{r} A_k(t,\eta_t)x(t)dw_k(t)$$

is ESMS.

The next result provides a set of conditions which guarantees the existence of a θ-periodic solution $\tilde{\mathbf{X}}(\cdot)$ of the SGRDE (1) which satisfies the sign condition (10).

Theorem 3.1. *Assume: a) the Assumptions 1.1 and 1.2 are fulfilled;*

b) the set $\mathcal{A}^{\Sigma}$ *is not empty;*

c) the auxiliary system (34) is stochastic detectable.

Under these conditions, the SGRDE (1) has a θ*-periodic solution* $\tilde{\mathbf{X}} : \mathbb{R} \to \mathcal{S}_n^N$ *having the properties:*

(i) if $\tilde{\mathbf{X}}(t) = (\tilde{X}(t,1), ..., \tilde{X}(t,N))$*, then* $\tilde{X}(t,i) \geq 0$, $\forall (t,i) \in \mathbb{R} \times \mathfrak{N}$;

(ii) $\tilde{\mathbf{X}}(\cdot)$ *satisfies the sign condition (10);*

(iii) the system

$$\begin{aligned} dx(t) &= (A_0(t,\eta_t) + B_{02}(t,\eta_t)\tilde{F}_2(t,\eta_t))x(t)dt \\ &+ \sum_{k=1}^{r} (A_k(t,\eta_t) + B_{k2}(t,\eta_t)\tilde{F}_2(t,\eta_t))x(t)dw_k(t) \end{aligned} \tag{35}$$

is ESMS, where $\tilde{F}_2(t,i) = \begin{pmatrix} 0 & I_{m_2} \end{pmatrix} \tilde{F}(t,i)$, $\tilde{F}(t,i)$ *being computed as in (6).*

(iv) $\tilde{\mathbf{X}}(\cdot)$ *is the stabilizing and* θ*-periodic solution of (1).*

Proof. From Proposition 3.1 and Proposition 3.2, one obtains that the functions $\tilde{X}(\cdot, i)$ are well defined by

$$\tilde{X}(t,i) = \lim_{\tau \to \infty} X_\tau(t,i) \tag{36}$$

$t \in \mathbb{R}$, $i \in \mathfrak{N}$. Employing the Lebesque's dominate convergence theorem one shows in a standard way, that $t \to \tilde{\mathbf{X}}(t) = (\tilde{X}(t,1), ..., \tilde{X}(t,N))$ is a global solution of (1). Proceeding as in the proof of Theorem 5.5.1 from[1], one shows that $\tilde{X}(\cdot)$ defined via (36) is a periodic function with period θ. Since $\tilde{X}(t,i) \geq X_\tau(t,i)$, $\forall \tau > 0$ we obtain that $\tilde{X}(t,i) \geq 0$, $\forall (t,i) \in \mathbb{R} \times \mathfrak{N}$. Thus, we have obtained that (i) holds and additionally $\tilde{X}(\cdot)$ satisfies (12). Further, combining Proposition 3.1 and (19) we deduce that $R_{11}(t,i) + \sum_{k=1}^{r} B_{k1}(t,i)X_\tau(t,i)B_{k1}(t,i) \leq -\mu^2 I_{m_1}$ with μ not depending upon τ, t, i. Taking the limit for $\tau \to \infty$ we obtain

$$R_{11}(t,i) + \sum_{k=1}^{r} B_{k1}^T(t,i)\tilde{X}(t,i)B_{k1}(t,i) \leq -\mu^2 I_{m_1}. \tag{37}$$

From (37) it follows immediately that the solution $\tilde{\mathbf{X}}(\cdot)$ satisfies the sign condition (13). So, we may conclude via Corollary 2.1 that the solution $\tilde{\mathbf{X}}(\cdot)$ defined via (36) satisfies the sign condition (10). Thus, we have proved that the assertion (ii) in the statement holds. It remains to check the validity of (iii). First, let us remark that if the solution $\tilde{\mathbf{X}}(\cdot)$ satisfies the sign condition (10) the corresponding feedback gains $\tilde{F}(t,i)$ are well defined by (6). By a standard procedure, involving detectability of the system (34) (see Proposition 6.2 in[11]) one obtains that the system (35) is ESMS. The details are omitted for shortness. To prove (iv) one can proceed as in the proof of Theorem 5.8 from[8]. □

Remark 3.1. To prove that the solution $\tilde{\mathbf{X}}(\cdot)$ of (1) defined via (36) satisfies the sign conditions (13), we have proved firstly that it satisfies some stronger sign conditions, namely (37). This fact is due to the way in which the set $\mathcal{A}^\Sigma$ was defined. Changing the definition of the set $\mathcal{A}^\Sigma$ one can prove a result similar to the one from Theorem 3.1, but with the difference that the solution $\tilde{\mathbf{X}}(\cdot)$ satisfies a condition of type (13) instead of the stronger condition (37). This fact will be presented into a future paper, where also the detailed proofs of the results included in the present work will be given.

References

1. V. Dragan, T. Morozan and A.M. Stoica, *Mathematical Methods in Robust Control of Linear Stochastic Systems*, 2nd edn. (New York: Springer, 2013).
2. H. Abou-Kaudil, G. Freiling, V. Ionescu and G. Jank, *Matrix Riccati equations in Control Systems Theory,* (Birhauser, Basel, 2003).
3. V.M. Ungureanu and V. Dragan, *Nonlinear differential equations of Riccati type on ordered Banach spaces*, Electronic Journal of Qualitative Theory of Differential Equations, Proc. 9th Coll. QTDE, (2012), 17, 1–22.
4. T. Morozan, *Parametrized Riccati equation and inputoutput operators for timevarying stochastic differential equations with state dependent noise*, Stud. Cercet. Mat.,1, (1999), pp. 99-115.
5. D. Hinrichsen and A.J. Pritchard, *Stochastic H_∞*, SIAM J. Control Optim., 36,(1998), pp. 1504-1538.
6. I.R. Petersen, V.A. Ugrinovskii and A.V. Savkin, *Robust Control Design using H_∞ Methods*, (Springer,Berlin, 2000).
7. T. Morozan, *Parametrized Riccati equations for controlled linear differential systems with jump Markov perturbations*, Stoch. Anal. Appl., 14, 4, (1998), pp.661–682.
8. V. Dragan and T. Morozan, *Global solutions to a game-theoretic Riccati equation of stochastic control*, J. Differ. Equ., 138, 2, (1997), pp. 328-350.
9. Y. Feng, B.D.O. Anderson, *An iterative algorithm to solve state-perturbed stochastic algebraic Riccati equations in LQ zero-sum games*, Systems and Control Letters, 59, (2010), pp.50–56.
10. D. Vrabie, F. Lewis, *Adaptive dynamic programming for online solution of a zero-sum differential game*, Journal Contr. Theory and Appl., 9, 3, (2011), pp. 353–360.
11. V. Dragan and T. Morozan, *Game theoretic coupled Riccati equations associated to controlled linear differential systems with jump Markov perturbations*, Stochastic Analysis and Appl., 19, 5, (2001), pp. 715–751.
12. V. Dragan, A. Halanay and V. Ionescu, *Infinite horizon disturbance attenuation for discrete time systems. A Popov-Yakubovich approach*, Integr. Equat. Oper. Th., 19, (1994), pp. 153–215.

A maximum principle for a class of first order differential operators

Maria Fărcăşeanu[a], Mihai Mihăilescu[b]

Department of Mathematics, University of Craiova, 200585 Craiova, Romania
E-mail: [a]farcaseanu.maria@yahoo.com, [b]mmihailes@yahoo.com

Denisa Stancu-Dumitru

Department of Mathematics and Computer Science, University Politehnica of Bucharest, 060042 Bucharest, Romania
E-mail: denisa.stancu@yahoo.com

In this paper we derive a maximum-type principle for a certain class of differential operators from some earlier results related to an eigenvalue problem involving the $p(\cdot)$-Laplacian.

Keywords: maximum principle; differential equation; differential inequality; elliptic operators; eigenvalue problem.

1. Introduction

The goal of this paper is to present how a series of results obtained in connection with an eigenvalue problem involving variable exponents can be used in order to obtain a maximum principle which complements the classical maximum principle for elliptic operators. In particular, the advances presented here improve earlier results obtained by Fan, Zhang & Zhao[1] in Theorem 3.1, Mihăilescu, Rădulescu & Stancu-Dumitru[2] in Theorem 1 and Mihăilescu, Moroşanu & Stancu-Dumitru[3] in Proposition 1. The paper is organized as follows: in Subsections 1.1 and 1 we recall the statement of the maximum principle for elliptic operators and some known results concerning an eigenvalue problem for the $p(\cdot)$-Laplace operator; in Subsection 2 we present the main results of our paper; Section 2 is devoted to the proofs of the main results.

Throughout this paper we assume $\Omega \subset \mathbb{R}^N$ ($N \geq 2$) is a bounded domain with smooth boundary.

1.1. *The maximum principle for elliptic operators*

One of the basic tools involved in the study of PDE's is the *maximum principle.* The importance of the maximum principles is mainly due to the fact that it *enables us to obtain information about differential equations and inequalities without any explicit knowledge of the solutions themselves* (according to Pucci & Serrin[4] p. IX). The most common statement of the maximum principle leads us to think back to the *classical* maximum principle for *elliptic* operators. Recall that operator L in Ω,

$$Lp=\sum_{i,j=1}^{N} a_{ij}(x)\frac{\partial^2 p}{\partial x_i \partial x_j}+\sum_{i=1}^{N} b_i(x)\frac{\partial p}{\partial x_i}\,,$$

for $p\in C^2(\Omega)\cap C(\overline{\Omega})$, is called *elliptic* if the matrix $[a_{ij}(x)]$ is positive, this means that for any compact subset K of Ω, the quadratic form

$$\sum_{i,j=1}^{N} a_{ij}(x)\xi_i\xi_j\,,$$

is positive and uniformly bounded from 0 for all $x\in K$ and all $\xi\in\mathbb{R}^N$ with $|\xi|=1$, and the coefficients a_{ij}, b_i are locally bounded in Ω.

The classical maximum principle for elliptic operators reads as follows (see, e.g. Theorem 3.1 in Gilbarg & Trudinger[5], Theorem 2.1.1 in Pucci & Serrin[4] or Chapter 2 in Protter & Weinberger[6]):

Theorem 1.1. *Let L be a linear elliptic differential operator in Ω. Suppose that*

$$Lp\geq 0\ \ (\leq 0),\qquad \text{in}\ \Omega\,, \tag{1}$$

with $p\in C^2(\Omega)\cap C(\overline{\Omega})$. Then the maximum (minimum) of p in $\overline{\Omega}$ is achieved on $\partial\Omega$, that is

$$\max_{x\in\overline{\Omega}} p(x)=\max_{x\in\partial\Omega} p(x)\ \ (\min_{x\in\overline{\Omega}} p(x)=\min_{x\in\partial\Omega} p(x))\,.$$

Remark 1. a. As it is pointed out by Pucci & Serrin[4] *the hypothesis that p is twice differentiable is essential for Theorem 1.1.*
b. The prototype differential operator for which Theorem 1.1 holds true is the *Laplace operator*, i.e.

$$\Delta p=\sum_{i=1}^{N}\frac{\partial^2 p}{\partial x_i^2}\,.$$

c. If we replace condition (1) by

$$Lp + c(x)p \geq 0 \ \ (\leq 0), \qquad \text{in } \Omega\,, \tag{2}$$

with $c(x) \leq 0$ in Ω then the conclusion of Theorem 1.1 remains valid if we assume that $p \geq 0$ in Ω ($p \leq 0$ in Ω).

A natural question can be considered in connection with the *classical maximum principle for elliptic operators*: **Does there exist another class of operators (different from elliptic operators) for which the conclusion of Theorem 1.1 holds true?**

In this paper we will give a *positive* answer to the above question.

1.2. *An eigenvalue problem for the* $p(\cdot)$*-Laplace operator*

In this section we allow $N \geq 1$. Let $p : \Omega \to (1, \infty)$ be a continuous function satisfying $\inf_{x\in\Omega} p(x) > 1$ and $\sup_{x\in\Omega} p(x) < \infty$. Define the $p(\cdot)$*-Laplace operator* by

$$\Delta_{p(\cdot)} u := \operatorname{div}(|\nabla u|^{p(\cdot)-2}\nabla u)\,.$$

Consider the eigenvalue problem

$$\begin{cases} -\Delta_{p(x)} u(x) = \lambda |u(x)|^{p(x)-2} u(x), & \text{for } x \in \Omega \\ u(x) = 0, & \text{for } x \in \partial\Omega\,. \end{cases} \tag{3}$$

We say that $\lambda \in \mathbb{R}$ is an *eigenvalue* of problem (3) if there exists a *nontrivial* solution u of the problem (such a solution is defined usually in a weak sense for a function u belonging to a suitable Sobolev-type space). Let us denote by Λ the set of all eigenvalues of problem (3) and define

$$\lambda_\star := \inf \Lambda\,.$$

Define also the quotient

$$\lambda_1 := \inf_{u \in C_0^1(\Omega)\setminus\{0\}} \frac{\int_\Omega |\nabla u|^{p(x)}\, dx}{\int_\Omega |u|^{p(x)}\, dx}\,.$$

Fan, Zhang & Zhao[1] proved that $\lambda_\star \geq 0$, the case $\lambda_\star = 0$ can occur and $\lambda_\star > 0$ if and only if $\lambda_1 > 0$ (see, Lemma 3.1 in Ref.[1]). Moreover, in Theorem 3.2 Ref.[1] has been showed that *if* $N = 1$ *then* $\lambda_1 > 0$ *if and only if* $p(\cdot)$ *is a monotone function.* Unfortunately, in high-dimensional spaces a necessary and sufficient condition relating the positivity of λ_1 and the properties of function $p(\cdot)$ is not available in the literature. However, some advances can be also found in Theorem 3.1 Ref.[1] where it was proved that *if there is an open subset* $U \subset \Omega$ *and a point* $x_0 \in U$ *such that* $p(x_0) <$ *(or*

>) $p(x)$ for all $x \in \partial\Omega$, then $\lambda_\star = 0$. On the other hand, by Theorem 3.3 Ref.[1] we know that *if $N \geq 2$ and there exists a vector $l \in \mathbb{R}^N \setminus \{0\}$ such that for any $x \in \Omega$ function $f(t) := p(x + tl)$ is monotone for $t \in I_x := \{t : x + tl \in \Omega\}$ then $\lambda_1 > 0$.* In a similar context, Mihăilescu, Rădulescu & Stancu-Dumitru showed that *if $N \geq 2$, $p \in C^1(\Omega)$ and there exists a vectorial function $\overrightarrow{a} : \Omega \to \mathbb{R}^N$ such that $\overrightarrow{a} \in C^1(\Omega; \mathbb{R}^N)$ and*

$$\exists\ a_0 > 0 \quad \text{s.t. } \operatorname{div} \overrightarrow{a}(x) \geq a_0 > 0, \quad \forall\, x \in \overline{\Omega}$$

and

$$\overrightarrow{a}(x) \cdot \nabla p(x) = 0, \quad \forall\, x \in \Omega\,,$$

then $\lambda_1 > 0$ (see, Theorem 1 Ref.[2]). Note that similar results as those given by Theorem 1 Ref.[2] were also obtained in a more general context by Allegretto[7] and by Mihăilescu, Moroşanu & Stancu-Dumitru[3], Theorem 1.

1.3. *The main results*

Using as starting point the results presented in Section 1 we will obtain some maximum principles which complement the result of Theorem 1.1 from Section 1.1. The first main result of this paper is given by the following theorem.

Theorem 1.2. *Let $\overrightarrow{a} : \Omega \to \mathbb{R}^N$ be a vectorial function such that $\overrightarrow{a} \in C^1(\Omega; \mathbb{R}^N) \cap C(\overline{\Omega}; \mathbb{R}^N)$. Assume that there exists a positive constant $a_0 > 0$ such that*

$$\operatorname{div} \overrightarrow{a}(x) \geq a_0 > 0, \quad \forall\, x \in \overline{\Omega}\,. \tag{4}$$

If $p \in C^1(\Omega) \cap C(\overline{\Omega})$ is a solution of the differential inequality

$$\overrightarrow{a}(x) \cdot \nabla p(x) \leq 0, \quad \forall\, x \in \Omega\,, \tag{5}$$

then for each open set $U \subset \Omega$ the minimum of p in $\overline{U}$ is achieved on ∂U.

If $p \in C^1(\Omega) \cap C(\overline{\Omega})$ is a solution of the differential inequality

$$\overrightarrow{a}(x) \cdot \nabla p(x) \geq 0, \quad \forall\, x \in \Omega\,, \tag{6}$$

then for each open set $U \subset \Omega$ the maximum of p in $\overline{U}$ is achieved on ∂U.

If $p \in C^1(\Omega) \cap C(\overline{\Omega})$ is a solution of the PDE's

$$\overrightarrow{a}(x) \cdot \nabla p(x) = 0, \quad \forall\, x \in \Omega\,, \tag{7}$$

then for each open set $U \subset \Omega$ the maximum and minimum of p in $\overline{U}$ are achieved on ∂U.

Corollary 1.1. *Assume that condition (4) is satisfied. Assume* p_1, $p_2 \in C^1(\Omega) \cap C(\overline{\Omega})$ *are such that*

$$\vec{a}(x) \cdot \nabla p_1(x) \ \le \ (\ge, \ =) \ \vec{a}(x) \cdot \nabla p_2(x), \quad \forall\, x \in \Omega\,.$$

Let $U \subset \Omega$ *be an open set such that*

$$p_1(x) \ge \ (\le, \ =) \ p_2(x), \quad \forall\, x \in \partial U\,.$$

Then

$$p_1(x) \ge \ (\le, \ =) \ p_2(x), \quad \forall\, x \in U\,.$$

Remark 2. a. In the case when condition (7) in Theorem 1.2 is satisfied the conclusion of the theorem remains true if we replace assumption (4) by

$$\operatorname{div} \vec{a}(x) \le -a_0 < 0, \quad \forall\, x \in \overline{\Omega}\,.$$

b. Note that in order to obtain the conclusion of Theorem 1.2 it is enough to have $p \in C^1(\Omega)$ (instead of $p \in C^2(\Omega)$ as in the case of Theorem 1.1).
c. Theorem 1.2 complements some earlier results obtained by Fan, Zhang & Zhao[1] in Theorem 3.1, Mihăilescu, Rădulescu & Stancu-Dumitru[2] in Theorem 1 and Mihăilescu, Moroşanu & Stancu-Dumitru[3] in Proposition 1.
d. The analytic conclusion of Theorem 1.2 has a geometrical explanation. Indeed, from a geometrical point of view the assumption (5) means that $\vec{a}$ and ∇p form at every point of the domain an obtuse angle which leads to the minimum of p on the boundary of the domain.
Example. We point out an example of functions p and $\vec{a}$ for which Theorem 1.2 can be applied. Let $N = 2$. Taking $\Omega = B_{\frac{1}{3^{1/3}}}(0)$, $\vec{a}(x) = (-x_1, 2x_2)$ and $p(x) = x_1^2 x_2 + \frac{3}{2}$ it is easy to check that relations (4) and (7) are fulfilled. Thus, p achieves its minimum and maximum on $\partial\Omega$. Note also that function p given above is not harmonic since $\Delta p(x_1, x_2) = 2x_2$ and consequently the above conclusion can not be obtained by applying Theorem 1.1. More examples of functions satisfying conditions (4) and (7) can be found in Ref.[2] and Ref.[3].

Next, we reformulate the maximum principle in Theorem 1.2 by relaxing condition (4). To be more specific we prove.

Theorem 1.3. *Let* $\vec{a} : \Omega \to \mathbb{R}^N$ *be a vectorial function such that* $\vec{a} \in C^1(\Omega; \mathbb{R}^N) \cap C(\overline{\Omega}; \mathbb{R}^N)$*. Assume*

$$\operatorname{div} \vec{a}(x) = 0, \quad \forall\, x \in \overline{\Omega}, \tag{8}$$

and there exists a function $A : \overline{\Omega} \to (0, \infty)$ such that $A \in C^1(\Omega) \cap C(\overline{\Omega})$ and for which there exists a positive constant $A_0 > 0$ such that

$$\vec{a}(x) \cdot \nabla A(x) \geq A_0 > 0, \quad \forall\, x \in \overline{\Omega}. \tag{9}$$

If $p \in C^1(\Omega) \cap C(\overline{\Omega})$ is a solution of the differential inequality

$$\vec{a}(x) \cdot \nabla p(x) \leq 0, \quad \forall\, x \in \Omega, \tag{10}$$

then for each open set $U \subset \Omega$ the minimum of p in $\overline{U}$ is achieved on ∂U.

If $p \in C^1(\Omega) \cap C(\overline{\Omega})$ is a solution of the differential inequality

$$\vec{a}(x) \cdot \nabla p(x) \geq 0, \quad \forall\, x \in \Omega, \tag{11}$$

then for each open set $U \subset \Omega$ the maximum of p in $\overline{U}$ is achieved on ∂U.

If $p \in C^1(\Omega) \cap C(\overline{\Omega})$ is a solution of the PDE's

$$\vec{a}(x) \cdot \nabla p(x) = 0, \quad \forall\, x \in \Omega \tag{12}$$

then for each open set $U \subset \Omega$ the maximum and minimum of p in $\overline{U}$ are achieved on ∂U.

Example. a. We point out an example of functions $\vec{a}$, p and A for which Theorem 1.3 can be applied. Indeed, it is enough to take $N = 2$, $\Omega \subset \{(x_1, x_2) \in \mathbb{R}^2 : \exists\, c > 0, \text{ s.t. } x_1 \geq c,\ x_2 \leq -c\}$, $\vec{a}(x_1, x_2) = (-x_1, x_2)$, $p(x_1, x_2) = x_1^2 x_2^2$ and $A(x_1, x_2) = x_1^2 x_2 + C$, where $C > 0$ is a constant sufficiently large such that $A(x_1, x_2) > 0$ for each $(x_1, x_2) \in \overline{\Omega}$. Then conditions (8), (9) and (12) are satisfied and consequently the maximum and minimum of p in $\overline{\Omega}$ are achieved on $\partial\Omega$.

b. On the other hand, we note that if in the above example we take Ω to be a ball centered in the origin then $\min_{\Omega} p = 0$ is achieved in $(0, 0)$ which does not belong to $\partial\Omega$ (actually, this represents an example which shows that if condition (4) is replaced by (8) then the conclusion of Theorem 1.2 does not hold true).

2. Proof of the main results

2.1. *Proof of Theorem 1.2*

First, we note that we can always assume that $\min_{x \in \overline{\Omega}} p(x) > 1$. Indeed, it is easy to remark that if $p(x)$ satisfies one of the relations (5), (6), (7) then $C + p(x)$ satisfies the corresponding relation too, for any real constant C. Thus, we may assume $\min_{x \in \overline{\Omega}} p(x) > 1$ eventually by choosing C to be positive and large enough and replacing $p(x)$ by $C + p(x)$.

In the sequel we will use the following notations

$$p^+ := \max_{x\in\overline{\Omega}} p(x), \qquad p^- := \min_{x\in\overline{\Omega}} p(x).$$

Using the above remark we can assume that $p^- > 1$.

We will prove Theorem 1.2 only in the case when conditions (4) and (5) are satisfied. The case when conditions (4) and (6) hold true can be treated similarly in the sense that we can replace $p(x)$ by $C - p(x)$, where C is a positive constant, large enough such that $\min_{x\in\overline{\Omega}}(C - p(x)) > 1$. Thus, $C - p(x)$ will satisfy condition (5) and consequently we will infer that $\min_{x\in\overline{U}}(C - p(x)) = \min_{x\in\partial U}(C - p(x))$ or $\max_{x\in\overline{U}} p(x) = \max_{x\in\partial U} p(x)$.

Finally, we note that in the case when conditions (4) and (7) are satisfied the conclusion of Theorem 1.2 is obvious as a consequence of the above two cases.

In the following we prove Theorem 1.2 under assumptions (4) and (5). The conclusion of Theorem 1.2 will follow as a consequence of two lemmas that will be presented below.

Denote

$$\mu_1 := \inf_{u\in C_0^1(\Omega)\setminus\{0\},\ 0\le u\le 1} \frac{\displaystyle\int_\Omega |\nabla u|^{p(x)}\, dx}{\displaystyle\int_\Omega |u|^{p(x)}\, dx}.$$

Lemma 2.1. *If there exists an open set $U \subset \Omega$ and a point $x_0 \in U$ such that $p(x_0) < p(x)$ for all $x \in \partial U$ then $\mu_1 = 0$.*

Remark 3. The result of Lemma 2.1 represents a slight improvement of Theorem 3.1 Ref. [1]. The idea of the proof below follows the lines of Theorem 3.1 Ref. [1].

Proof. Without loss of generality we may assume that $\overline{U} \subset \Omega$ and there exist two positive constants ϵ_0 and ϵ_1 such that

$$p(x_0) < p(x) - 4\epsilon_0, \quad \forall\, x \in \partial U, \tag{13}$$

$$p(x_0) < p(x) - 2\epsilon_0, \quad \forall\, x \in B(\partial U, \epsilon_1), \tag{14}$$

where

$$B(\partial U, \epsilon_1) := \{x \in \Omega :\ \exists\, y \in \partial U \text{ s.t. } |x - y| < \epsilon_1\}.$$

Indeed, since p is continuous on the compact set $\overline{\Omega}$ it is clear that it is uniformly continuous. Consequently, there exists $\epsilon_1 > 0$ such that for each

x, $z \in \Omega$ with $|x-z| < \epsilon_1$ we have $|p(x)-p(z)| < 2\epsilon_0$. It follows that for each $z \in B(\partial U, \epsilon_1)$ there exists $x_z \in \partial U$ such that $|z-x_z| < \epsilon_1$ and we have $|p(x_z)-p(z)| < 2\epsilon_0$. Then using (13) it is clear that for each $z \in B(\partial U, \epsilon_1)$ we have $p(x_0) < p(x_z) - 4\epsilon_0 \leq |p(x_z)-p(z)| + p(z) - 4\epsilon_0 < p(z) - 2\epsilon_0$. Thus, (14) is satisfied.

Next, we note that there exists a positive constant $\epsilon_2 > 0$ such that

$$B(x_0, \epsilon_2) \subset U \setminus B(\partial U, \epsilon_1)\,,$$

and

$$|p(x_0) - p(x)| < \epsilon_0, \quad \forall\, x \in B(x_0, \epsilon_2)\,. \tag{15}$$

Obviously, there exists a function $u \in C_0^1(\Omega)$ such that

$$|\nabla u(x)| \leq C \quad \text{and} \quad 0 \leq u(x) \leq 1, \quad \forall\, x \in \Omega\,,$$

and

$$u(x) = \begin{cases} 0 & \text{for } x \notin U \cup B(\partial U, \epsilon_1) \\ 1 & \text{for } x \in U \setminus B(\partial U, \epsilon_1)\,. \end{cases}$$

Next, it is clear that for each $t \in (0,1)$ we have

$$\begin{aligned} \mu_1 &\leq \frac{\int_\Omega |\nabla(tu)|^{p(x)}\,dx}{\int_\Omega |tu|^{p(x)}\,dx} \leq \frac{\int_{B(\partial U,\epsilon_1)} |\nabla(tu)|^{p(x)}\,dx}{\int_{B(x_0,\epsilon_2)} |tu|^{p(x)}\,dx} \\ &\leq \frac{\int_{B(\partial U,\epsilon_1)} |tC|^{p(x)}\,dx}{\int_{B(x_0,\epsilon_2)} |t|^{p(x)}\,dx} \leq \frac{C_1|tC|^{p(\eta_1)}}{C_2|t|^{p(\eta_2)}}\,, \end{aligned}$$

where $C_1 := |B(\partial U, \epsilon_1)|$, $C_2 := |B(x_0, \epsilon_2)|$, $\eta_1 \in B(\partial U, \epsilon_1)$ and $\eta_2 \in B(x_0, \epsilon_2)$.

Using relations (14) and (15) we find

$$p(\eta_1) - p(\eta_2) > \epsilon_0\,,$$

and we get $\mu_1 \leq C_3 t^{\epsilon_0}$ for all $t \in (0,1)$ and therefore $\mu_1 = 0$.

The proof of Lemma 2.1 is complete. □

Lemma 2.2. *Assume that conditions (4) and (5) are fulfilled. Then* $\mu_1 > 0$.

Remark 4. The proof of Lemma 2.2 will follow the lines of Theorem 1 Ref.[2] and is inspired by the ideas in Théorème 20.7 Ref.[8].

Proof. Simple computations show that for each $u \in C_0^1(\Omega)$ with $0 \leq u \leq 1$ the following equality holds true

$$\begin{aligned}\operatorname{div}(u(x)^{p(x)}\overrightarrow{a}(x)) &= u(x)^{p(x)}\operatorname{div}\overrightarrow{a}(x) + p(x)u(x)^{p(x)-1}\nabla u(x)\cdot\overrightarrow{a}(x)\\ &\quad + u(x)^{p(x)}\log(u(x))\nabla p(x)\cdot\overrightarrow{a}(x).\end{aligned}$$

On the other hand, the flux-divergence theorem implies that for each $u \in C_0^1(\Omega)$ with $0 \leq u \leq 1$ we have

$$\int_\Omega \operatorname{div}(u(x)^{p(x)}\overrightarrow{a}(x))\,dx = \int_{\partial\Omega} u(x)^{p(x)}\overrightarrow{a}(x)\cdot\overrightarrow{n}\,d\sigma(x) = 0.$$

Using the above pieces of information and condition (5) we infer that for each $u \in C_0^1(\Omega)$ with $0 \leq u \leq 1$ it holds true

$$\begin{aligned}\int_\Omega u(x)^{p(x)}\operatorname{div}\overrightarrow{a}(x)\,dx &= -\int_\Omega p(x)u(x)^{p(x)-1}\nabla u(x)\cdot\overrightarrow{a}(x)\,dx\\ &\quad -\int_\Omega u(x)^{p(x)}\log(u(x))\nabla p(x)\cdot\overrightarrow{a}(x)\\ &\leq p^+\int_\Omega u(x)^{p(x)-1}|\nabla u(x)|\,|\overrightarrow{a}(x)|\,dx.\end{aligned}$$

Next, we recall that for each $\epsilon > 0$, each $x \in \Omega$ and each $A,\ B \geq 0$ the following Young type inequality holds true (see, e.g. the footnote on p. 56 from Ref. [9])

$$AB \leq \epsilon A^{\frac{p(x)}{p(x)-1}} + \frac{1}{\epsilon^{p(x)-1}}B^{p(x)}.$$

We fix $\epsilon > 0$ such that

$$p^+\epsilon < a_0,$$

where a_0 is given by relation (4).

The above facts and relation (4) yield

$$\begin{aligned}a_0\int_\Omega u(x)^{p(x)}\,dx &\leq p^+\epsilon\int_\Omega u(x)^{p(x)}\,dx\\ &\quad + p^+\int_\Omega\left(\frac{1}{\epsilon}\right)^{p(x)-1}|\overrightarrow{a}(x)|^{p(x)}|\nabla u(x)|^{p(x)}\,dx,\end{aligned}$$

for any $u \in C_0^1(\Omega)$ with $0 \leq u \leq 1$, or

$$\begin{aligned}(a_0 - \epsilon p^+)\int_\Omega u(x)^{p(x)}\,dx &\leq p^+\left(\frac{1}{\epsilon}\right)^{p^- -1}\int_\Omega|\overrightarrow{a}(x)|^{p(x)}|\nabla u(x)|^{p(x)}\,dx\\ &\quad + p^+\left(\frac{1}{\epsilon}\right)^{p^+ -1}\int_\Omega|\overrightarrow{a}(x)|^{p(x)}|\nabla u(x)|^{p(x)}\,dx,\end{aligned}$$

for any $u \in C_0^1(\Omega)$ with $0 \leq u \leq 1$.

The proof of Lemma 2.2 is complete. □

2.2. *Proof of Theorem 1.3*

We use the same notations introduced in Section 3. The same arguments as in the proof of Theorem 1.2 can be used in order to consider that $p^- > 1$. On the other hand, similar arguments as in the proof of Theorem 1.2 assure that it is enough to give the proof of Theorem 1.3 in the case when conditions (8), (9) and (10) are satisfied. Undoubtedly, Lemma 2.1 still holds true in the hypothesis of Theorem 1.3. We prove the following auxiliary result:

Lemma 2.3. *Assume that conditions (8), (9) and (10) are fulfilled. Then* $\mu_1 > 0$.

Proof. By relation (8) it follows that

$$\int_\Omega \overrightarrow{a}(x) \cdot \nabla\phi(x)\, dx = 0, \quad \forall\, \phi \in C_0^1(\Omega)\,.$$

Next, note that for each $u \in C_0^1(\Omega)$ with $0 \leq u \leq 1$ we have $\phi = u(x)^{p(x)}A(x) \in C_0^1(\Omega)$ and since

$$\begin{aligned}\nabla\phi(x) = {} & u(x)^{p(x)}\nabla A(x) + A(x)u(x)^{p(x)}\log(u(x))\nabla p(x) \\ & + A(x)u(x)^{p(x)-1}p(x)\nabla u(x)\,,\end{aligned}$$

we get by using relation (10) that

$$\begin{aligned}\int_\Omega u(x)^{p(x)}\nabla A(x) \cdot \overrightarrow{a}(x)\, dx = {} & -\int_\Omega A(x)u(x)^{p(x)-1}p(x)\nabla u(x) \cdot \overrightarrow{a}(x)\, dx \\ & - \int_\Omega A(x)u(x)^{p(x)}\log(u(x))\nabla p(x) \cdot \overrightarrow{a}(x)\, dx\,,\end{aligned}$$

for any $u \in C_0^1(\Omega)$ with $0 \leq u \leq 1$.

The proof continues as in the case of Lemma 2.2. □

Obviously, Theorem 1.3 is a consequence of Lemmas 2.1 and 2.3.

Acknowledgments

The research of the three authors has been partially supported by CNCS-UEFISCDI Grant No. PN-II-ID-PCE-2012-4-0021 "Variable Exponent Analysis: Partial Differential Equations and Calculus of Variations".

References

1. X. Fan, Q. Zhang, & D. Zhao, Eigenvalues of $p(x)$-Laplacian Dirichlet problem, *Journal of Mathematical Analysis and Applications* **302**, 306-317 (2005).
2. M. Mihăilescu, V. Rădulescu, & D. Stancu-Dumitru, On a Caffarelli-Kohn-Nirenberg type inequality in bounded domains involving variable exponent growth conditions and applications to PDE's, *Complex Variables-Elliptic Equations* **56**, 659-669 (2011).
3. M. Mihăilescu, G. Moroşanu, & D. Stancu-Dumitru, Equations involving a variable exponent Grushin-type operator, *Nonlinearity* **24**, 2663-2680 (2011).
4. P. Pucci & J. Serrin, *The maximum principle* (Birkhäuser, Berlin, 2007).
5. D. Gilbarg & N. S. Trudinger, *Elliptic partial differential equations of second order* (Springer Berlin, 1998).
6. M. H. Protter & H. F. Weinberger, *Maximum Principles in Differential Equations* (Prentice-Hall, Englewood Cliffs, N.J., 1967).
7. W. Allegretto, Form estimates for the $p(x)$-Laplacean, *Proceedings of the American Mathematical Society* **135**, 2177-2185 (2007).
8. M. Willem, *Analyse Fonctionnelle Elémentaire*, (Cassini, Paris, 2003).
9. H. Brezis, *Analyse fonctionnelle. Théorie, Méthodes et Applications* (Masson, Paris, 1992).

Differentiability and integrability properties for solutions to nonlocal equations

Mikil Foss* and Petronela Radu**

Department of Mathematics, University of Nebraska-Lincoln, Lincoln, NE 68588, USA

E-mail: *mfoss@math.unl.edu

**pradu@math.unl.edu, www.math.unl.edu/∼ pradu3

The main results of this paper concern integrability and differentiability properties for solutions for a nonlocal system. The boundary value problem is associated with an integro-differential equation that exhibits a weakly singular kernel. The main result provides a nonlocal equivalent of classical regularity theorems established for elliptic systems. In addition, we show that solutions inherit additional integrability properties of the forcing term, thus providing a simple proof to well-posedness of solutions in L^p for $p > 2$.

Keywords: nonlocal operators, classical differentiability, higher-integrability, weakly integrable kernels, peridynamics.

1. Introduction

Nonlocal theories have emerged as successful tools in investigations of singular phenomena (such as fracture) and also in phenomena with nonlocal features in biology [1,2], thermal diffusion [3], image processing [4], sand-pile formation and more (see the monography [5] and the references therein). In particular, the theory of peridynamics introduced by Silling [6] introduces a model for tracking cracks by using a nonlocal operator which sums pairwise interactions within a neighborhood of radius δ called horizon. This framework can also be used to model elastic deformations with no breakage of bonds; in that case, the solution $u : \mathbb{R}^n \to \mathbb{R}$ will satisfy the steady state system

$$\begin{cases} \mathcal{L}_\mu u(x) = f(x), & x \in \Omega \\ u(x) = 0, & x \in \mathcal{B}\Omega. \end{cases} \tag{1}$$

Above, the nonlocal Laplacian is defined on a set $\Omega \subset \mathbb{R}^n$ with smooth boundary and its collar $\mathcal{B}\Omega$ (an open set surrounding Ω of positive measure, which will be precisely defined below) as

$$\mathcal{L}_\mu u(x) := \int_{\Omega\cup\mathcal{B}\Omega} [u(y) - u(x)]\mu(x,y)\,\mathrm{d}y, \tag{2}$$

for a measurable kernel $\mu : (\Omega \cup \mathcal{B}\Omega) \times (\Omega \cup \mathcal{B}\Omega) \to \mathbb{R}$. The source term satisfies $f \in L^2(\Omega)$; as such, the equalities in (11) and in the sequel are seen a.e.

Recently, many papers investigated the system (11) when the kernel μ is weakly integrable (i.e. $\mu(x,y) = \mu(|x-y|) \in L^1(\mathbb{R})$) as well as when it has a non-integrable singularity ([7,8]). The main contribution of this paper is to provide a general and versatile roadmap for the study of regularity of solutions to nonlocal systems. More precisely, as end results, we

- Establish additional integrability properties of solutions and prove well-posedness of (11) in L^p. The higher-integrability property makes up for a weaker version of Poincareé's inequality that is available in the nonlocal framework.
- Show Sobolev differentiability of solutions of the nonlocal problem corresponding to a C^1 forcing term. The result holds only up to the boundary, the investigation of regularity on the full domain is still ongoing.

1.1. *Connections with classical elliptic problems*

Nonlocal operators are natural generalizations of classical differential operators. This idea is enforced by several results, of which we mention:

- If the kernel μ is given by a combination of derivatives of the Dirac mass, then for this particular choice of kernel we have $\mathcal{L}_\mu = \Delta u$ in the sense of distributions (see[9]);
- As the radius of the horizon δ goes to zero, the nonlocal solutions converge to their classical counterparts. This fact has been proven for the Laplacian[10], as well as for the biharmonic operator[11];
- Dirichlet's principle and variational arguments apply for nonlocal systems as proven in[12];
- A weighted mean-value theorem is available; indeed, one can easily see that a nonlocal harmonic function for which $\mathcal{L}_\mu u(x) = 0$ satisfies

$$u(x) = \frac{\int_{\Omega\cup\mathcal{B}\Omega} \mu(|x-y|)u(y)\,\mathrm{d}y}{\int_{\Omega\cup\mathcal{B}\Omega} \mu(|x-y|)\,\mathrm{d}y}.$$

Using this result, one can prove

- $\Delta u = 0$ if and only if $\mathcal{L}_\mu u(x) = 0$ for every integrable symmetric kernel $\mu(x,y) = \mu(|x-y|) \in L^1(\mathbb{R})$ ([13]).

1.2. *Motivation and main result*

It is known that solutions to the classical Laplace equation are harmonic functions which are infinitely differentiable (in the classical sense). We establish here a counterpart to this result in the nonlocal setting. From the regularity point of view, it is clear that the nonlocal Laplacian with μ integrable will not provide a gain in differentiability as the classical Laplacian does; we will prove, however, that nonlocal solutions enjoy interior differentiability at the same level as the differentiability of the forcing term f.

It is known[10] that the solutions u_δ to the problem

$$\begin{cases} \mathcal{L}_\delta u_\delta(x) = f(x), & x \in \Omega \\ u_\delta(x) = 0, & x \in \mathcal{B}\Omega, \end{cases} \tag{3}$$

obtained for each $\delta > 0$, converge strongly in L^2 to u the solution to the classical Laplace equation

$$\begin{cases} \Delta u(x) = f, & x \in \Omega \\ u(x) = 0, & x \in \partial\Omega. \end{cases}$$

However, it has not been established under what conditions on f the nonlocal solutions u_δ enjoy a certain level of regularity. Our main result reads:

Theorem 1.1 (Regularity of nonlocal solutions). *Let $f \in C^1(\bar{\Omega})$. For a nonnegative, integrable kernel μ the unique solution u_δ of (3) satisfies*

$$u_\delta \in W^{1,2}(\Omega).$$

Finally, we mention that the results of this paper can be easily extended to vector-valued functions, but for the sake of simplicity we keep the presentation in the scalar case.

The paper is organized as follows. In the next section we introduce notation and assumptions for our results, after which we derive the scaling needed for obtaining pointwise error estimates between nonlocal and local Laplacians applied to a smooth function. A nonlocal version of Poincaré's inequality that provides essential estimates for our main result will conclude the section of preliminary results. After proving the higher integrability property of solutions and wellposedness of (11) in L^p in section 3 we proceed to prove the regularity theorem in section 4.

1.3. *Definitions and setup*

Let $\mu : \mathbb{R}^n \times \mathbb{R}^n \to \mathbb{R}$ be a measurable nonnegative function such that $\mu(x,y) = \mu(|x-y|)$ [a], and let $\Omega \subset \mathbb{R}^n$ be an open and bounded domain. Define

$$\mathcal{B}\Omega := \overline{\cup_{x\in\Omega}\mathrm{supp}(k(x,\cdot)) \setminus \Omega}.$$

Denote by ω_n the volume of the unit ball in n dimensions, so $n\omega_{n-1}$ will be its surface area. The n dimensional ball of radius δ centered at x will be denoted by $B_\delta(x)$ and ∇u is the spatial gradient.

We also use the standard notation for the average value of a function

$$f_\Omega := \fint_\Omega f(x)\,dx = \frac{1}{|\Omega|}\int\limits_\Omega f(x)\,dx. \tag{4}$$

For $\varepsilon > 0$ denote by

$$\Omega_\varepsilon := \{x \in \Omega : B_\varepsilon(x) \subseteq \Omega\}.$$

If $\phi \in C^\infty(\mathbb{R}^n)$ and $u \in L^2(\Omega \cup \mathcal{B}\Omega)$, then we define the convolution product

$$(\phi * u)(x) = \int\limits_{\mathbb{R}^n} \phi(y-x)\bar{u}(y)\,\mathrm{d}y$$

where

$$\bar{u}(y) := \begin{cases} u(y), & y \in \Omega \cup \mathcal{B}\Omega \\ 0, & y \notin \Omega \cup \mathcal{B}\Omega. \end{cases}$$

Assumptions for the kernel μ. Let μ be a nonnegative kernel that satisfies

(A1) μ is weakly singular, more precisely, there exist $0 \le \beta < n$ and $c > 0$ such that for every $r > 0$ we have

$$0 < \mu(r) \le cr^{-\beta}.$$

(A2) There exists $C_0 > 0$ such that $\mu(r) > C_0$ for every $r < \delta$.

---

[a] Throughout the paper we will allow the abuse of notation that denotes a radial function and its radial representative by the same letter.

(A3) For every $\delta > 0$ we let

$$\mu_\delta(r) = \begin{cases} \mu(r), & 0 < r \le \delta \\ 0, & \delta < r. \end{cases}$$

For every $\delta > 0$ we introduce the function $\pi : (0, \infty) \to [0, \infty)$

$$\pi(r) := \int_\delta^r \rho\, \mu_\delta(\rho)\, \mathrm{d}\rho. \tag{5}$$

Thus, if $\mu(r) = r^{-\beta}$ then we find that

$$\pi(r) = \begin{cases} \frac{1}{2-\beta}[r^{2-\beta} - \delta^{2-\beta}], & 0 < r \le \delta, \beta \ne 2 \\ \ln r - \ln \delta, & 0 < r \le \delta, \beta = 2 \\ 0, & r > \delta. \end{cases}$$

Under the above assumptions we will use a slightly different notation for the nonlocal Laplacian which keeps track of the size of the support of μ and incorporates the scaling factor, namely

$$\mathcal{L}_\delta u(x) = \sigma(\delta) \int_{\Omega \cup \mathcal{B}\Omega} [u(y) - u(x)] \mu_\delta(|x-y|)\, \mathrm{d}y. \tag{6}$$

The horizon-dependent scaling σ will be determined later.

2. Scaling of the nonlocal Laplacian and pointwise estimates

In this section we will select an appropriate scaling $\sigma(\delta)$ which normalizes the nonlocal operator. In fact, we will show that the scaled $\mathcal{L}_\delta u$ converges to Δu at the rate δ^2. This error bounds between the local and nonlocal Laplacian have been derived before (see for example[14]). Below we will present a different argument that can be easily generalized to obtain error estimates between nonlocal solutions and their local equivalents.

Assume that $u \in C^4(\Omega \cup \mathcal{B}\Omega)$. By using the fundamental theorem of calculus we have that

$$\begin{aligned} \mathcal{L}_\delta u(x) &= \sigma(\delta) \int_{B_\delta(x)} [u(y) - u(x)] \mu_\delta(|x-y|)\, \mathrm{d}y \\ &= \sigma(\delta) \int_{B_\delta(x)} \int_0^1 \nabla u(x + s(y-x)) \cdot (y-x) \mu_\delta(|y-x|)\, \mathrm{d}s\ \mathrm{d}y. \end{aligned}$$

After changing the order of integration in the double integral and changing variables $z = y - x$ we obtain that

$$\mathcal{L}_\delta u(x) = \sigma(\delta) \int_0^1 \int_{B_\delta(0)} \nabla u(x+sz) \cdot [z\mu_\delta(|z|)] \, \mathrm{d}z \, \mathrm{d}s.$$

With π given by (5) we have that

$$\nabla_z \pi(|z|) = \pi'(z)\frac{z}{|z|} = z\mu_\delta(|z|),$$

so

$$\mathcal{L}_\delta u(x) = \sigma(\delta) \int_0^1 \int_{B_\delta(0)} \nabla u(x+sz) \cdot \nabla_z \pi(|z|) \, \mathrm{d}z \, \mathrm{d}s.$$

After performing an integration by parts we obtain

$$\mathcal{L}_\delta u(x) = \sigma(\delta) \int_0^1 \int_{\partial B_\delta(0)} \left[\nabla u(x+sz) \cdot \frac{z}{\delta}\right] \pi(|z|) \, \mathrm{d}\omega(z) \, \mathrm{d}s$$

$$- \sigma(\delta) \int_0^1 \int_{B_\delta(0)} \mathrm{div}_z[\nabla u(x+sz)]\pi(|z|) \, \mathrm{d}z \, \mathrm{d}s.$$

For $z \in \partial B_\delta(0)$ we have $\pi(|z|) = \pi(\delta) = 0$, thus we can write

$$\mathcal{L}_\delta u(x) = -\sigma(\delta) \int_0^1 \int_{B_\delta(0)} s\Delta u(x+sz)\pi(|z|) \, \mathrm{d}z \, \mathrm{d}s$$

which it further becomes

$$\mathcal{L}_\delta u = -\sigma(\delta) \int_0^1 \int_{B_\delta(0)} s[\Delta u(x+sz) - \Delta u(x)]\pi(|z|) \, \mathrm{d}z \, \mathrm{d}s \tag{7}$$

$$-\Delta u(x)\sigma(\delta) \int_0^1 \int_0^\delta \int_{\partial B_\rho(0)} s\pi(|\rho|) d\omega(z) \, \mathrm{d}\rho \, \mathrm{d}s.$$

We compute

$$\int_0^1 \int_0^\delta \int_{\partial B_\rho(0)} s\pi(\rho) \, \mathrm{d}\omega(z) \, \mathrm{d}\rho \, \mathrm{d}s = \frac{1}{2} n\omega_{n-1} \int_0^\delta \pi(\rho)\rho^{n-1} \, \mathrm{d}\rho.$$

This suggests that we use the scaling

$$\sigma(\delta) = \frac{-2}{n\omega_{n-1} \int\limits_0^\delta \pi(\rho)\rho^{n-1}\,\mathrm{d}\rho} \tag{8}$$

in order to have the coefficient of the Laplacian to be 1 on the RHS of (1) above.

In particular, if $n = 1$ and $\mu(r) = \begin{cases} r^{-\beta}, & 0 < r \le \delta \\ 0, & \delta < r \end{cases}$ then

$$\sigma(\delta) = \frac{3-\beta}{\delta^{3-\beta}}.$$

With the choice of scaling given in (8) above we write

$$\mathcal{L}_\delta u(x) - \Delta u(x) = -\sigma(\delta) \int\limits_{B_\delta(0)} \int\limits_0^1 s[\Delta u(x+sz) - \Delta u(x)]\pi(|z|)\,\mathrm{d}s\,\mathrm{d}z. \tag{9}$$

By employing the fundamental theorem of calculus and simplifying we obtain

$$\begin{aligned}\mathcal{L}_\delta u(x) - \Delta u(x) &= -\sigma(\delta) \int\limits_{B_\delta(0)} \int\limits_0^1 \frac{(1-s^2)}{2}[\Delta\nabla u(x+sz)]\cdot z\pi(|z|)\,\mathrm{d}s\,\mathrm{d}z \\ &= -\sigma(\delta) \int\limits_{B_\delta(0)} \int\limits_0^1 \frac{(1-s^2)}{2}[\Delta\nabla u(x+sz) \\ &\qquad -\Delta\nabla u(x)]\cdot z\pi(|z|)\,\mathrm{d}s\,\mathrm{d}z,\end{aligned}$$

where we used the fact that

$$\int\limits_{B_\delta(0)} z\pi(|z|)\,\mathrm{d}z = 0.$$

Further we have

$$\begin{aligned}\mathcal{L}_\delta u(x) - \Delta u(x) = -\sigma(\delta) \int\limits_{B_\delta(0)} \int\limits_0^1 \left(\frac{1}{3} - \frac{s}{2} + \frac{s^3}{6}\right) \big[\Delta\nabla^2 u(x \\ +sz)z\big]\cdot z\pi(|z|)\,\mathrm{d}s\,\mathrm{d}z.\end{aligned}$$

With

$$M_4 = \sup_{x\in\Omega\cup\mathcal{B}\Omega} |D^4 u(x)|$$

we estimate

$$|\mathcal{L}_\delta u(x) - \Delta u(x)| \leq \sigma(\delta) M_4 \int\limits_{B_\delta(0)} \int\limits_0^1 \left(\frac{1}{3} - \frac{s}{2} + \frac{s^3}{6}\right) |z|^2 |\pi(|z|)| \,\mathrm{d}s\,\mathrm{d}z$$

$$= \frac{\sigma(\delta) M_4}{8} n\omega_{n-1} \int\limits_0^\delta \rho^{n+1} |\pi(\rho)| \,\mathrm{d}\rho.$$

Finally, we use the fact that $\rho^{n+1} \leq \rho^{n-1}\delta^2$ to obtain

$$|\mathcal{L}_\delta u(x) - \Delta u(x)| \leq \frac{M_4}{4}\delta^2. \tag{10}$$

This estimate shows that the nonlocal Laplacian approaches the classical Laplacian as the horizon goes to zero at a rate δ^2, independently of the dimension. The convergence also demonstrates that the scaling selected in (8) is the correct one.

We conclude this section by stating the following nonlocal Poincaré-type inequality, versions of which which can be found in [15] or [12]:

Theorem 2.1 (Nonlocal Poincaré inequality). *Let Ω be an open, bounded domain and $\mu \in L^1(\mathbb{R}^n)$ a nonnegative kernel that satisfies (A2) and (A3). If $u \in L^2(\Omega \cup \mathcal{B}\Omega)$ then there is a constant $\lambda_p(\delta, n, \Omega) > 0$ s.t.*

$$\lambda_p \int\limits_\Omega |u(x)|^2 \,\mathrm{d}x \leq \int\limits_\Omega \int\limits_{\Omega\cup\mathcal{B}\Omega} \mu_\delta(|x-y|)|u(y) - u(x)|^2 \,\mathrm{d}y\,\mathrm{d}x + \int\limits_{\mathcal{B}\Omega} |u(x)|^2 \,\mathrm{d}x.$$

3. Higher integrability

In this section we prove some regularizing properties for solutions of

$$\begin{cases} \mathcal{L}_\mu u(x) = f(x), & x \in \Omega \\ u(x) = 0, & x \in \mathcal{B}\Omega. \end{cases} \tag{11}$$

More specifically we show that a solution increases its integrability properties to match those of the function f. Thus, this result addresses one of the deficiencies of Poincaré's inequality, mainly, the fact the the integrability of a function cannot be improved based on bounds on its nonlocal gradient, as in the classical framework. The main theorem of this section, Theorem 3.1, states that whenever the nonlocal Laplacian has an integrable kernel[b],

[b]Note that if the kernel is highly singular then there are compact embedding results and Poincaré-type inequalities that do increase the integrability by using estimates on the gradient.

solutions to (11) will be bounded if $f \in L^\infty$.

Theorem 3.1. *Let $\mu \geq 0$, $\mu \in L^1(\mathbb{R}^n)$ and w.l.o.g. assume that $\|\mu\|_{L^1(\mathbb{R}^n)} = 1$. If u is a solution of (11) with a priori regularity $u \in L^p(\Omega)$, $p > 1$ and we assume that $f \in L^r(\Omega)$ with $r > p$, then u lifts its regularity so that $u \in L^r(\Omega)$.*

Proof. In light if our assumptions we write the integro-differential equation in (11) as

$$\int_{\mathbb{R}^n} u(y)\mu(x,y)\,\mathrm{d}y - u(x)\int_{\mathbb{R}^n} \mu(x,y)\,\mathrm{d}y = (u * \mu)(x) - u(x) = f(x). \quad (12)$$

For a function u that satisfies the above equation, consider its extension by zero outside Ω as given by

$$\bar{u}(x) = \begin{cases} u(x), & x \in \Omega, \\ 0, & x \in \mathbb{R}^n \setminus \Omega. \end{cases}$$

In the last equality of (12) we apply Young's inequality to obtain

$$\|u\|_r \leq \|u * \mu\|_r + \|f\|_r \leq \|u\|_p \|\mu\|_q + \|f\|_r,$$

where $\frac{1}{r} = \frac{1}{p} + \frac{1}{q} - 1$ with q chosen depending on the following cases:

(i) $\beta < n\dfrac{p-1}{p}$. In this case take $q = \frac{p}{p-1}$ to obtain $r = \infty$. This means that we can increase the integrability of the function u from the $p-$level of integrability given to match the integrability of the function f. In particular, if $f \in L^\infty$, then $u \in L^\infty$.

(ii) $\beta \geq n\dfrac{p-1}{p}$. Then let $q = \dfrac{n-\varepsilon}{\beta}$ for ε arbitrarily small. Then, the degree of integrability could be raised up to (but not including) $r_0 = \dfrac{pn}{n + p(\beta - n)}$. Now, assuming that $f \in L^r$ with $r > r_0\dfrac{pn}{n + p(\beta - n)}$, then we apply again Young's inequality, by using the fact that $u \in L^{r_0}$ so we have $\dfrac{1}{r_1} = \dfrac{1}{r_0} + \dfrac{1}{q_1} - 1$. Take $q_1 = \dfrac{p(n-\varepsilon)}{(n-\varepsilon)(2p-1) - p\beta}$ to obtain $r_1 = \infty$, so again we are able to lift the order of integrability of u to match f's order.

□

Remark 3.1. It is clear that the above result is not only dimension independent, also it is applicable for Ω finite domain, as well as for $\Omega = \mathbb{R}^n$.

Finally, note that the above theorem allows us to prove

Theorem 3.2 (Wellposedness of solutions in L^p). *If $f \in L^p(\Omega)$ for $p > 2$ and Ω is a finite measurable domain in $\mathbb{R}^n$ then there exists $u \in L^p(\Omega)$ a solution to* (11).

Proof. Since $f \in L^p(\Omega)$ and Ω is a finite measurable domain we have that $f \in L^2(\Omega)$. We have then that there exists a unique $u \in L^2(\Omega)$ that solves (11); see[9] or[12] when $p = 2$. By Theorem 3.1 we have that since $f \in L^p(\Omega)$ and $u \in L^2(\Omega)$ with $p > 2$ the solution u belongs to $L^p(\Omega)$ which concludes the proof. □

4. Regularity estimates and main results

We begin this section with the following theorem, which shows an improvement in differentiability for solutions of the system (11) when the kernel is smooth. Our final goal is to establish regularity of solutions when μ is only integrable.

Theorem 4.1. *If $\mu \in C^1(\mathbb{R}^n), f \in C^1(\mathbb{R}^n)$ and $u \in L^1(\mathbb{R}^n)$ is a solution to $\mathcal{L}_\mu u = f$, i.e.*

$$\int_{\mathbb{R}^n} (u(y) - u(x))\mu(x - y)dy = f(x),$$

then $u \in C^1(\mathbb{R}^n)$.

Proof. The theorem follows easily from the equality

$$u(x) = (u * \mu)(x) - f(x)$$

and the smoothing properties of the convolution operator. □

In the absence of differentiability assumptions for μ we will need the following lemmas for the proof of the main result.

Lemma 4.1. *If $u \in C^1(\Omega \cup \mathcal{B}\Omega)$ satisfies*

$$\mathcal{L}_\delta u(x) = f(x), \text{ for all } x \in \Omega$$

for $f \in C^1(\Omega)$ then $\mathcal{L}_\delta(\nabla u)(x) = \nabla f(x)$ for all $x \in \Omega$.

Proof. The linearity of the $\mathcal{L}_\delta$ operator trivially gives the result. □

Lemma 4.2. *If $u \in L^2(\Omega \cup \mathcal{B}\Omega)$ satisfies*

$$\mathcal{L}_\delta u(x) = f(x), \text{ for all } x \in \Omega,$$

then

$$\mathcal{L}_\delta[\phi * u](x) = (\phi * f)(x)$$

for all $x \in \Omega_\varepsilon$ where $\phi \in C^\infty(\mathbb{R}^n)$ with $supp(\phi) \subset B_\varepsilon(0)$.

Proof. Again, employ the linearity of $\mathcal{L}_\delta$ and of the convolution product. Note though that the equality holds in $\Omega_\varepsilon \subset \Omega$ since the equality $\mathcal{L}_\delta u(x) = f(x)$ holds only inside Ω and the convolution function will capture values of u from outside Ω at ε distance away from $\partial\Omega$. □

The above lemmas yield will be employed in the proof of our main result which we state and prove below:

Theorem 4.2 (Regularity for solutions of nonlocal equations). *Let $f \in C^1(\bar{\Omega})$, μ a nonnegative and integrable kernel that satisfies Assumptions (A2) and (A3). Then there exists a unique solution u to the system in* (11) *from section 1 such that $u \in W^{1,2}(\Omega)$.*

Proof. The existence and uniqueness of $u \in L^2(\Omega \cup \mathcal{B}\Omega)$ when $f \in L^2(\Omega)$ was proven before (see[12] for a variational argument). Below we will establish uniform estimates in the gradient which yield C^1 regularity for u up to the boundary.

Let $\{\phi_\varepsilon\}_{\varepsilon>0}$ be a family of mollifiers such that $\phi_\varepsilon := \frac{1}{\varepsilon^n}\phi(\frac{x}{\varepsilon})$ and $\text{supp}(\phi_\varepsilon) \subset B_\varepsilon(0)$. By Lemma 4.2, for each $\varepsilon > 0$

$$\mathcal{L}_\delta[\phi_\varepsilon * u](x) = (\phi_\varepsilon * f)(x)$$

for each $x \in \Omega_\varepsilon$. Put $u_\varepsilon := \phi_\varepsilon * u$ and $f_\varepsilon := \phi_\varepsilon * f$. Since $u_\varepsilon \in C^\infty(\Omega \cup \mathcal{B}\Omega)$, Lemma 4.1 implies

$$\mathcal{L}_\delta[\nabla u_\varepsilon](x) = \nabla f_\varepsilon(x) \tag{13}$$

for each $x \in \Omega_\varepsilon$. Since $u = 0$ on $\Omega \cup \mathcal{B}\Omega$ we have that $u_\varepsilon = 0$ on

$$\Omega^\varepsilon := \{x \in \mathbb{R}^n | \text{dist}(x, \Omega) < \varepsilon\},$$

hence $\nabla u_\varepsilon = 0$ on $\Omega \cup \mathcal{B}\Omega \setminus \Omega^\varepsilon$. We then have from (13) above that for all $x \in \Omega_\varepsilon$

$$\int_{\Omega^\varepsilon} [\nabla u_\varepsilon(y) - \nabla u_\varepsilon(x)]\mu(x-y)dy = \nabla f_\varepsilon(x).$$

We multiply the equation by ∇u_ε and integrate on Ω_ε to obtain

$$\int\limits_{\Omega_\varepsilon}\int\limits_{\Omega^\varepsilon}[(\nabla u_\varepsilon(y)-\nabla u_\varepsilon(x))\mu(|y-x|)]\cdot\nabla u_\varepsilon(x)\,\mathrm{d}y\,\mathrm{d}x = \int\limits_{\Omega_\varepsilon}\nabla f_\varepsilon(x)\cdot\nabla u_\varepsilon(x)\,\mathrm{d}x. \quad (14)$$

We estimate the LHS above as follows

$$\begin{aligned}&\int\limits_{\Omega_\varepsilon}\int\limits_{\Omega_\varepsilon}[(\nabla u_\varepsilon(y)-\nabla u_\varepsilon(x))\mu(|y-x|)]\cdot\nabla u_\varepsilon(x)\,\mathrm{d}y\,\mathrm{d}x\\ &+\int\limits_{\Omega_\varepsilon}\int\limits_{\Omega^\varepsilon\setminus\Omega_\varepsilon}[(\nabla u_\varepsilon(y)-\nabla u_\varepsilon(x))\mu(|y-x|)]\cdot\nabla u_\varepsilon(x)\,\mathrm{d}y\,\mathrm{d}x\\ &=\frac{1}{2}\int\limits_{\Omega_\varepsilon}\int\limits_{\Omega_\varepsilon}[(\nabla u_\varepsilon(y)-\nabla u_\varepsilon(x))^2\mu(|y-x|)]\,\mathrm{d}y\,\mathrm{d}x\\ &+\int\limits_{\Omega_\varepsilon}\int\limits_{\Omega^\varepsilon\setminus\Omega_\varepsilon}[(\nabla u_\varepsilon(y)-\nabla u_\varepsilon(x))\mu(|y-x|)]\cdot\nabla u_\varepsilon(x)\,\mathrm{d}y\,\mathrm{d}x.\end{aligned}$$

From (14) and above equality we obtain

$$\begin{aligned}\int\limits_{\Omega_\varepsilon}\int\limits_{\Omega_\varepsilon}[(\nabla u_\varepsilon(y)-\nabla u_\varepsilon(x))^2\mu(|y-x|)]\,\mathrm{d}y\,\mathrm{d}x &= 2\int\limits_{\Omega_\varepsilon}\nabla f_\varepsilon(x)\cdot\nabla u_\varepsilon(x)\,\mathrm{d}x\\ &-2\int\limits_{\Omega_\varepsilon}\int\limits_{\Omega^\varepsilon\setminus\Omega_\varepsilon}[(\nabla u_\varepsilon(y)-\nabla u_\varepsilon(x))\mu(|y-x|)]\cdot\nabla u_\varepsilon(x)\,\mathrm{d}y\,\mathrm{d}x.\end{aligned} \quad (15)$$

By Poincaré's inequality as given by Theorem 2.1 applied on the domain Ω_ε with collar $\Omega^\varepsilon\setminus\Omega_\varepsilon$ we have

$$\begin{aligned}\|\nabla u_\varepsilon\|^2_{L^2(\Omega_\varepsilon)} &\le C\int\limits_{\Omega_\varepsilon}\int\limits_{\Omega_\varepsilon}[(\nabla u_\varepsilon(y)-\nabla u_\varepsilon(x))^2\mu(|y-x|)]\,\mathrm{d}y\,\mathrm{d}x \quad (16)\\ &+C\int\limits_{\Omega_\varepsilon}\int\limits_{\Omega^\varepsilon\setminus\Omega_\varepsilon}[(\nabla u_\varepsilon(y)-\nabla u_\varepsilon(x))^2\mu(|y-x|)]\,\mathrm{d}y\,\mathrm{d}x+\|\nabla u_\varepsilon\|^2_{L^2(\Omega^\varepsilon\setminus\Omega_\varepsilon)}.\end{aligned}$$

By using Hölder's inequality in (15) and combining it with (16) we obtain

$$\|\nabla u_\varepsilon\|^2_{L^2(\Omega_\varepsilon)} \le C\Big\{ \|\nabla f_\varepsilon\|^2_{L^2(\Omega_\varepsilon)} + \|\nabla u_\varepsilon\|^2_{L^2(\Omega^\varepsilon\setminus\Omega_\varepsilon)} + \int_{\Omega_\varepsilon}\int_{\Omega^\varepsilon\setminus\Omega_\varepsilon} [(\nabla u_\varepsilon(y) - \nabla u_\varepsilon(x))\mu(|y-x|)]\cdot\nabla u_\varepsilon(x)\,\mathrm{d}y\,\mathrm{d}x\Big\}. \tag{17}$$

Since $\nabla u_\varepsilon(x) = (u * \nabla\phi_\varepsilon)(x)$ with $\nabla\phi_\varepsilon(x) = \frac{1}{\varepsilon^{n+1}}\nabla\phi\left(\frac{x}{\varepsilon}\right)$ we have by Young's inequality for convolutions that

$$\|\nabla u_\varepsilon\|^2_{L^2(\Omega^\varepsilon\setminus\Omega_\varepsilon)} \le \|u\|^2_{L^1(\Omega^\varepsilon\setminus\Omega_\varepsilon)}\|\nabla\phi_\varepsilon\|^2_{L^2(\Omega^\varepsilon\setminus\Omega_\varepsilon)}.$$

For $f \in C^1(\bar\Omega) \subset L^\infty(\Omega)$ we have $u \in C^1(\bar\Omega) \subset L^\infty(\Omega)$ by Theorem 3.1, and since $|\Omega^\varepsilon\setminus\Omega_\varepsilon| < C\varepsilon$ [c] the above estimate becomes

$$\|\nabla u_\varepsilon\|^2_{L^2(\Omega^\varepsilon\setminus\Omega_\varepsilon)} \le C\varepsilon^2\|u\|^2_{L^\infty(\Omega)}\frac{\|\nabla\phi\|^2_{L^1(\mathbb{R}^n)}}{\varepsilon^2} < C.$$

Finally, we estimate the last term on the RHS of (17) by using Young's inequality for convolutions.

$$\begin{aligned}
&\left|\int_{\Omega_\varepsilon}\int_{\Omega^\varepsilon\setminus\Omega_\varepsilon} [(\nabla u_\varepsilon(y) - \nabla u_\varepsilon(x))\mu(|y-x|)]\cdot\nabla u_\varepsilon(x)\,\mathrm{d}y\,\mathrm{d}x\right| \\
&\le \left|\int_{\Omega_\varepsilon}\int_{\Omega^\varepsilon\setminus\Omega_\varepsilon} \nabla u_\varepsilon(y)\cdot\nabla u_\varepsilon(x)\mu(|y-x|)\,\mathrm{d}y\,\mathrm{d}x\right| + \|\nabla u_\varepsilon\|_{L^2(\Omega_\varepsilon)}\int_{\Omega^\varepsilon\setminus\Omega_\varepsilon}\mu(|y-x|)\,\mathrm{d}y \\
&\le \int_{\Omega_\varepsilon} |[(\chi_{\Omega^\varepsilon\setminus\Omega_\varepsilon}(y)\nabla u_\varepsilon) * \mu](x)|dx + \|\nabla u_\varepsilon\|_{L^2(\Omega_\varepsilon)}\int_{\Omega^\varepsilon\setminus\Omega_\varepsilon}\mu(|y-x|)\,\mathrm{d}y \\
&\le \eta\|\nabla u_\varepsilon\|_{L^2(\Omega^\varepsilon)} + C(\eta)\|\mu\|^2_{L^1((\Omega_\varepsilon))}\|\nabla u_\varepsilon\|^2_{L^2(\Omega^\varepsilon\setminus\Omega_\varepsilon)} \\
&\qquad + \|\nabla u_\varepsilon\|_{L^2(\Omega_\varepsilon)}\int_{\Omega^\varepsilon\setminus\Omega_\varepsilon}\mu(|y-x|)\,\mathrm{d}y
\end{aligned}$$

for $\eta > 0$, and where χ_A denotes the characteristic function for the set A. Since $\mu \in L^1(\mathbb{R}^n)$ we can choose ε sufficiently small to absorb the third term above in the LHS of (17); similarly, by choosing η sufficiently small we obtain

$$\|\nabla u_\varepsilon\|^2_{L^2(\Omega_\varepsilon)} \le C. \tag{18}$$

[c]Since the boundary of Ω is smooth

The above uniform estimate shows that for an open set $\Omega' \subset\subset \Omega$ we can extract a subsequence ∇u_{ε_k} that converges weakly to some $h \in L^2(\Omega')$. We have that h is the distributional gradient of u since for $\phi \in C_c(\Omega')$ the following holds:

$$(h, \phi)_{L^2(\Omega')} = \lim_{k\to\infty} (\nabla u_{\varepsilon_k}, \phi)_{L^2(\Omega')} = -\lim_{k\to\infty} (u_{\varepsilon_k}, \nabla\phi)_{L^2(\Omega')} = -(u, \nabla\phi)_{L^2(\Omega')}.$$

This concludes the proof of the theorem. □

Remark 4.1. Note that a boot-strapping argument shows that $f \in C^\infty(\bar{\Omega})$ implies $u \in C^\infty(\bar{\Omega})$.

References

1. H. Sun, D. Uminsky and A. L. Bertozzi, Stability and clustering of self-similar solutions of aggregation equations, *J. Math. Phys.* **53**, 115610, 18 (2012).
2. A. Mogilner and L. Edelstein-Keshet, A non-local model for a swarm, *Journal of Mathematical Biology* **38**, 534 (1999).
3. S. Oterkus, E. Madenci and A. Agwai, Peridynamic thermal diffusion, *J. Comput. Phys.* **265**, 71 (2014).
4. G. Gilboa and S. Osher, Nonlocal Operators with Applications to Image Processing , *Multiscale Modeling & Simulation* **7**, 1005 (2008).
5. F. Andreu-Vaillo, J. M. Mazón, J. D. Rossi and J. J. Toledo-Melero, *Nonlocal diffusion problems*, Mathematical Surveys and Monographs, Vol. 165 (American Mathematical Society, Providence, RI; Real Sociedad Matemática Española, Madrid, 2010).
6. S. Silling, Reformulation of elasticity theory for discontinuities and long-range forces, *Journal of the Mechanics and Physics of Solids* **48**, 175 (2000).
7. M. Foss and J. Geisbauer, Partial regularity for subquadratic parabolic systems with continuous coefficients, *Manuscripta Math.* **139**, 1 (2012).
8. L. A. Caffarelli, R. Leitão and J. M. Urbano, Regularity for anisotropic fully nonlinear integro-differential equations, *Math. Ann.* **360**, 681 (2014).
9. Q. Du, M. Gunzburger, R. B. Lehoucq and K. Zhou, Analysis and approximation of nonlocal diffusion problems with volume constraints, *SIAM Rev.* **54**, 667 (2012).
10. T. Mengesha and Q. Du, The bond-based peridynamic system with Dirichlet-type volume constraint, *Proc. Roy. Soc. Edinburgh Sect. A* **144**, 161 (2014).

11. P. Radu, D. Toundykov and J. Trageser, A nonlocal biharmonic operator and its connection with the classical analogue (submitted).
12. B. Hinds and P. Radu, Dirichlet's principle and wellposedness of solutions for a nonlocal p-Laplacian system, *Appl. Math. Comput.* **219**, 1411 (2012).
13. J. Bracewell, A. Brucker, M. Foss, C. Hoffman, P. Radu and T. Vossberg, Steady state models for nonlocal heat conduction in peridynamics (2012), (preprint).
14. Q. Du, M. Gunzburger, R. B. Lehoucq and K. Zhou, A nonlocal vector calculus, nonlocal volume-constrained problems, and nonlocal balance laws, *Mathematical Models and Methods in Applied Sciences* **23**, 493 (2013).
15. B. Aksoylu and M. L. Parks, Variational theory and domain decomposition for nonlocal problems, *Appl. Math. Comput.* **217**, 6498 (2011).

Ferroelectric thin structures

Antonio Gaudiello
DIEI, Università degli Studi di Cassino e del Lazio Meridionale, via G. Di Biasio 43, 03043 Cassino (FR), Italia
E-mail: gaudiell@unina.it

Kamel Hamdache
Leonard de Vinci Pôle Universitaire, Technology Lab, 92916 Paris La Défense Cedex, France
E-mail: kamel.hamdache@devinci.fr

In this note we summarize some results obtained in [19] and in [20], where starting from a non-convex and nonlocal 3D-variational model for the electric polarization in a ferroelectric material, via an asymptotic process we obtain $2D$-variational models for a thin film and $1D$-variational models for a thin wire, respectively.

Keywords: electric polarization, thin film, thin wire, nonlocal problems

1. Ferroelectric thin films and ferroelectric thin wires

The ferroelectricity is a property of some materials to have a spontaneous electrical polarization that can be reversed by the application of an external electric field. Ferroelectricity was demonstrated for first time in 1920 in Rochelle salt by J. Valasek. The chemical and crystallographic complexity of this salt restrained the research and suggested that the ferroelectricity was an exotic property requiring very specific conditions. Moreover, this property did not present any practical interest. A leap in the study of the ferroelectricity happened in the early 1950s with the discovery of ferroelectric oxides with perovskite structure: barium titanate (BaTiO3), lead titanate (PbTiO3), etc. These simpler materials enabled the development of theories of ferroelectricity and opened the way for industrial use of ferroelectric materials. Dielectric and piezoelectric properties of ferroelectric materials are now used in a wide variety of contexts. In particular, thin ferroelectric materials are receiving great interest due to their various applications to memory and storage devices, electronic circuits with miniaturized and integrated forms, etc. as, for instance, radio frequency identification cards (RFID) and ferroelectric tunnel junction (FTJ) (see[10] and[29]).

In[19] and in[20], we start from a non-convex and nonlocal 3D-variational model for the electric polarization in a ferroelectric material. Via an asymptotic process based on dimensional reduction, we obtain $2D$-variational models for a thin film and $1D$-variational models for a thin wire, respectively. The reduced models are justified by reasons of simplicity and economy, especially by a numerical point of view. In these two papers, we do not consider any deformation of the ferroelectric material.

Let $h_n \in]0,1[$, $n \in \mathbb{N}$, be a parameter tending to zero and $\omega \subset \mathbb{R}^2$ be an open polygonal set. Let

$$\Omega_n = \omega \times \left]-\frac{h_n}{2}, \frac{h_n}{2}\right[, \quad n \in \mathbb{N}, \tag{1}$$

be a $3D$ ferroelectric cylindrical device with cross-section $\omega \subset \mathbb{R}^2$ and small thickness h_n in the case of modeling a thin film, or

$$\Omega_n = h_n\omega \times \left]-\frac{1}{2}, \frac{1}{2}\right[, \quad n \in \mathbb{N}, \tag{2}$$

be a 3D ferroelectric cylindrical device with small cross-section $h_n\omega$ and thickness 1 in the case of modeling a thin wire.

In a ferroelectric material, the response to an applied electric field changes the electric displacement as $\mathbf{D} = \varepsilon E + 4\pi p$, where $\varepsilon > 0$ is the dielectric permeability, p is the spontaneous electric polarization field and $E = -D\varphi$ is the electric field which satisfies the electrostatic equation

$$\begin{cases} \operatorname{div}(-\varepsilon D\varphi + 4\pi p) = 0 \text{ in } \Omega_n, \\ \\ (-\varepsilon D\varphi + 4\pi p) \cdot \nu = 0 \text{ on } \partial\Omega_n, \end{cases} \tag{3}$$

with ν denoting the unit outer normal on $\partial\Omega_n$. The free energy associated with Ω_n is non-convex, nonlocal and it is given by (see[7,28,30] for an explanation of the model)

$$\begin{aligned} \mathcal{E}_n : p = (p_1, p_2, p_3) \in (H^1(\Omega_n))^3 \to \beta \int_{\Omega_n} |rotp|^2 dx + \int_{\Omega_n} |divp|^2 dx \\ + \alpha \int_{\Omega_n} (|p|^2 - 1)^2 dx + \int_{\Omega_n} |D\varphi|^2 dx - \int_{\Omega_n} (g_n \cdot p) dx, \end{aligned} \tag{4}$$

where α and β are positive constants, $g_n \in \left(L^2(\Omega_n)\right)^3$ is an external electric field, $x = (x_1, x_2, x_3) = (x', x_3)$ denotes the generic point of $\mathbb{R}^3$, φ is the unique solution of (3), and $\cdot$ denotes the inner product in $\mathbb{R}^3$.

Remark that $\beta \int_{\Omega_n} |rotp|^2 dx + \int_{\Omega_n} |divp|^2 dx$ reduces to the classical energy $\int_{\Omega_n} |Dp|^2 dx$ when $\beta = 1$, so roughly speaking this term penalizes the spatial variation of p, while the term $\alpha \int_{\Omega_n} (|p|^2 - 1)^2 dx$ induces a phase transition of p. So the body is driven to have regions of uniform polarization separated by thin transition layers. The external energy $\int_{\Omega_n} g_n \cdot p\, dx$ favors the polarization parallel to an externally applied field.

Imposing suitable convergence assumptions on the rescaled exterior field in $\Omega = \omega \times \left]-\frac{1}{2}, \frac{1}{2}\right[$, the goal is to study the asymptotic behavior, as h_n vanishes, of minimum of energy (4) with suitable boundary conditions. Precisely, we consider two types of boundary conditions:

$$p \cdot \nu = 0 \text{ on } \partial\Omega_n, \tag{5}$$

or

$$p \wedge \nu = 0 \text{ on } \partial\Omega_n, \tag{6}$$

with $\wedge$ denoting the cross product in $\mathbb{R}^3$. We prove the following four results.

I) In the case of a thin wire, i.e. when Ω_n is defined by (2), we prove that

$$\begin{aligned} &\lim_n \min \left\{ \frac{\mathcal{E}_n(p)}{h_n^2} : p \in (H^1(\Omega_n))^3,\ p \cdot \nu = 0 \text{ on } \partial\Omega_n \right\} \\ &= \min \left\{ E_\infty(q) : q \in H_0^1\left(\left]-\tfrac{1}{2}, \tfrac{1}{2}\right[\right) \right\}, \end{aligned} \tag{7}$$

where

$$E_\infty : q \in H_0^1\left(\left]-\frac{1}{2}, \frac{1}{2}\right[\right) \to \beta|\omega| \int_{-\frac{1}{2}}^{\frac{1}{2}} \left|\frac{dq}{dx_3}\right|^2 dx_3 + \alpha|\omega| \int_{-\frac{1}{2}}^{\frac{1}{2}} \left(|q|^2 - 1\right)^2 dx_3$$

$$+ \left(\frac{4\pi}{\varepsilon}\right)^2 |\omega| \int_{-\frac{1}{2}}^{\frac{1}{2}} |q|^2 dx_3 - \int_{-\frac{1}{2}}^{-\frac{1}{2}} \overline{g}_3 q dx_3,$$

with $\overline{g} = (\overline{g}_1, \overline{g}_2, \overline{g}_3)$ denoting the integral over x' on cross section ω of the $\left(L^2(\Omega)\right)^3$-weak limit of the rescaled external field. More precisely, we obtain $(H^1(\Omega))^3$-strong convergence for a subsequence of rescaled polarizations. The limit polarization depends only on $x_3 \in \left]-\frac{1}{2}, \frac{1}{2}\right[$ and it is parallel to the wire, i.e. the first two components are zero, while its third component solves the limit problem in (7). We remark that the nonlocal term transforms into $\left(\frac{4\pi}{\varepsilon}\right)^2 |\omega| \int_{-\frac{1}{2}}^{\frac{1}{2}} |q|^2 dx_3$. So the limit problem becomes local.

II) In the case of a thin wire with boundary condition (6), we prove that the rescaled polarization converges strongly in $(H^1(\Omega))^3$ to $(0, 0, 0)$.

III) In the case of a thin film, i.e. when Ω_n is defined by (1), we prove that

$$\lim_n \min\left\{\frac{\mathcal{E}_n(p)}{h_n} : p \in (H^1(\Omega_n))^3,\ p\cdot\nu = 0 \text{ on } \partial\Omega_n\right\}$$
$$= \min\left\{F_\infty(q) : q \in \left(H^1(\omega)\right)^2 : q\cdot\nu' = 0 \text{ on } \partial\omega\right\}, \tag{8}$$

where

$$F_\infty : q \in \left(H^1(\omega)\right)^2 \to \beta\int_\omega |rotq|^2 dx' + \int_\omega |divq|^2 dx'$$
$$+\alpha\int_\omega (|q|^2-1)^2 dx' + \int_\omega |D\psi|^2 dx' - \int_\omega (\underline{g_1}, \underline{g_2})\cdot q\, dx',$$

with ν' denoting the unit outer normal on $\partial\omega$, $\underline{g} = (\underline{g}_1, \underline{g}_2, \underline{g}_3)$ the integral over x_3 on $\left]-\frac{1}{2}, \frac{1}{2}\right[$ of the $\left(L^2(\Omega)\right)^3$-weak limit of the rescaled external field, and ψ the unique solution of

$$\begin{cases} \operatorname{div}(-\varepsilon D\psi + 4\pi q) = 0 \text{ in } \omega, \\ (-\varepsilon D\psi + 4\pi q)\cdot\nu' = 0 \text{ on } \partial\omega. \end{cases}$$

More precisely, also in this case we obtain $(H^1(\Omega))^3$-strong convergence for a subsequence of rescaled polarizations. The limit is independent of x_3 and parallel to cross-section ω, i.e. the third component is zero, while the in-plane components solve the limit problem in (8). This limit problem is similar to the original problem.

IV) In the case of a thin film with boundary condition (6), we prove that

$$\lim_n \min\left\{\frac{\mathcal{E}_n(p)}{h_n} : p \in (H^1(\Omega_n))^3,\ p\cdot\nu = 0 \text{ on } \partial\Omega_n\right\}$$
$$= \min\left\{G_\infty(q) : q \in H_0^1(\omega)\right\} \tag{9}$$

where

$$G_\infty : q \in H^1(\omega) \to \int_\omega \left(\beta|Dq|^2 + \alpha(|q|^2-1)^2 + \left(\frac{4\pi}{\varepsilon}\right)^2 |q|^2 - \underline{g_3} q\right) dx'.$$

Also in this case we obtain $(H^1(\Omega))^3$-strong convergence for a subsequence of rescaled polarizations with a limit independent of x_3, but now it is orthogonal to ω, i.e. the first two components are zero while the out-plane component solves the limit problem in (9). The nonlocal term transforms into $\left(\frac{4\pi}{\varepsilon}\right)^2 \int_\omega |q|^2 dx'$. So the limit problem becomes local.

In these two papers, both boundary conditions (5) and (6) allow to estimate $\|Dp\|_{(L^2(\Omega_n))^9}$ in terms of $\|rotp\|_{(L^2(\Omega_n))^3}$ and $\|divp\|_{L^2(\Omega_n)}$, independently of $n \in N$. To this aim, we assume that ω is a polygon (see[9]). We reformulate our problem on a fixed domain through appropriate rescalings of the kind proposed in[8] and impose appropriate convergence assumptions on the rescaled exterior fields. The proofs of our results are based on accurate *a priori* estimates and on the given boundary conditions which provide a first characterization of the $H^1(\Omega)$-weak limit of the rescaled polarization. These information allows us to identify the limit of the nonlocal term, i.e. the limit of the electric field satisfying the electrostatic equation (3). Then, using the main idea of Γ-convergence method introduced in[11], we characterize completely the limit polarization as solution of a minimization problem. Finally, we prove that the convergences are $H^1(\Omega)$-strong.

For the study of ferromagnetic thin structures, whose 3D model is close to our 3D model, we refer to[1], [2], [3], [4], [5], [6], [12], [13], [14], [15], [16], [17], [18], [23], [24], [25], [26], and[27]. For other recent problems with thin wires, see[21] and[22].

References

1. R. Alicandro, C. Leone, *3D-2D Asymptotic Analysis for Micromagnetic Thin Films*, ESAIM Control Optim. Calc. Var. **6** (2001), 489-498.
2. F. Alouges, S. Labbé, *Convergence of a ferromagnetic film model*, C. R. Math. Acad. Sci. Paris **344** (2007), 2, 77-82.
3. H. Ammari, L. Halpern, K. Hamdache, *Asymptotic behavior of thin ferromagnetic films*, Asymptot. Anal. **24** (2000), 277-294.
4. M. Baía, E. Zappale, *A note on the 3D-2D dimensional reduction of a micromagnetic thin film with nonhomogeneous profile*, Appl. Anal. **86** (2007), n. 5, 555-575.
5. G. Carbou, *Thin Layers in Micromagnetism*, M^3AS: Math. Models Methods Appl. Sci. **11** (2001), n.9, 1529-1546.
6. P. G. Carbou, S. Labbé, E. Trélat, *Control of travelling walls in a ferromagnetic nanowire*, Discrete Contin. Dyn. Syst. Ser. S **1** (2008), 1, 5159.
7. P. Chandra, P. B. Littlewood, *A Landau primer for ferroelectrics*, The Physics of ferroelectrics: A modern perspective, Topics Appl. Physics **105** (2007), 69-116 (K. Rabe, C. H. Ahn, J.-M. Triscone Eds).
8. P. G. Ciarlet, P. Destuynder, *A Justification of the two-dimensional linear plate model*, J. Mècanique **18** (1979), 2, 315-344.

9. M. Costabel, M. Dauge, S. Nicaise, *Singularities of Maxwell interface problems*, M2AN Math. Model. Numer. Anal. **33** (1999), 3, 627-649.
10. L. E. Cross, R. E. Newnham, *History of Ferroelectrics*, Reprinted from the Ceramics and Civilization, Volume II High-Technology Ceramics-Past, Present, and Future. Copyright 1987 by The American Ceramic Society. Inc.
11. E. De Giorgi, T. Franzoni, *Su un tipo di convergenza variazionale*, Atti Accad. Naz. Lincei Rend. Cl. Sci. Fis. Mat. Natur. **58** (8) (1975), 6, 842-850.
12. U. De Maio, L. Faella, C. Perugia, *Quasy-stationary ferromagnetic thin films in degenerated cases,* Ricerche Mat. (2014), **63** (2014), 225-237.
13. U. De Maio, L. Faella, S. Soueid, *Junction of quasi-stationary ferromagnetic thin films,* Asymptot. Anal. **94** (2015), 211-240.
14. A. DeSimone, R. V. Kohn, S. Muller, F. Otto, *A reduced theory for thin-film micromagnetics*, Commun. Pure Appl. Math. **55** (2002), 11, 1408-1460.
15. A. Gaudiello, R. Hadiji, *Junction of one-dimensional minimization problems involving S^2 valued maps*, Adv. Differ. Equ. **13** (2008), 9-10, 935-958.
16. A. Gaudiello, R. Hadiji, *Asymptotic analysis, in a thin multidomain, of minimizing maps with values in S^2*, Ann. Inst. Henri Poincaré, Anal. Non Linéaire **26** (2009), 1, 59-80.
17. A. Gaudiello, R. Hadiji, *Junction of ferromagnetic thin films*, Calc. Var. Partial Differential Equations **39** (2010), 3, 593-619.
18. A. Gaudiello, R. Hadiji, *Ferromagnetic thin multi-structures*, J. Differential Equations **257** (2014), 1591-1622.
19. A. Gaudiello, K. Hamdache, *The polarization in a ferroelectric thin film: local and nonlocal limit problems*, ESAIM Control Optim. Calc. Var. **19** (2013), 657-667.
20. A. Gaudiello, K. Hamdache, *A reduced model for the polarization in a ferroelectric thin wire*, NoDEA Nonlinear Differential Equations Appl. **22** (2015), 6, 1883-1896.
21. A. Gaudiello, G. Panasenko, A. Piatnitski, *Asymptotic analysis and domain decomposition for a biharmonic problem in a thin multi-structure*, Commun. Contemp. Math. DOI: 10.1142/S0219199715500571 (2015), 27 pages.
22. A. Gaudiello, A. Sili, Homogenization of highly oscillating boundaries with strongly contrasting diffusivity, SIAM J. Math. Anal. **47** (2015), 3, 1671-1692.

23. G. Gioia, R.D. James, *Micromagnetism of very thin films,* Proc. R. Lond. A **453** (1997), 213-223.
24. R. Hadiji, K. Shirakawa, *Asymptotic analysis of micromagnetics on thin films governed by indefinite material coefficients*, Commun. Pure Appl. Anal. **9** (2010), 5, 1345-1361.
25. R. V. Kohn, V. V. Slastikov, *Another thin-film limit of micromagnetics,* Arch. Rational Mech. Anal. **178** (2005), 227-245.
26. D. Sanchez, *Behaviour of the Landau-Lifschitz equation in a ferromagnetic wire,* Math. Methods Appl. Sci. **32** (2009), 2, 167-205.
27. V. Slastikov, C. Sonnenberg, *Reduced models for ferromagnetic nanowires*, IMA J. Appl. Math **77** (2012) 2, 220-235.
28. Y. Su, C. M. Landis, *Continuum thermodynamics of ferroelectric domain evolution: theory, finite element implementation, and application to domain wall pinning*, J. Mech. Phys. Solids **55** (2007), 2 280-305.
29. J. Ballato and M. C. Gupta (Editors), *The Handbook of Photonics, Second Edition*, CRC Press 2006. V. Gopalan, K. L. Schepler, V. Dierolf, I. Biaggio, *Chapter 6. Ferroelectric Materials.*
30. W. Zhang, K. Bhattacharya, *A computational model of ferroelectric domains. Part I. Model formulation and domain switching*, Acta Mater. **53** (2005), 185-198.

Sliding modes for a phase-field system

Gianni Gilardi
Dipartimento di Matematica "F. Casorati", University of Pavia,
and IMATI-CNR of Pavia
via Ferrata 1, 20100 Pavia, Italy
E-mail: gianni.gilardi@unipv.it

The present contribution regards a sliding mode control problem for a phase-field model of Caginalp type and summarizes the results of a joint work with Viorel Barbu, Pierluigi Colli, Gabriela Marinoschi and Elisabetta Rocca obtained in the framework of the three year Italian-Romanian project on *Nonlinear PDEs with applications in modeling cell growth, chemotaxis and phase transition.*

Keywords: phase field system, nonlinear boundary value problems, phase transition, sliding mode control, state-feedback control law.

1. Introduction

Sliding mode control is one of the fundamental approaches for the systematic design of robust controllers for nonlinear complex dynamic systems and it is considered a classical tool for the regulation of continuous - or discrete - time systems in finite-dimensional settings. The main advantage is that it reduces complexity, as we show very roughly. Assume that a Cauchy problem for an ODE system is given. Namely

$$y'(t) = f(t, y(t)) \quad \text{for } t \geq 0 \quad \text{and} \quad y(0) = y_0 \tag{1}$$

where $f : [0, +\infty) \times \mathbb{R}^n \to \mathbb{R}^n$ is a suitably smooth function, $y_0 \in \mathbb{R}^n$ is a prescribed initial datum, and $y : [0, +\infty) \to \mathbb{R}^n$ is the solution. Then, the main idea is first to identify a manifold of lower dimension

$$\mathcal{M} = \{x \in \mathbb{R}^n : \; g(x) = 0\}$$

called the sliding manifold, and then to design a feed-back control law $y \mapsto u(y)$ that forces the trajectories of the system to reach the sliding manifold and maintain them on it. The modified system looks like

$$y'(t) = f(t, y(t)) + u(y(t)) \quad \text{for } t \geq 0 \quad \text{and} \quad y(0) = y_0\,. \tag{2}$$

Therefore, $u : \mathbb{R}^n \to \mathbb{R}^n$ has to be chosen in order that the solution y to the modified problem satisfies $y(t) \in \mathcal{M}$, i.e., $g(y(t)) = 0$, for t large enough. Thus, for large times the complexity is strongly reduced. For this reason, sliding mode control is very popular in the engineering world (automatic control) and the literature on this subject is very wide. We confine ourselves to quote the books[6,8,10,11] and the state-of-the-art monographs[1,5,7,12]. On the contrary, the literature for analogous problems in the framework of partial differential equations is very poor. In this respect, we can cite the papers[4,9], and[13].

In the paper[2] the authors employ – to the best of our knowledge for the first time in the literature – a sliding mode contol technique for a nonlinear PDE system of phase-field type. In this contribution, we summarize the results of[2] and start considering the following rather simple version of the phase-field system of Caginalp type (see[3]):

$$\partial_t\big(\theta + \ell\phi\big) - \kappa\Delta\theta = f \quad \text{and} \quad \partial_t\phi - \nu\Delta\phi + F'(\phi) = \gamma\theta \tag{3}$$

complemented with the Neumann homogeneous boundary conditions and some initial conditions for both θ and ϕ. The above equations model two-phase transition processes and involve the relative temperature θ around some critical value and the so-called order parameter ϕ, which is an index of how much of one of the phases is present at every point and every time. Moreover, ℓ, κ, ν and γ are physical constants (the first one is related to the latent heat), f is a source term and F' is the derivative of a double-well potential F (i.e., a function with two minimum points, termed wells). Typical and important examples of such potentials are

$$F_{reg}(r) = \tfrac{1}{4}\,(r^2-1)^2 \tag{4}$$

$$F_{log}(r) = \big((1+r)\ln(1+r) + (1-r)\ln(1-r)\big) - (1+c)\,r^2 \tag{5}$$

$$F_{obs}(r) = I(r) - c\,r^2 \tag{6}$$

where r runs over $(-1,1)$ in (5) and $c > 0$ in both (5) and (6). The function I in (6) is the indicator function of $[-1,1]$, i.e., it takes the values 0 or $+\infty$ according to whether or not r belongs to $[-1,1]$. The potential (4) and (5) are the usual classical regular potential and the so-called logarithmic potential, respectively, while the third potential (6) takes the value $+\infty$ somewhere. All the above examples fit the following more general setting: the potential F is the sum $F = \hat\beta + \Pi$, where $\hat\beta$ is a (possibly infinite) proper and lower semicontinuous convex function and Π is a smooth perturbation. In such a case, the derivative F' is replaced by the sum $\beta + \pi$, where $\beta := \partial\hat\beta$ is the subdifferential of $\hat\beta$ and π is the derivative of Π, and the second equation in (3) becomes a differential inclusion. Thus, (6) is some kind of prototype of the worst situation.

Our aim is to find out some state-feedback control laws $(\theta, \phi) \mapsto u(\theta, \phi)$ that can be inserted in one of the equations in order that the dynamics of the system modified in this way forces the value $(\theta(t), \phi(t))$ of the solution to reach some manifold of the phase space in a finite time and then lie there with a sliding mode. We study three cases. In the first one, a feedback control is added to the first equation in (3) in order to force a linear relationship between θ and ϕ; in the second and third cases, a prescribed distribution ϕ^* of the order parameter is forced by means of a feedback control added to the second equations.

The third problem uses the sign function. On the contrary, in the first two situations we present, the feedback control law is nonlocal, since it is related to the function Sign defined as the subdifferential of the L^2-norm.

While the shape of the feedback control laws is chosen in advance, their size depends on a parameter $\rho > 0$. The sliding mode is ensured if ρ is large enough, without significant restrictions in the nonlocal case, with a proper restriction in the local one.

2. Results

In this section, we describe the problem more precisely and present the main results of[2]. We assume that

$$\ell,\ \kappa,\ \nu,\ \gamma \in (0, +\infty) \tag{7}$$

$$\hat{\beta} : \mathbb{R} \to [0, +\infty] \quad \text{is convex, proper and l.s.c.} \quad \text{with} \quad \hat{\beta}(0) = 0 \tag{8}$$

$$\Pi : \mathbb{R} \to \mathbb{R} \quad \text{is a } C^1 \text{ function} \quad \text{and} \quad \Pi' \quad \text{is Lipschitz continuous} \tag{9}$$

We set for brevity

$$\beta := \partial\hat{\beta} \quad \text{and} \quad \pi := \Pi'. \tag{10}$$

Next, we assume $\Omega \subset \mathbb{R}^3$ to be open, bounded, connected, and smooth, and set for convenience

$$V := H^1(\Omega), \quad H := L^2(\Omega), \quad W := \{v \in H^2(\Omega) : \partial_n v = 0\} \tag{11}$$

where the symbol ∂_n stands for the (say, outer) normal derivative on the boundary of Ω. Furthermore, we introduce the operators Sign: $H \to 2^H$ and sign : $\mathbb{R} \to 2^{\mathbb{R}}$ as the subdifferentials of the maps $\|\ \|_H \cdot : H \to \mathbb{R}$ and

$|\cdot| : \mathbb{R} \to \mathbb{R}$, respectively. Thus, more explicitly, we have

$$\text{Sign } v = \frac{v}{\|v\|_H} \quad \text{if } v \neq 0 \quad \text{and} \quad \text{Sign } 0 = \text{closed unit ball of } H \tag{12}$$

$$\text{sign } r = \frac{r}{|r|} \quad \text{if } r \neq 0 \quad \text{and} \quad \text{sign } 0 = [-1, 1] \tag{13}$$

for $v \in H$ and $r \in \mathbb{R}$. All the operators β, Sign and sign are maximal monotone graphs. As they are in general multi-valued, the equations become differential inclusions and the selections from the graphs (thus, new unknown functions) enter the state systems modified by the state-feedback control laws. We set for convenience $Q := \Omega \times (0, T)$, where T is some fixed final time. As said in the Introduction, we study three cases: in the first one, a control term is added to the first equation (3), while in the other ones, it is added to the second equation. In principle, for the data, we require that

$$f \in L^2(Q), \quad \theta_0 \in V, \quad \phi_0 \in V \quad \text{and} \quad \hat{\beta}(\phi_0) \in L^1(\Omega) \tag{14}$$

and we look for a quadruplet $(\theta, \phi, \xi, \sigma)$ satisfying the regularity requirements and the boundary and initial conditions

$$\theta, \phi \in H^1(0, T; H) \cap L^\infty(0, T; V) \cap L^2(0, T; W) \tag{15}$$

$$\xi \in L^2(0, T; H) \quad \text{and} \quad \sigma \in L^\infty(0, T; H) \tag{16}$$

$$\theta(0) = \theta_0 \quad \text{and} \quad \phi(0) = \phi_0 \tag{17}$$

and solving the system depending on the parameter $\rho > 0$ we introduce at once. Notice that (15) actually contains the Neumann homogeneous boundary conditions for both θ and ϕ, due to the definition (11) of W. Given $\eta^* \in W$ and $\alpha \in \mathbb{R}$, the first system is

$$\partial_t(\theta + \ell\phi) - \kappa\Delta\theta = f - \rho\sigma \tag{18}$$

$$\partial_t\phi - \nu\Delta\phi + \xi + \pi(\phi) = \gamma\theta \quad \text{and} \quad \xi \in \beta(\phi) \tag{19}$$

$$\sigma(t) \in \text{Sign}(\theta(t) + \alpha\phi(t) - \eta^*) \quad \text{for a.a. } t \in (0, \mathrm{T}) \tag{20}$$

where (18)–(19) are understood a.e. in Q. In the sequel, we also term such problem *Problem (A)*. The second problem, which we call *Problem (B)*, depends on a given $\phi^* \in W$ and consists in the equations

$$\partial_t(\theta + \ell\phi) - \kappa\Delta\theta = f \tag{21}$$

$$\partial_t\phi - \nu\Delta\phi + \xi + \pi(\phi) = \gamma\theta - \rho\sigma \quad \text{and} \quad \xi \in \beta(\phi) \tag{22}$$

$$\sigma(t) \in \text{Sign}(\phi(t) - \phi^*) \quad \text{for a.a. } t \in (0, \mathrm{T}) \tag{23}$$

where (21)–(22) are understood a.e. in Q. The last case, termed *Problem (C)*, is the same as the previous one with the following difference: the non-local operator Sign is replaced by the local sign : $\mathbb{R} \to 2^{\mathbb{R}}$. For the sake of clarity, we write Problem (C), explicitly. Given $\phi^* \in W$, the quadruplet $(\theta, \phi, \xi, \sigma)$ has to satisfy the equations and conditions, all of them being understood a.e. in Q

$$\partial_t\big(\theta + \ell\phi\big) - \kappa\Delta\theta = f \tag{24}$$

$$\partial_t\phi - \nu\Delta\phi + \xi + \pi(\phi) = \gamma\theta - \rho\sigma \quad \text{and} \quad \xi \in \beta(\phi) \tag{25}$$

$$\sigma \in \text{sign}\,(\phi - \phi^*). \tag{26}$$

Here are the results of[2] on the well-posedness of the above problems.

Theorem 2.1. *Assume (7)–(9), (14)*

$$\eta^* \in W \quad \textit{and} \quad \alpha \in \mathbb{R}. \tag{27}$$

Then, for every $\rho > 0$, Problem (A) has at least a solution $(\theta, \phi, \xi, \sigma)$ satisfying (15)–(17) and the estimates

$$\|\theta\|_{L^\infty(0,T;H)\cap L^2(0,T;V)} + \|\phi\|_{H^1(0,T;H)\cap L^2(0,T;W)} \\ + \|\xi\|_{L^2(0,T;H)} + \|\sigma\|_{L^\infty(0,T;H)} \le C_1 \tag{28}$$

$$\|\theta\|_{H^1(0,T;H)\cap L^2(0,T;W)} \le C_2\big(\rho^{1/2} + 1\big) \tag{29}$$

where C_1 and C_2 depend only on the quantities involved in assumptions (7)–(9), (14) and (27). Moreover, the solution is unique if $\alpha = \ell$. Furthermore, if in addition

$$\phi_0 \in W \quad \textit{and} \quad \beta^0(\phi_0) \in H \tag{30}$$

then, there exists a solution that also satisfies

$$\phi \in W^{1,\infty}(0,T;H) \cap H^1(0,T;V) \cap L^\infty(0,T;W) \\ \textit{and} \quad \xi \in L^\infty(0,T;H) \tag{31}$$

$$\|\phi\|_{W^{1,\infty}(0,T;H)\cap H^1(0,T;V)\cap L^\infty(0,T;W)} \le C_3\big(\rho^{1/2} + 1\big) \tag{32}$$

where C_3 depends on the norms related to (30) as well. In particular, ϕ is bounded. Finally, the component θ of any solution satisfying all the above regularity requirements is bounded whenever $\theta_0 \in V \cap L^\infty(\Omega)$ and $f \in L^\infty(0,T;H)$.

Theorem 2.2. *Assume (7)–(9), (14), as well as*

$$\phi^* \in W \quad \textit{and} \quad \beta^0(\phi^*) \in H. \tag{33}$$

Then, for every $\rho > 0$, Problem (B) has at least a solution $(\theta, \phi, \xi, \sigma)$ satisfying (15)–(17) and the estimates

$$\begin{aligned} &\|\theta\|_{L^\infty(0,T;H)\cap L^2(0,T;V)} + \|\phi\|_{L^\infty(0,T;H)\cap L^2(0,T;V)} \\ &\quad + \|\sigma\|_{L^\infty(0,T;H)} \le C_4 \end{aligned} \tag{34}$$

$$\begin{aligned} &\|\theta\|_{H^1(0,T;H)\cap L^2(0,T;W)} + \|\phi\|_{H^1(0,T;H)\cap L^2(0,T;W)} \\ &\quad + \|\xi + \rho\sigma\|_{L^2(0,T;H)} \le C_5\big(\rho^{1/2} + 1\big) \end{aligned} \tag{35}$$

where C_4 and C_5 depend only on the quantities involved in assumptions (7)–(9), (14) and (33). Furthermore, the components θ and ϕ of the solution are uniquely determined, and ξ and σ are uniquely determined as well if β is single-valued.

A similar result holds for Problem (C). We present the corresponding statement in a more accurate form for a reason that will be clear later on. To this concern, a remark is needed. In all the above estimates, one can give the usual meaning to the subscripts concerning norms and the norms in the spaces involved are the standard ones, but for the space W defined in (11). Indeed, following[2], it is more convenient to endow W with the norm defined by

$$\|v\|_W^2 = \|v\|_H^2 + |\Omega|^{4/3}\|\Delta v\|_H^2 \quad \text{for every } v \in W. \tag{36}$$

With the above notation, we term C_Ω the best constant realizing the inequality

$$\|v\|_\infty \le C_\Omega \, \|v\|_W \quad \text{for every } v \in W. \tag{37}$$

The reason of this choice will be explained later on (see the forthcoming Remark 2.4). Now, we just notice that $\|\ \|_W$. is equivalent to the norm induced on W by the standard one in $H^2(\Omega)$ (thanks to the regularity theory of elliptic equations). Notice that the constant C_Ω actually exists due to the continuous embedding $H^2(\Omega) \subset L^\infty(\Omega)$ (since $\Omega \subset \mathbb{R}^3$ is bounded and smooth) and only depends on Ω. At this point, we can state the following well-posedness result:

Theorem 2.3. *Assume (7)–(9), (14) and (33). Then, for every $\rho > 0$, Problem (C) has at least a solution $(\theta, \phi, \xi, \sigma)$ satisfying (15)–(17). Furthermore, the components θ and ϕ of the solution are uniquely determined, and ξ and σ are uniquely determined as well if β is single-valued. Finally, if the conditions*

$$f \in H^1(0,T;H), \quad \theta_0 \in W, \quad \phi_0 \in W \quad \textit{and} \quad \beta^0(\phi_0) \in H \tag{38}$$

are assumed in addition, then (31) holds as well as

$$\theta \in W^{1,\infty}(0,T;H) \cap H^1(0,T;V) \cap L^\infty(0,T;W). \tag{39}$$

In particular, both ϕ *and* θ *are bounded. Moreover, the estimates*

$$\|\phi - \phi^*\|_\infty \le \rho\, C_{str}\, C_\Omega |\Omega|^{7/6} + C_6 \tag{40}$$

$$\|\theta\|_\infty \le \rho\, C_{str}\, C_\Omega |\Omega|^{7/6} + C_7 \tag{41}$$

hold true with a structural constant C_{str} *depending only on the physical parameters* ℓ, κ, ν *and* γ, *the constant* C_Ω *given by (37) and some constants* C_6 *and* C_7 *depending on the structure of the systems,* Ω, T *and on the norms of the data involved.*

Remark 2.1. The above results are quite general. In particular, both potentials (4) and (5) are certainly allowed and the multi-valued potential (6) has to be excluded just in the parts of Theorems 2.2 and 2.3 regarding uniqueness for the pair (ξ, σ), which might be not uniquely determined, in general. Concerning the constant C_{str} of Theorem 2.3, it is proved in[2] that one can take

$$C_{str} = 2 \max\Big\{ \frac{6^{1/2}}{\nu},\ \frac{\ell}{\kappa^{1/2}\nu^{1/2}} + \frac{4\ell}{\kappa} \Big\}. \tag{42}$$

However, no sharpness is guaranteed at all.

For each of the first two problems, the existence of the desired sliding mode is ensured for ρ large enough. We have indeed

Theorem 2.4. *Assume (7)–(9), (14), (27), (30) and* $f \in L^\infty(0,T;H)$. *Then, for some* $\rho^* > 0$ *and for every* $\rho > \rho^*$, *there exist a solution* $(\theta, \phi, \xi, \sigma)$ *to Problem (A) and a time* $T^* \in [0,T)$ *such that*

$$\theta(t) + \alpha\phi(t) = \eta^* \quad \text{for a.a. } x \in \Omega \quad \textit{for every } t \in [T^*, T]. \tag{43}$$

Theorem 2.5. *Assume (7)–(9), (14) and (33). Then, for some* $\rho^* > 0$ *and for every* $\rho > \rho^*$, *there exist a solution* $(\theta, \phi, \xi, \sigma)$ *to Problem (B) and a time* $T^* \in [0,T)$ *such that*

$$\phi(t) = \phi^* \quad \text{for a.a. } x \in \Omega \quad \textit{for every } t \in [T^*, T]. \tag{44}$$

Remark 2.2. In the proof given in[2], possible values of ρ^* and T^* that fit the conclusions of these results are computed, explicitly. For Problems (A) and (B), one can take respectively

$$\rho^* := C_A^2 + 2C_A + \frac{2}{T}\,\|\theta_0 + \alpha\phi_0 - \eta^*\|_H \quad \text{and} \quad T^* := \frac{2\|\theta_0 + \alpha\phi_0 - \eta^*\|_H}{\rho - C_A^2 - 2C_A}$$

$$\rho^* := 2C_B + \frac{2}{T}\,\|\phi_0 - \phi^*\|_H \quad \text{and} \quad T^* := \frac{2\|\phi_0 - \phi^*\|_H}{\rho - 2C_B}$$

where the constants C_A and C_B are constructed in order that

$$\|f - (\ell - \alpha)\partial_t\phi - \kappa\alpha\Delta\phi - \kappa\Delta\eta^*\|_{L^\infty(0,T;H)} \le C_A\big(\rho^{1/2} + 1\big)$$
$$\|\gamma\theta + \nu\Delta\phi^* - \beta^z(\phi^*)\|_{L^\infty(0,T;H)} \le C_B\,.$$

We observe that the time T^* is roughly proportional to $1/\rho$ in both cases, whence it tends to zero as ρ tends to infinity. Therefore, the sliding mode can be forced to start as soon as one desires by prescribing a sufficiently big factor ρ in front of the feedback control.

The situation for Problem (C) is different, since we can ensure the existence of the desired sliding mode for ρ large enough only if further conditions are fulfilled. Namely, we need a restriction involving the structure of the system and the domain Ω (that is why we have written the statement of Theorem 2.3 in that form). Our result only involves the component ϕ of the solution, and we recall that ϕ is uniquely determined.

Theorem 2.6. *Assume (7)–(9), (14), (33), (38) and*

$$\Delta\phi^* \in L^\infty(\Omega) \quad \textit{and} \quad \beta^0(\phi^*) \in L^\infty(\Omega). \tag{45}$$

Let C_{str} and C_Ω be the constants appearing in (41) and in (37), and assume that

$$\gamma\, C_{str} C_\Omega |\Omega|^{7/6} < 1. \tag{46}$$

Then, for some $\rho^* > 0$ and for every $\rho > \rho^*$, the following is true: if $(\theta, \phi, \xi, \sigma)$ is a solution to Problem (C), there exists a time $T^* \in [0, T)$ such that

$$\phi(t) = \phi^* \quad \text{for a.a. } x \in \Omega \quad \textit{for every } t \in [T^*, T]. \tag{47}$$

Remark 2.3. Assume that the constants C_{str}, C_Ω and C_7 realize the inequalities (41) and (46). Then, possible values of ρ^* and T^* that fit the conclusion of the above theorem are given by (here L is the Lipschitz constant of π)

$$\rho^* := \frac{\gamma C_7 + \nu\|\Delta\phi^*\|_\infty + \|\xi\|_\infty^* + M_\pi^* + M_0/T}{1 - \gamma\, C_{str} C_\Omega |\Omega|^{7/6}} \quad \text{and} \quad T^* := \frac{M_0}{\rho - A(\rho)}$$

where $M_\pi^* := L(M_0 + \|\phi^*\|_\infty) + |\pi(0)|\,, \quad M_0 := \|\phi_0 - \phi^*\|_\infty$ and $A(\rho) := \gamma\big(C_{str} C_\Omega |\Omega|^{7/6}\rho + C_7\big) + \nu\|\Delta\phi^*\|_\infty + \|\xi\|_\infty^* + M_\pi^*\,.$

Note that $\rho^* > 0$ by (46) and that $T^* \in [0, T)$ if $\rho > \rho^*$. Moreover, the last sentence of Remark 2.2 also applies to the present case.

Remark 2.4. In order to understand the meaning of (46), let us assume that the structure of the system is chosen, so that the physical constants are fixed, and let us think of a class of open sets having the same shape. Precisely, we fix an open set Ω_0 of measure 1 and assume that $\Omega = x_0 + \lambda R\,\Omega_0$ for some $x_0 \in \mathbb{R}^3$, $\lambda > 0$ and some rotation $R \in SO(3)$. Then $\lambda = |\Omega|^{1/3}$ and one easily checks that our definition (36) of $\|\ \|_{W^*}$ yields $C_\Omega = C_{\Omega_0}\,|\Omega|^{-1/2}$, since the H-norms of v and of Δv are properly balanced in the norm $\|v\|_W$ under a rescaling of a function v. Then, the smallness condition (46) means that $|\Omega|$ is small enough. Indeed, the left-hand side of (46) becomes $\gamma C_{str} C_{\Omega_0} |\Omega|^{2/3}$ in the chosen class of domains.

The technique used for Theorems 2.1-2.3 on well-posedess relies on the theory of evolution equations of monotone type, essentially. Here we confine ourselves to give just an idea of the proof of the existence of the sliding modes. The following lemma is crucial.

Lemma 2.1. *Let $a_0, b_0, \psi_0, \rho \in \mathbb{R}$ be such that*

$$a_0,\, b_0,\, \psi_0 \geq 0 \quad \textit{and} \quad \rho > a_0^2 + 2b_0 + 2\,\frac{\psi_0}{T} \tag{48}$$

and let $\psi : [0, T] \to [0, +\infty)$ be an absolutely continuous function satisfying $\psi(0) = \psi_0$ and

$$\psi' + \rho \leq a_0\,\rho^{1/2} + b_0 \quad \textit{a.e. in the set} \quad P := \{t \in (0, T) :\ \psi(t) > 0\}. \tag{49}$$

Then, the following conclusions hold true.
i) If $\psi_0 = 0$, then ψ vanishes identically.
ii) If $\psi_0 > 0$, there exists $T^* \in (0, T)$ satisfying $T^* \leq 2\psi_0/(\rho - a_0^2 - 2b_0)$ such that ψ is strictly decreasing in $(0, T^*)$ and ψ vanishes in $[T^*, T]$.

The above lemma is suitably applied after changing the unknown functions of Problem (A). We keep ϕ, replace θ by $\eta := \theta + \alpha\phi - \eta^*$ and rewrite equation (18) and condition (20) in the form

$$\partial_t \eta - \kappa\Delta\eta + \rho\sigma = g := f - (\ell - \alpha)\partial_t\phi - \kappa\alpha\Delta\phi + \kappa\Delta\eta^* \tag{50}$$

$$\sigma(t) \in \mathrm{Sign}(\eta(t)) \quad \text{for a.a. } t \in (0,T). \tag{51}$$

In fact, the argument should be developed for a suitable approximating problem. Here, we are forced to proceed formally. By the a priori estimates given in Theorem 2.1, we have

$$\|g\|_{L^\infty(0,T;H)} \le C\big(\rho^{1/2} + 1\big) \tag{52}$$

where C depends only on the structure and the data involved in the statement. At this point, we set

$$\rho^* := C^2 + 2C + \frac{2}{T}\,\|\theta_0 + \alpha\phi_0 - \eta^*\|_H \tag{53}$$

and assume $\rho > \rho^*$. By multiplying (50) by σ and observing that the term involving the laplacian produces a nonnegative contribution, we obtain

$$\psi'(t) + \rho\|\sigma(t)\|_H^2 \le C\big(\rho^{1/2} + 1\big) \quad \text{for a.a. } t \in (0,T) \tag{54}$$

where $\psi(t) := \|\eta(t)\|_H$. As $\|\sigma(t)\|_H = 1$ if $\psi(t) > 0$ by (12), we can apply the lemma with $a_0 = b_0 = C$ and we observe that our condition $\rho > \rho^*$ completely fits the assumptions by (53). Thus, we find $T^* \in [0,T)$ such that $\eta(t) = 0$ for every $t \in [T^*, T]$, i.e., (43).

The idea for Theorem 2.5 is similar. In this case we also replace ϕ by $\chi := \phi - \phi^*$ and formally multiply the equation for χ derived from (22) by σ. We obtain an inequality like (54) where now $\psi(t) := \|\chi(t)\|_H$. Also in this case, we have $\|\sigma(t)\|_H = 1$ if $\psi(t) > 0$. Thus, we apply the lemma and conclude. In both situations, the power $\rho^{1/2}$ in the a priori estimates given by Theorems 2.1 and 2.2 is crucial. This is not the case for Problem (C) (see estimates (40)–(41)): for that reason, the existence of the sliding mode requires the condition (46). Moreover, by multiplying the equation for $\chi := \phi - \phi^*$ by the right selection $\sigma \in \mathrm{sign}\,\chi$, the left-hand side we obtain is the same as in (54) where now $\psi(t) := \|\chi(t)\|_{L^1(\Omega)}$. Thus, $\psi(t) > 0$ does not imply $\|\sigma(t)\|_H = 1$ (nor $\|\sigma(t)\|_H \ge \alpha$ for some constant $\alpha > 0$), i.e., the local operator sign does not guarantee the same as the nonlocal Sign does. For these reasons, Lemma 2.1 cannot be applied and the argument that can prove our result must be different. The method used in[2] relies on a comparison technique. Namely, one proves that $|\chi| \le w$ a.e., where w solves an ordinary Cauchy problem constructed in order that w vanishes in a finite time. Such a problem involves the quantities mentioned in Remark 2.3 and it turns out that $w(t) = 0$ for $t \geq T^*$ if $\rho > \rho^*$, by construction. However, the details are complicated and cannot be summarized here.

References

1. G. Bartolini, L. Fridman, A. Pisano, E. Usai (eds.), *Modern Sliding Mode Control TheoryNew Perspectives and Applications*, Lecture Notes in Control and Information Sciences **375**, Springer, 2008.
2. V. Barbu, P. Colli, G. Gilardi, G. Marinoschi, E. Rocca, *Sliding mode control for a nonlinear phase-field system*, preprint arXiv:1506.01665 [math.AP] (2015) 1-28.
3. G. Caginalp, *An analysis of a phase field model of a free boundary*, Arch. Rational Mech. Anal **92** (1986), 205-245.
4. M.-Bi Cheng, V. Radisavljevic, W.-C. Su, *Sliding mode boundary control of a parabolic PDE system with parameter variations and boundary uncertainties*, Automatica J. IFAC **47** (2011), 381–387.
5. C. Edwards, E. Fossas Colet, L. Fridman (eds.), *Advances in Variable Structure and Sliding Mode Control*, Lecture Notes in Control and Information Sciences **334**, Springer-Verlag, 2006.
6. C. Edwards, S. Spurgeon, *Sliding Mode Control: Theory and Applications*, Taylor and Francis, London, 1999.
7. L. Fridman, J. Moreno, R. Iriarte (eds.), *Sliding Modes After the First Decade of the 21st Century: State of the Art*, Lecture Notes in Control and Information Sciences **412**, Springer, 2011.
8. U. Itkis, *Control systems of variable structure*, Wiley, 1976.
9. L. Levaggi, *Existence of sliding motions for nonlinear evolution equations in Banach spaces*, Discrete Contin. Dynam. Systems, Supplement 2013, 477–487.
10. V. Utkin, *Sliding Modes in Control and Optimization*, Communications and Control Engineering Series, Springer-Verlag, 1992.
11. V. Utkin, J. Guldner, J. Shi, *Sliding Mode Control in Electro-Mechanical Systems*, 2nd Edition CRC Press, Automation and Control Engineering, 2009.
12. K.D. Young, Ü. Özgüner (eds.), *Variable Structure Systems*, Sliding Mode and Nonlinear Control, Springer-Verlag, 1999.
13. H. Xing, D. Li, C. Gao, Y. Kao, *Delay-independent sliding mode control for a class of quasi-linear parabolic distributed parameter systems with time-varying delay*, J. Franklin Inst. **350** (2013), 397–418.

Uniformly hyperbolic viable sets in affine IFS

Vasile Glavan* and Valeriu Guţu**

*,** *Department of Mathematics, Moldova State University, Chişinău, Republic of Moldova,*
* *Siedlce University of Natural Sciences and Humanities, Siedlce, Poland*
E-mail: * *vglavan@uph.edu.pl,*
** *vgutu@yahoo.com*
www.usm.md

Viable sets in multi-valued dynamics, generated by finite families of mappings, are considered. In the case of affine Iterated Function Systems, satisfying a generalized cone condition, the existence of compact viable sets is stated, and some topological characteristics of these sets are given.

Keywords: Iterated Function System, viability, cone condition, Shadowing property, linear skew-product flow.

1. Introduction

Contractive Iterated Functions Systems (IFS for short) are very well studied, their attractors represent classical fractals (and not only) and various algorithms for constructing these fractals are elaborated (see, e.g., Ref. 1). Generalizations for (weakly) contracting relations have been purposed in Refs. 2 and 3, where existence of the attractors has been stated and the dynamics on the attractor has been studied. The attractor is (forwardly) invariant and the dynamics on it characterizes the dynamics on the whole phase space. If the IFS is not contractive, the dynamic itself is much more complex and proper invariant subsets may not exist. In control theory a weaker notion is considered - viability.

In our paper we are concerned with viable compact sets in affine IFS. We adapt the well known concept of cone condition from ordinary dynamics to set-valued cases and give sufficient conditions for a compact viable set to exist. We also prove that under these conditions the compact viable set is the closure of periodic points. The property of shadowing pseudo-orbits near the viable set under the generalized cone condition is stated as well.

2. Viability and invariance in IFS

Let $F = \{X; f_1, f_2, \ldots, f_m\}$ be a collection of continuous functions on the complete metric space (X, d), called also as an *Iterated Function System* (IFS).

Let $T = \mathbb{Z}$, or $T = \mathbb{Z}_+ = \mathbb{N} = \{n \in \mathbb{Z} \,|\, n \geq 0\}$, or $T = \mathbb{Z}_- = \{n \in \mathbb{Z} \,|\, n < 0\}$.

A sequence $(x_n)_{n \in T}$ in X is called a *chain* of the IFS F if for every $n \in T$ there exists $t_n \in \{1, 2, \ldots, m\}$ such that $x_{n+1} = f_{t_n}(x_n)$. Given $\delta \geq 0$ a sequence $(x_n)_{n \in T}$ in X is called a *δ-chain* (*pseudo-chain*) if for every $n \in T$ there exists $t_n \in \{1, 2, \ldots, m\}$ such that $d(x_{n+1}, f_{t_n}(x_n)) \leq \delta$. Denote $\Theta = (t_n)_{n \in T}$ and call it a *control sequence* for the *δ-chain* $(x_n)_{n \in T}$, including the case of chains as 0-chains.

Two chains (or pseudo-chains) are called *concordant*, if they admit a common control sequence.

Remark 2.1. The control sequence need not be uniquely determined by the chain.

Given the IFS F, a finitely valued mapping,

$$x \mapsto \{f_1(x)\} \cup \{f_2(x)\} \cup \ldots \cup \{f_m(x)\}$$

(which we will denote also by F), together with a discrete inclusion $x_{n+1} \in F(x_n)$, are also defined. A chain $(x_n)_{n \in T}$ of the IFS is called a *solution* of the inclusion. If the IFS is contracting, then there is a compact global attractor, which attracts every compact non-empty subset. Moreover, every solution of the corresponding discrete inclusion is asymptotic with a uniquely determined chain from the attractor (the so-called *"asymptotic phase"*) (see Refs. 2 and 3, where the "asymptotic phase property" is stated for any contracting closed relation).

Let $\mathcal{P}(X)$ denote the space of all nonempty compact subsets of X, endowed with Pompeiu-Hausdorff metrics. Associated with this IFS there is a continuous mapping, called the *Nadler-Hutchinson operator*

$$F_* : \mathcal{P}(X) \to \mathcal{P}(X), \quad F_*(C) = \bigcup_{j=1}^{m} f_j[C],$$

where $f_j[C] = \{f_j(x) \,|\, x \in C\}$ for any $C \in \mathcal{P}(X)$.

A closed subset Λ is said to be *F_*+-invariant* if $F_*(\Lambda) \subseteq \Lambda$, and *$F_*$-invariant* if the inclusion becomes a equality. In particular, $f_j[\Lambda] \subseteq \Lambda$, for all $j = 1, 2, \ldots, m$, i.e. every chain, starting at a point $x \in \Lambda$, is contained in Λ.

Invariance is a cornerstone concept in ordinary dynamics, fixed points and periodic orbits being the simplest examples, although, if the function is not invertible, a fixed point need not be f^{-1}-invariant. In that case the double equality $f^{-1}(\Lambda) = \Lambda = f(\Lambda)$ is coined as *complete invariance.* In set-valued dynamics, as well as in control theory, a weaker counterpart exists. More precisely, given $\alpha \in \mathbb{Z}_- \cup \{-\infty\}$ and $\beta \in \mathbb{Z}_+ \cup \{+\infty\}$, the nonempty subset $A \subseteq X$ is called *viable* on $S = [\alpha, \beta]$ if for every $a \in A$ there exists at least one chain $(x_n)_{n \in S}$ such that $x_0 = a$ and $x_n \in A$ for all $n \in S$. For more details on invariance and viability in difference, or differential inclusions, evolutions, governed by PDE, see Refs. 4 and 5.

Remark 2.2. Thus, invariance requires that every chain "starting" at a point from this set, must belong to it, while viability requires only existence of at least one such chain.

Obviously, every (completely) invariant set is viable on ($\mathbb{Z}$) $\mathbb{Z}_+$, as well as each fixed point or periodic chain does. A viable set need not be closed, nor bounded.

Lemma 2.1. *A set A is viable on $\mathbb{Z}$ if and only if it is viable on $\mathbb{Z}_+$ and on $\mathbb{Z}_-$.*

Proposition 2.1. *A set A is viable on $\mathbb{Z}_+$ if and only if $A \subset F_*^{-1}(A)$.*

Proof. Assume that A is viable on $\mathbb{Z}_+$. Given $x \in A$ let $(x, x_1, x_2, \ldots) \subset A$ be the corresponding chain, guaranteed by viability. For $x_1 \in A$ there is j such that $x_1 = f_j(x) \in A$, which means that $x \in f_j^{-1}(A)$, and, as a consequence, $x \in F_*^{-1}(A) = f_1^{-1}(A) \cup \ldots \cup f_m^{-1}(A)$.

Conversely, assume that $A \subset F_*^{-1}(A)$ and let $x \in A$. By assumption, this implies that $x \in f_j^{-1}(A)$ for some $j = 1, 2, \ldots m$, i.e. there exists $x_1 \in A$, such that $x_1 = f_j(x)$. Repeating the procedure inductively, we obtain a chain $(x, x_1, x_2, \ldots)$ in A. □

For a homeomorphism $f : X \to X$ and a closed subset $\Lambda \subset X$ the following properties are equivalent:

(1) $f(\Lambda) \subset \Lambda$;
(2) $f^{-1}(\Lambda) = \Lambda = f(\Lambda)$;
(3) $\Lambda \subset f(\Lambda) \cap f^{-1}(\Lambda)$.

For a continuous function $f : X \to X$ the last inclusion is equivalent to the fact that Λ is $f+$-invariant and viable on $\mathbb{Z}_-$. For a general closed relation, and in particular, for an IFS, this inclusion means only viability on $\mathbb{Z}$. More precisely, the following result holds true, as a consequence of Lemma 2.1 and Proposition 2.1.

Theorem 2.1. *A closed set A is viable on $\mathbb{Z}$ if and only if $A \subset F_*(A) \cap F_*^{-1}(A)$.*

Remark 2.3. E. Akin[6,7], while dealing with general closed relations in compact metric spaces, calls a relation, that satisfies the last inclusion, surjective on the subset A.

Theorem 2.2. *Let $F = \{X; f_1, f_2, \ldots, f_m\}$ be an IFS, consisting of homeomorphisms. If the set A is viable on $S \subset \mathbb{Z}$, then its closure $\bar{A}$ is also viable on S.*

Proof. Suppose the set A is viable on S.

Let $y \in \bar{A}$. It follows that there exists a sequence $(a_n)_{n\in\mathbb{N}} \subset A$, which converges to y, i.e. $\lim_{n\to+\infty} a_n = y$.

For every $n \in \mathbb{N}$ there exists a chain $(a_{k,n})_{k\in S} \subset A$ such that $a_{0,n} = a_n$ and $a_{k+1,n} = f_{t_{k,n}}(a_{k,n})$, where $t_{k,n} \in \{1, \ldots, m\}$.

Since the range of all values $(t_{k,n})$ is finite, for $k = 0$ there is a value t_{n_0}, which occurs infinitely times among $(t_{0,n})_{n\in\mathbb{N}}$. It follows that there exists a subsequence $(a_{n_p})_{p\in\mathbb{N}}$ such that $a_{1,n_p} = f_{t_0}(a_{0,n_p})$ for all $p \in \mathbb{N}$.

Consider this subsequence as the initial sequence $(a_n)_{n\in\mathbb{N}}$ and repeat the reasons consecutively for $k = 1, -1, 2, -2, \ldots$.

By induction, one can show that there exists a new sequence $(b_n)_{n\in\mathbb{N}}$, which converges to y, and for every $n \in \mathbb{N}$ there exists a chain $(b_{k,n})_{k\in S}$ such that $b_{0,n} = b_n$ and $b_{k+1,n} = f_{t_k}(b_{k,n})$. This means that all the chains $(b_{k,n})_{k\in S}$ are concordant for all $n \in \mathbb{Z}$.

Denote $y_0 = y = \lim_{n\to+\infty} b_n$, $y_1 = f_{t_0}(y_0)$ and $y_{-1} = f_{t_{-1}}^{-1}(y_0)$. Then there exists the limits

$$y_1 = f_{t_0}(\lim_{n\to+\infty} b_n) = f_{t_0}(\lim_{n\to+\infty} b_{0,n}) = \lim_{n\to+\infty} f_{t_0}(b_{0,n}) = \lim_{n\to+\infty} b_{1,n} \in \bar{A},$$

$$y_{-1} = f_{t_{-1}}^{-1}(\lim_{n\to+\infty} b_n) = f_{t_{-1}}^{-1}(\lim_{n\to+\infty} b_{0,n}) = \lim_{n\to+\infty} f_{t_{-1}}^{-1}(b_{0,n}) =$$

$$\lim_{n\to+\infty} b_{-1,n} \in \bar{A}.$$

By induction one can show that the chain $(y_k)_{k\in S}$, defined by $y_0 = y$, and $y_{k+1} = f_{t_k}(y_k)$ for $k \geq 0$ and $y_{k-1} = f_{t_{k-1}}^{-1}(y_k)$ for $k \leq 0$, lies in $\bar{A}$.

This completes the proof. □

3. IFS and extensions of Bernoulli shifts

Let $\Sigma_m = \{1, 2, \ldots, m\}^{\mathbb{Z}}$ denote the set of bi-infinite symbolic sequences with values in $\{1, 2, \ldots, m\}$. Endowed with product topology it becomes a compact Hausdorff space.

Consider the IFS $F = \{X; f_1, f_2, \ldots, f_m\}$ and assume that each function is a homeomorphism. Given a chain $(x_n)_{n\in\mathbb{Z}}$, there exists a control sequence

$$\Theta = (\ldots, t_{-k}, \ldots, t_{-1}, t_0, t_1, \ldots, t_k, t_{k+1}, \ldots) \in \Sigma_m,$$

such that $x_{n+1} = f_{t_n}(x_n)$ $(n \in \mathbb{Z})$.

Let $\sigma : \Sigma_m \to \Sigma_m$ denote the *Bernoulli shift*: $(\sigma(\Theta))_n := t_{n+1}$ $(n \in \mathbb{Z})$ and consider the *extension* $\hat{F} : \Sigma_m \times X \to \Sigma_m \times X$, defined by

$$\hat{F}(\Theta, z) := \big(\sigma(\Theta), f_{t_0}(z)\big).$$

Thus, $\hat{F}^n(\Theta, z) = \big(\sigma^n(\Theta), (f_{t_{n-1}} \circ \cdots \circ f_{t_1} \circ f_{t_0})(z)\big)$. Denote $F^n(\Theta) = f_{t_{n-1}} \circ \cdots \circ f_{j_1} \circ f_{t_0}$.

By definition, every orbit $\big(\hat{F}^n(\Theta, z)\big)_{n\in\mathbb{Z}}$ of the extension $\hat{F} : \Sigma_m \times X \to \Sigma_m \times X$ covers the orbit $\big(\sigma^n(\Theta)\big)_{n\in\mathbb{Z}}$ of the shift transformation σ. In other words, the sequence $(z_n)_{n\in\mathbb{Z}}$, with $z_{n+1} = F^n(\Theta)(z_0)$, represents an orbit of $\hat{F}$, starting at (Θ, z_0). We call such orbit a Θ-*chain* (or a $(\Theta, 0)$-*chain*). In this setting, we will say that the sequence $(y_n)_{n\in\mathbb{Z}}$ is a (Θ, δ)-*chain* if

$$d(y_{n+1}, F\big(\sigma^n(\Theta)\big)(y_n)) \le \delta \quad (n \in \mathbb{Z}).$$

Recall (see, e.g., Ref. 2), that the IFS has the *Shadowing Property* (*on* T) if for every $\varepsilon > 0$ there exists $\delta > 0$ such that for every δ-chain $(x_n)_{n\in T}$ there is a chain $(y_n)_{n\in T}$, satisfying $d(x_n, y_n) \le \varepsilon$ for all $n \in T$. In this case we say that $(y_n)_{n\in T}$ ε-*shadows* $(x_n)_{n\in T}$. If the chain $(y_n)_{n\in T}$ can be chosen to be concordant with $(x_n)_{n\in T}$, then we will speak about the *concordant Shadowing Property.*

We say that the extension $\hat{F}$ has the *Shadowing Property* if, given $\varepsilon > 0$ and $\Theta \in \Sigma_m$, there exists $\delta = \delta(\varepsilon, \Theta) > 0$ such that every (Θ, δ)-chain is ε-shadowed by a $(\Theta, 0)$-chain. In case δ does not depend on $\Theta \in \Sigma_m$ we say that the extension $\hat{F}$ has the *uniform Shadowing Property.*

Theorem 3.1. *An IFS $F = \{X; f_1, f_2, \ldots, f_m\}$ has the concordant Shadowing Property if and only if the associated extension has the uniform Shadowing Property.*

Proof. Let F have the concordant Shadowing Property: given $\varepsilon > 0$ there exists $\delta > 0$ such that for every δ-chain $(z_n)_{n\in\mathbb{Z}}$ there is a chain $(w_n)_{n\in\mathbb{Z}}$, concordant with $(z_n)_{n\in\mathbb{Z}}$, and verifying $d(z_n, w_n) \le \varepsilon$ $(n \in \mathbb{Z})$. Let Θ denote their common control sequence, i.e.

$$d(z_{n+1}, f_{t_n}(z_n)) \le \delta \qquad (n \in \mathbb{Z}) \tag{1}$$

and $w_{n+1} = f_{t_n}(w_n)$ for all $n \in \mathbb{Z}$, where $(t_n)_{n\in\mathbb{Z}} = \Theta$.

Fix $\varepsilon > 0$ and take δ from (1). Hence, $(z_n)_{n\in\mathbb{Z}}$ is a (Θ, δ)-chain for the extension $\hat{F}$ and $(w_n)_{n\in\mathbb{Z}}$ is a $(\Theta, 0)$-chain for $\hat{F}$. Since δ does not depend on $\Theta \in \Sigma_m$, this means that the uniform Shadowing Property for the extension $\hat{F}$ is stated.

The converse is obtained similarly. Assume that the extension $\hat{F}$ has the uniform Shadowing Property: given $\varepsilon > 0$ there exists $\delta = \delta(\varepsilon) > 0$ such that for every $\Theta \in \Sigma_m$ and every (Θ, δ)-chain $(z_n)_{n\in\mathbb{Z}}$ there is a $(\Theta, 0)$-chain $(w_n)_{n\in\mathbb{Z}}$, which ε-shadows $(z_n)_{n\in\mathbb{Z}}$.

By construction, $\Theta \in \Sigma_m$ represents a control sequence, the same for the δ-chain $(z_n)_{n\in\mathbb{N}}$ and the chain $(w_n)_{n\in\mathbb{N}}$. □

4. Generalized cone condition in linear IFS'

In what follows we are concerned with a class of expanding affine IFS' with "coherent hyperbolic structure". Roughly speaking, this means that each affine component of such IFS is hyperbolic, i.e. the spectrum of its linear part does not intersect the unit circle, the inside parts of the spectra have the same number of elements, counted with multiplicities. In particular, all stable subspaces have the same dimension. This, and other concordance properties are guaranteed by the so-called "cone-condition", introduced by V. M. Alexeev (see Refs. 8–10).

In treating cone condition we follow Ref. 11.

Let $\mathbb{R}^n = \mathbb{R}^k \oplus \mathbb{R}^{n-k}$ and let $x = (u, v) \in \mathbb{R}^n$. Given a real number γ, one calls *standard horizontal γ-cone* at the origin of coordinates the subset

$$H_\gamma = \{(u, v) \in \mathbb{R}^n : \|v\| \le \gamma \|u\|\}.$$

The corresponding *standard vertical cone* is defined analogically. More generally, by a *cone* at the origin of coordinates $K \subset \mathbb{R}^n$ we mean the image of a standard cone under a linear invertible application.

Let $F = \{\mathbb{R}^k; f_1, f_2, \ldots, f_m\}$, with $f_i(x) = A_i x + b_i$, be an affine IFS, and denote by F_0 the corresponding linear IFS.

In what follows by the inclusion $A \Subset B$ we mean that the closed subset A is contained in the interior of the set B.

Definition 4.1. One says that a linear IFS satisfies the *generalized cone condition*, if there are two cone-fields $\mathcal{H}, \mathcal{V} \subset \Sigma_m \times \mathbb{R}^k$ such that:

(1) for any $t = (\dots, t_{-1}, t_0, t_1, \dots) \in \Sigma_m$, the interior of each cone $\mathcal{H}_t$ and $\mathcal{V}_t$ is nonempty, these cones intersect only in zero, and $A_{t_0}\mathcal{H}_t \Subset \mathcal{H}_{\sigma(t)}$ and $A_t^{-1}\mathcal{V}_t \Subset \mathcal{V}_{\sigma^{-1}(t)}$;
(2) there exists $\mu > 1$ such that $\|A_{t_0}\xi\| \geq \mu\|\xi\|$ for any $\xi \in \mathcal{H}_t$, $t \in \Sigma_m$, and $\|A_{t_0}^{-1}\eta\| \geq \mu\|\eta\|$ for any $\eta \in \mathcal{V}_t$, $t \in \Sigma_m$.

Remark 4.1. In Ref. 12 a multi-cone condition is formulated for linear IFS', consisting of matrices from $SL(2;\mathbb{R})$, and this multi-cone is common for all matrices. In our opinion, although the associated skew-product is of step type, in terminology of Yu. S. Ilyashenko (see, e.g., Ref. 13), i.e. the cocycle depends only on the zero-coordinate, as in our case, the spectral sub-bundles, being "solutions" of some cohomological equations, need not be as "smooth" as the right hand side of the equation does, or, in other words, they need not be constant on the cylinders, defined by zero coordinate. The same is valuable for the cone fields, that localize these spectral sub-bundles. We don't know whether the "generalized cone condition" can be substituted, without loss of generality, by a locally constant (multi) cone condition.

Easily seen that the cone fields $\mathcal{H}_t$ and $\mathcal{V}_t$, restricted over the fixed points of the Bernoulli shift, force each matrix to be hyperbolic with stable subspaces of the same dimension. A finite collection of hyperbolic matrices with common stable and unstable spaces, as well as their small enough perturbations, represent linear IFS', which satisfy the generalized cone-condition.

In what follows by $\pi_0 : \Sigma_m \to \{1, 2, \dots, m\}, \pi_0(t) = t_0$, we denote the projection on zero coordinate.

Theorem 4.1. *Assume that the linear IFS $\{\mathbb{R}^n; A_1, A_2, \dots, A_m\}$ consists of invertible matrices and satisfies the generalized cone condition. Then there exist two families of linear subspaces $E_t^u \subset \mathcal{H}_t$, $E_t^s \subset \mathcal{V}_t$ and a real number $\mu > 1$ such that for any $t \in \Sigma_m$ the following holds true:*

$$E_t^s \oplus E_t^u = \mathbb{R}^n, \tag{2}$$

$$A_{t_0}(E_t^s) = E_{\sigma(t)}^s, \quad A_{t_0}(E_t^u) = E_{\sigma(t)}^u, \quad A_{t_0}^{-1}(E_{\sigma(t)}^u) = E_t^u, \tag{3}$$

$$\|A_{t_0}\xi\| \leq \frac{1}{\mu}\|\xi\| \ (\xi \in E_t^s), \quad \|A_{t_0}^{-1}\eta\| \leq \frac{1}{\mu}\|\eta\| \ (\eta \in E_{\sigma(t)}^u). \tag{4}$$

Moreover, the vector spaces E_t^s, E_t^u depend continuously on $t \in \Sigma_m$, and thus, $E^s := \bigcap_{t\in\Sigma_m} E_t^s, E^u := \bigcap_{t\in\Sigma_m} E_t^u$ are vector sub-bundles of $\Sigma_m \times \mathbb{R}^n$.

Proof. Given $t = (\ldots, t_{-1}, t_0, t_1, \ldots) \in \Sigma_m$, define

$$\begin{aligned} E_t^s &= \bigcap_{n\geq 0} \left(A_{t_0}^{-1} \circ A_{t_1}^{-1} \circ \cdots \circ A_{t_n}^{-1}\right) \mathcal{V}_{\sigma^n(t)} \\ &= \bigcap_{n\geq 0} \left(A_{\pi_0(t)}^{-1} \circ A_{\pi_0(\sigma(t))}^{-1} \circ \cdots \circ A_{\pi_0(\sigma^n(t))}^{-1}\right) \mathcal{V}_{\sigma^n(t)} \end{aligned}$$

and

$$\begin{aligned} E_t^u &= \bigcap_{n>0} \left(A_{t_{-1}} \circ \cdots \circ A_{t_{-n}}\right) \mathcal{H}_{\sigma^{-n}(t)} \\ &= \bigcap_{n>0} \left(A_{\pi_0(\sigma^{-1}(t))} \circ \cdots \circ A_{\pi_0(\sigma^{-n}(t))}\right) \mathcal{H}_{\sigma^{-n}(t)}. \end{aligned}$$

To prove invariance of closed subsets E^s and E^u we make use of the following diagrams:

$$\begin{array}{ccccccc} \mathcal{V}_t & \xrightarrow{A_{\pi_0(\sigma(t))}} & \mathcal{V}_{\sigma(t)} \longrightarrow \cdots \longrightarrow & \mathcal{V}_{\sigma^{n-1}(t)} & \xrightarrow{A_{\pi_0(\sigma^n(t))}} & \mathcal{V}_{\sigma^n(t)} \\ \downarrow & & \downarrow & \downarrow & & \downarrow \\ t & \xrightarrow{\sigma} & \sigma(t) \longrightarrow \cdots \longrightarrow & \sigma^{n-1}(t) & \xrightarrow{\sigma} & \sigma^n(t) \end{array}$$

and

$$\begin{array}{ccccccc} \mathcal{H}_{\sigma^{-n}(t)} & \xrightarrow{A_{\pi_0(\sigma^{-n}(t))}} & \mathcal{H}_{\sigma^{-n+1}(t)} \longrightarrow \cdots \longrightarrow & \mathcal{H}_{\sigma^{-1}(t)} & \xrightarrow{A_{\pi_0(\sigma^{-1}(t))}} & \mathcal{H}_t \\ \downarrow & & \downarrow & \downarrow & & \downarrow . \\ \sigma^{-n}(t) & \xrightarrow{\sigma} & \sigma^{-n+1}(t) \longrightarrow \cdots \longrightarrow & \sigma^{-1}(t) & \xrightarrow{\sigma} & t \end{array}$$

Next we proceed as follows:

$$\begin{aligned} A_{t_0}(E_t^s) &= A_{t_0}\left(\bigcap_{n\geq 0} \left(A_{\pi_0(t)}^{-1} \circ A_{\pi_0(\sigma(t))}^{-1} \circ \cdots \circ A_{\pi_0(\sigma^n(t))}^{-1}\right) \mathcal{V}_{\sigma^n(t)}\right) \\ &= \bigcap_{n\geq 0} \left(A_{t_0} \circ A_{\pi_0(t)}^{-1} \circ A_{\pi_0(\sigma(t))}^{-1} \circ \cdots \circ A_{\pi_0(\sigma^n(t))}^{-1}\right) \mathcal{V}_{\sigma^n(t)} \\ &= \bigcap_{n\geq 0} \left(A_{\pi_0(t)} \circ A_{\pi_0(t)}^{-1} \circ A_{\pi_0(\sigma(t))}^{-1} \circ \cdots \circ A_{\pi_0(\sigma^{n+1}(t))}^{-1}\right) \circ A_{\pi_0(\sigma^{n+1}(t))} \mathcal{V}_{\sigma^n(t)} \\ &= \bigcap_{n\geq 0} \left(A_{\pi_0(\sigma(t))}^{-1} \circ \cdots \circ A_{\pi_0(\sigma^n(\sigma(t)))}^{-1}\right) \mathcal{V}_{\sigma^n(\sigma(t))} = E_{\sigma(t)}^s \end{aligned}$$

and

$$A_{t_0}(E_t^u) = A_{t_0}\left(\bigcap_{n>0}\left(A_{\pi_0(\sigma^{-1}(t))}\circ\cdots\circ A_{\pi_0(\sigma^{-n}(t))}\right)\mathcal{H}_{\sigma^{-n}(t)}\right)$$
$$= \bigcap_{n>0}\left(A_{t_0}\circ A_{\pi_0(\sigma^{-1}(t))}\circ\cdots\circ A_{\pi_0(\sigma^{-n+1}(t))}\right)\circ A_{\pi_0(\sigma^{-n}(t))}\mathcal{H}_{\sigma^{-n}(t)}$$
$$= \bigcap_{n>0}\left(A_{\pi_0(\sigma^{-1}(\sigma(t)))}\circ A_{\pi_0(\sigma^{-2}(\sigma(t)))}\circ\cdots\circ A_{\pi_0(\sigma^{-n}(\sigma(t)))}\right)\mathcal{H}_{\sigma^{-n}(\sigma(t))} = E^u_{\sigma(t)}.$$

Following Ref. 11, one can easily show that E_t^s and E_t^u are linear subspaces, while the condition, that the vertical and horizontal cones intersect only in zero, implies the equality (2). Two families of subspaces E_t^s and E_u^s, satisfying (2)-(4), mean that the linear skew-product (extension) over the Bernoulli shift (Σ_m, σ) does not admit nontrivial bounded motions. The Bernoulli shift being chain-recurrent, the vectorial subsets E_t^s and E_u^s, in fact, are vector subbundles in $\Sigma_m \times \mathbb{R}^n$ and the linear extension is hyperbolic (exponential dichotomic) (see Ref. 14). □

5. Uniformly hyperbolic viable sets in affine IFS'

Our goal is to put in evidence a closed viable subset, which has to be nonempty and compact, and maximal with respect to these properties, and at the same time, not too small, to include the periodic chains, which, of course, are viable on $\mathbb{Z}$.

Given an IFS F, denote by $Per(F)$ the subset of all periodic points of F (recall, that $x \in Per(F)$ iff $x \in F^n(x)$ for some $n \in \mathbb{N}\setminus\{0\}$).

The following theorem represents the main result of the paper.

Theorem 5.1. *Assume that the linear part of the affine IFS F satisfies the generalized cone-condition. Define Λ as the subset of points $x \in \mathbb{R}^n$, for which there exists a bounded chain $(x_n)_{n\in\mathbb{Z}}$ with $x_0 = x$. Then:*

(1) The subset Λ is nonempty compact and viable on $\mathbb{Z}$. Moreover, Λ is the maximal subset with these properties.

(2) The set Λ is the smallest closed and viable on $\mathbb{Z}$ subset containing $Per(F)$, thus $\Lambda = \overline{Per(F)}$.

(3) The affine IFS F has the concordant Shadowing Property on Λ, i.e. given $\varepsilon > 0$, there exists $\delta > 0$ such that for any δ-chain $(x_n)_{n\in\mathbb{Z}}$ satisfying $\varrho(x_n, \Lambda) \le \delta$ (here $\varrho(a, \Lambda) = \inf\{\|a - z\| : z \in \Lambda\}$), there exists a chain $(y_n)_{n\in\mathbb{Z}} \subset \Lambda$, concordant with $(x_n)_{n\in\mathbb{Z}}$ and such that $\|x_n - y_n\| < \varepsilon$ for all $n \in \mathbb{Z}$.

Proof. Let Λ be as in the condition. Since the linear extension $\hat{F}_0$ is exponentially dichotomic (hyperbolic) the corresponding affine extension has a unique invariant continuous cross-section (see, e.g., Ref. 14). Let $\hat{\gamma} : \Sigma_m \to \Sigma_m \times \mathbb{R}^n$ denote this cross-section. Then $\hat{\gamma}(t) = (t, \gamma(t))$ for some continuous function $\gamma : \Sigma_m \to \mathbb{R}^n$. The image of $\hat{\gamma}$ is the only compact invariant subset of $\Sigma_m \times \mathbb{R}^n$.

We shall prove that $\Lambda = pr_2(\hat{\gamma}(\Sigma_m))$, where pr_2 means the projection on the second multiplier.

For $x \in \Lambda$ let $(x_n)_{n\in\mathbb{Z}}, x_0 = x$, be the corresponding bounded chain and let $t \in \Sigma_m$ be a control sequence of $(x_n)_{n\in\mathbb{Z}}$. The last two satisfy the following equality:

$$x_{n+1} = A_{\pi_0(\sigma^n(t))} x_n + b_{\pi_0(\sigma^n(t))} \qquad (n \in \mathbb{Z}), \tag{5}$$

which means that $(\sigma^n(t), x_n)_{n\in\mathbb{Z}}$ is a bounded trajectory of the affine extension $\hat{F}$. Since $(x_n)_{n\in\mathbb{Z}}$ is bounded, $(\sigma^n(t), x_n)_{n\in\mathbb{Z}}$ is the only bounded trajectory of the affine extension $\hat{F}$, starting at (t, x_0). At the same time, $(t, \gamma(t))$ is the only initial point, from which a bounded orbit of $\hat{F}$ starts. This implies the equality $\gamma(t) = x_0$, and we have proved the inclusion $\Lambda \subset pr_2(\hat{\gamma}(\Sigma_m))$.

Conversely, given $t \in \Sigma_m$, the trajectory, defined by the affine recurrence (5), with $x_0 = \gamma(t)$, is contained in the compact invariant subset $\hat{\gamma}(\Sigma_m) \subset \Sigma_m \times \mathbb{R}^n$. Hence, $(x_n)_{n\in\mathbb{Z}}$ represents a bounded chain of the affine IFS. All these imply the inclusion $\Lambda \supset pr_2(\hat{\gamma}(\Sigma_m))$.

Now we prove viability of Λ on $\mathbb{Z}$. Let $x \in \Lambda$. From the equality

$$\Lambda = pr_2(\hat{\gamma}(\Sigma_m))$$

it follows that there exists $t \in \Sigma_m$ such that $x = \gamma(t)$.

Define the chain $(x_n)_{n\in\mathbb{Z}}$, with $x_0 = x$, by recurrent formula (5). This formula shows that $(\sigma^n(t), x_n)_{n\in\mathbb{Z}}$ is contained in $\hat{\gamma}(\Sigma_m)$, so $(x_n)_{n\in\mathbb{Z}} \subset \Lambda$. The restriction of the extension $\hat{F}$ on the invariant cross-section $\hat{\gamma}(\Sigma_m)$ is topologically equivalent with the Bernoulli shift (Σ_m, σ), in which the periodic points are dense. Since every periodic chain of the IFS is the image of some periodic orbit of (Σ_m, σ), it follows that $\overline{Per(F)} = \Lambda$. Since $\hat{\gamma}(\Sigma_m) \subset \Sigma_m \times \mathbb{R}^n$ is the maximal compact invariant sunset under the action of the affine extension $\hat{F}$, then Λ is the maximal compact viable subset of the IFS.

Finally, due to Theorem 3.1, the affine IFS has the concordant Shadowing Property if and only if the corresponding affine extensions have the uniform Shadowing Property. By translation $(t, x) \mapsto (t, x - \gamma(t))$ the affine extension $\hat{F}$ is reduced to the corresponding linear extension $\hat{F}_0$. The last one is equal to the direct sum of two linear extensions on the corresponding vector sub-bundles E^s and E^u. The restriction of $\hat{F}$ on E^s, and of its inverse $\hat{F}^{-1}$, respectively, on E^u are contracting, so each of them has the Shadowing Property. This implies that their sum also has the Shadowing Property. As a consequence, the corresponding affine IFS also has the Shadowing Property. □

6. Linear horseshoes as viable sets in the plane

In the particular case of an affine IFS in the plane with the same linear part we give topological characteristics of the viable set in terms of the eigenvalues of this operator. More precisely, consider an affine IFS of the type $x \mapsto Ax + b_i$ with diagonal 2×2 matrix $A = (\lambda, \mu)$, where $|\mu| < 1 < |\lambda|$ and $b_1 = (0,0)$, $b_2 = (1-\lambda, 0)$, $b_3 = (0, 1-\mu)$, $b_4 = (1-\lambda, 1-\mu)$.

The following result was announced in Ref. 15.

Theorem 6.1. *Consider the IFS F consisting of four affine mappings $x \mapsto Ax + b_i$ with diagonal 2×2 matrix $A = (\lambda, \mu)$, where $|\mu| < 1 < |\lambda|$ and $b_1 = (0,0)$, $b_2 = (1-\lambda, 0)$, $b_3 = (0, 1-\mu)$, $b_4 = (1-\lambda, 1-\mu)$. Then the viable subset Λ possesses the following properties:*

(1) For $|\mu| < 1/2$ and $|\lambda| > 2$, Λ is a Cantor set.
(2) For $|\mu| \geq 1/2$ and $|\lambda| > 2$, or $|\mu| < 1/2$ and $|\lambda| \leq 2$, Λ is a direct product of a Cantor set and the interval $[0; 1]$.
(3) For all other cases Λ coincides with the unit square.

Proof. Easily seen that the IFS F is the direct sum of two scalar IFS', say F' and F'', respectively, the second being contracting, while the first is expanding (see, e.g., Ref. 16). Denote by Λ' the repellor of F' and by Λ'' the attractor of F''. Due to the equalities $F'_*(\Lambda') = \Lambda'$ and $F''_*(\Lambda'') = \Lambda''$ (see, e.g., Ref. 1), these sets are viable on $\mathbb{Z}$. Moreover, every point behind them generates only unbounded chains, so that $\Lambda := \Lambda' \times \Lambda''$ is the maximal compact subset, viable on $\mathbb{Z}$. The inequalities $|\mu| < 1/2$ and $|\lambda| > 2$ assure that the IFS' F' and F'' satisfy the separation condition (see Ref. 1) and, as a consequence, Λ' and Λ'' are Cantor sets. So does Λ, as the direct product of two Cantor sets. If one of the inequalities, say $|\mu| < 1/2$, is violated, then the attractor, in this case, coincides with the whole interval $[0; 1]$, and Λ equals the Cartesian product of the corresponding subsets. □

Acknowledgments

This work was partially supported by the SCSTD ASM Project 15.817.02.26F and the DAAD Project "Center of Excellence for Applications of Mathematics".

References

1. M. Barnsley, *Fractals Everywhere*, 2nd edn. (Acad. Press Profess., Boston, 1993).
2. V. Glavan and V. Guţu, *On the Dynamics of Contracting Relations*, in Analysis and Optimization of Differential Systems, eds. V. Barbu et al. (Kluwer Acad. Publ., Boston, MA, 2003).
3. V. Glavan and V. Guţu, *Attractors and Fixed Points of Weakly Contracting Relations*, Fixed Point Theory **5**, 2 (2004).
4. J. P. Aubin, A. M. Bayen and P. Saint-Pierre, *Viability Theory. New Directions*, 2nd edn. (Springer, Heidelberg, 2011).
5. O. Cârjă, M. Necula and I. Vrabie, *Viability, Invariance and Applications*, North-Holland Mathematics Studies, 207 (Elsevier Science B.V., Amsterdam, 2007).
6. E. Akin, *The General Topology of Dynamical Systems* (AMS, Providence, 1993).
7. E. Akin, *Simplicial Dynamical Systems*, *Mem. Amer. Math. Soc.* **140**, 667 (1999).
8. V. M. Alexeev, *Quasirandom Dynamical Systems. I. Quasirandom Diffeomorphisms*, Mat. Sb. (N.S.) **76**, 118 (1968) (Russian).
9. V. M. Alexeev, *Quasirandom Dynamical Systems. II. One-dimensional Nonlinear Vibrations in a Periodically Perturbed Field*, Mat. Sb. (N.S.) **77**, 119 (1968) (Russian).
10. V. M. Alexeev, *Quasirandom Dynamical Systems. III. Quasirandom Vibrations of One-dimensional Oscillators*, Mat. Sb. (N.S.) **78**, 120 (1969) (Russian).
11. A. Katok and B. Hasselblatt, *Introduction to the Modern Theory of Dynamical Systems*, Cambridge University Press, Cambridge, 1995.
12. A. Avila, J. Bochi and J. -C. Yoccoz, *Uniformly Hyperbolic Finite-Valued $SL(2,\mathbb{R})$-Cocycles*, Comment. Math. Helv. **85** (2010).
13. A. S. Gorodetski and Yu. S. Ilyashenko, *Certain New Robust Properties of Invariant Sets and Attractors of Dynamical Systems*, Funct. Anal. Appl. **33**, 2 (1999).
14. I. U. Bronshtein, *Nonautonomous Dynamical Systems*, Shtiintsa,

Kishinev, 1984 (Russian).
15. V. Glavan and V. Guţu, *Plane Fractals as Viable Sets in Multi-Valued Dynamics*, in Computer Algebra Systems in Teaching and Research, Volume IV, no. 1. (Wydawnictwo Collegium Mazovia, Siedlce, 2013).
16. V. Glavan and V. Guţu, *Shadowing in Affine Iterated Function Systems*, Fixed Point Theory **10**, 2 (2009).

Some support considerations in the asymptotic optimality of two-scale controlled PDMP

Dan Goreac
Université Paris-Est, LAMA (UMR 8050), UPEMLV, UPEC, CNRS, F-77454, Marne-la-Vallée, France
E-mail: Dan.Goreac@u-pem.fr

Oana Silvia Serea
Université de Perpignan Via Domitia, Laboratoire de Mathématiques et de Physique, EA 4217, F-66860, Perpignan, France
E-mail: oana-silvia.serea@univ-perp.fr

The aim of this short note is to give a linear programming approach to optimality conditions (expressed through support criteria) in control problems with piecewise deterministic Markov dynamics. Two classes are considered: classical, discounted control problems and asymptotic problems associated to two-scales (perturbed) systems.

Keywords: Optimality conditions, Singular perturbations, Occupational measures, Piecewise deterministic Markov processes (PDMP)

1. Introduction

In this short note, we wish to show how linear programming methods may be used to infer support conditions for optimality for classical control problems for piecewise deterministic Markov processes (PDMP) and asymptotic control problems for two-scales PDMP.

We start by recalling the definition of a controlled PDMPs and by stating standard assumptions in Section 2. In Section 3, we deal with classical control problems with discounted cost and PDMP dynamics. We begin with recalling the linearized formulations (primal and dual) of discounted, infinite-horizon control problems. In this context, the dual formulation gives a reliable intuition on the structure of the support for the optimal measures. We briefly explain the notion of optimal pair and, whenever such pair exists, we explicitly give a support condition. In the general case, when the optimal value is not given by a regular function, we work with the optimality candidates using as regular function, the approximating functions obtained by the "shaking of coefficients" method in[1]. The support will then involve inner or outer limits of such sets.

In Section 4, we deal with control problems associated to two-scales PDMP. This is very much inspired by the literature on singularly perturbed systems (e.g.[2–7]). This kind of singular-perturbation problem is particularly challenging because, in many cases, one fails to identify the limit dynamics. This is the case even for simpler systems without jumps. Thus, a fortiori, one has difficulties in characterizing the asymptotic optimal controls. However, in particular cases, the value functions associated to the perturbed problem uniformly converge to the same value as the one obtained for the limit system. Moreover, the speed of convergence can be made explicit. We do not aim in this short note at giving explicit conditions to have this type of convergence. This makes the object of an ongoing research exploiting non-expansive properties for temperate virus inspired by[8]. The conditions are intended in the spirit of[9] Corollaries 3.4 and 4.3 (for Brownian setting). For already existing conditions of convergence, the interested reader is invited to consult the paper[6] or[3] and references therein. We use a double net (regularizations of the value functions associated to fixed scaling parameter) first to give a convenient dual asymptotic formulation and, second, to infer support conditions on the optimal measures. This is done using the intuitions already mentioned in Section 3.

2. Controlled Piecewise Deterministic Processes and Standing Assumptions

We consider U to be a compact metric space (the control space) and $\mathbb{R}^N$ be the state space, for some $N \geq 1$.

Piecewise deterministic control processes have first been introduced by Davis[10] (see also[11,12]. They can be summarized by a characteristic triplet: a vector field $f : \mathbb{R}^N \times U \to \mathbb{R}^N$ that determines the motion between two consecutive jumps, a jump rate $\lambda : \mathbb{R}^N \times U \to \mathbb{R}_+$ and a transition measure $Q : \mathbb{R}^N \times U \times \mathcal{B}\left(\mathbb{R}^N\right) \to \mathcal{P}\left(\mathbb{R}^N\right)$. Here $\mathcal{B}\left(\mathbb{R}^N\right)$ is the Borel σ-field on $\mathbb{R}^N$ and $\mathcal{P}\left(\mathbb{R}^N\right)$ stands for the family of probability measures on $\mathbb{R}^N$. For every $A \in \mathcal{B}\left(\mathbb{R}^N\right)$, the function $(u, x) \mapsto Q(x, u, A)$ is assumed to be measurable and, for every $(x, u) \in \mathbb{R}^N \times U$, $Q(x, u, \{x\}) = 0$.

Whenever $u \in \mathbb{L}^0\left(\mathbb{R}^N \times \mathbb{R}_+; U\right)$ (u is a Borel measurable function) and

$(t_0, x_0) \in \mathbb{R}_+ \times \mathbb{R}^N$, we consider the ordinary differential equation

$$\begin{cases} d\Phi_t^{t_0,x_0,u} = f\left(\Phi_t^{t_0,x_0,u}, u\left(x_0, t - t_0\right)\right) dt, \ t \geq t_0, \\ \Phi_{t_0}^{t_0,x_0,u} = x_0. \end{cases}$$

We choose the first jump time T_1 such that the jump rate $\lambda\big(\Phi_t^{0,x_0,u}, u(x_0,t)\big)$ satisfies

$$\mathbb{P}\left(T_1 \geq t\right) = \exp\left(-\int_0^t \lambda\left(\Phi_s^{0,x_0,u}, u\left(x_0, s\right)\right) ds\right).$$

The controlled piecewise deterministic Markov processes (PDMP) is defined by

$$X_t^{x_0,u} = \Phi_t^{0,x_0,u}, \ \text{if } t \in [0, T_1).$$

The post-jump location Y_1 has $Q\left(\Phi_\tau^{0,x_0,u}, u\left(x_0, \tau\right), \cdot\right)$ as conditional distribution given $T_1 = \tau$. Starting from Y_1 at time T_1, we select the inter-jump time $T_2 - T_1$ such that

$$\mathbb{P}\left(T_2 - T_1 \geq t \ / \ T_1, Y_1\right) = \exp\left(-\int_{T_1}^{T_1+t} \lambda\left(\Phi_s^{T_1,Y_1,u}, u\left(Y_1, s - T_1\right)\right) ds\right).$$

We set

$$X_t^{x_0,u} = \Phi_t^{T_1,Y_1,u}, \ \text{if } t \in [T_1, T_2).$$

The post-jump location Y_2 satisfies

$$\mathbb{P}\left(Y_2 \in A \ / \ T_2, T_1, Y_1\right) = Q\left(\Phi_{T_2}^{T_1,Y_1,u}, u\left(Y_1, T_2 - T_1\right), A\right),$$

for all Borel set $A \subset \mathbb{R}^N$. And so on.

2.1. *Standing Assumptions*

Throughout the paper, unless stated otherwise, we assume the following:

(A1) The function $f : \mathbb{R}^N \times U \longrightarrow \mathbb{R}^N$ is uniformly continuous on $\mathbb{R}^N \times U$ and there exists a positive real constant $C > 0$ such that

$$\left|f\left(x,u\right) - f\left(y,u\right)\right| \leq C\left|x - y\right|, \ \text{and} \ \left|f\left(x,u\right)\right| \leq C, \tag{A1}$$

for all $x, y \in \mathbb{R}^N$ and all $u \in U$.

(A2) The function $\lambda : \mathbb{R}^N \times U \longrightarrow \mathbb{R}_+$ is uniformly continuous on $\mathbb{R}^N \times U$ and there exists a positive real constant $C > 0$ such that

$$\left|\lambda\left(x,u\right) - \lambda\left(y,u\right)\right| \leq C\left|x - y\right|, \ \text{and} \ \lambda\left(x,u\right) \leq C, \tag{A2}$$

for all $x, y \in \mathbb{R}^N$ and all $u \in U$.

(A3) For each bounded uniformly continuous function $h \in BUC\left(\mathbb{R}^N\right)$, there exists a continuous function $\eta_h : \mathbb{R} \longrightarrow \mathbb{R}$ such that $\eta_h(0) = 0$ and

$$\sup_{u \in U} \left| \int_{\mathbb{R}^N} h(z)\, Q(x,u,dz) - \int_{\mathbb{R}^N} h(z)\, Q(y,u,dz) \right| \leq \eta_h(|x-y|). \quad \text{(A3)}$$

We will ask that η should only depend on the continuity modulus and bound of h but not of h itself.

(A4) For every $x \in \mathbb{R}^N$ and every decreasing sequence $(\Gamma_n)_{n\geq 0}$ of subsets of $\mathbb{R}^N$,

$$\inf_{n\geq 0} \sup_{u\in U} Q(x,u,\Gamma_n) = \sup_{u\in U} Q\left(x,u,\underset{n}{\cap}\Gamma_n\right). \quad \text{(A4)}$$

Remark 2.1. The assumption A3 can be somewhat weakened by imposing

(A3') For each bounded uniformly continuous function $h \in BUC\left(\mathbb{R}^N\right)$, there exists a continuous function $\eta_h : \mathbb{R} \longrightarrow \mathbb{R}$ such that $\eta_h(0) = 0$ and

$$\sup_{u\in U} \left|\lambda(x,u) \int_{\mathbb{R}^N} h(z)\, Q(x,u,dz) - \lambda(y,u) \int_{\mathbb{R}^N} h(z)\, Q(y,u,dz)\right| \\ \leq \eta_h(|x-y|).$$

It is obvious that whenever one assumes (A3) and $\lambda(\cdot)$ is bounded, the assumption A3' holds true.

To simplify the arguments, we assume the following invariance condition to hold true. There exists a compact set $\mathbb{K}$ such that, for every $x \in \partial\mathbb{K}$, every $p \in N_{\mathbb{K}}(x)$ and every $u \in U$, one has

$$\langle f(x,u), p\rangle + \lambda(x,u)\, Q(x,u,\mathbb{K}^c) \leq 0. \quad \text{(I)}$$

Here, $\mathbb{K}^c$ stands for $\mathbb{R}^N \setminus \mathbb{K}$, while $N_{\mathbb{K}}(x)$ stands for the normal cone to $\mathbb{K}$ at x, denoted by $N_{\mathbb{K}}(x)$ and is defined as

$$N_{\mathbb{K}}(x) = \left\{ \begin{array}{l} \mathbb{R}^N : \forall\varepsilon > 0, \exists\eta > 0 \text{ such that} \\ \forall y \in \mathbb{K} \cap \overline{B}(x,\eta),\ \langle p, y-x\rangle \leq \varepsilon |y-x| \end{array} \right\}.$$

We recall that $\overline{B}(x,\eta) = \left\{ y \in \mathbb{R}^N : |y-x| \leq \eta \right\}$. This assumption implies that the set $\mathbb{K}$ is invariant with respect to the piecewise deterministic Markov dynamics (every trajectory starting from a point of $\mathbb{K}$ remains in $\mathbb{K}$ almost surely). For further details on invariance of piecewise deterministic Markov processes, the reader is referred to[13].

Remark 2.2. This condition implies that if $\mathbb{K}$ is invariant w.r.t. the PDMP with characteristic triplet (f, λ, Q), it also holds true for $\left(\delta f, \delta'\lambda, \frac{\delta}{\delta'}Q\right)$, for all $\delta, \delta' > 0$.

3. An LP Approach to Optimality in Classical Control Problems with PDMP Dynamics

3.1. *Linear Programming Tools for PDMP*

Given a Lipschitz-continuous, bounded cost functional $g : \mathbb{R}^N \longrightarrow \mathbb{R}$, we introduce the value function by setting

$$v_g(x) := \inf_u \mathbb{E}\left[\int_0^\infty e^{-t} g\left(X_t^{x,u}\right) dt\right], \text{ for all } x \in \mathbb{R}^N. \tag{1}$$

The following results are taken from[14].

We denote by $\mathcal{P}\left(\mathbb{R}^N \times U\right)$ the set of all probability measures on $\mathbb{R}^N \times U$ and by

$$\Theta(x) = \left\{ \begin{array}{l} \gamma \in \mathcal{P}\left(\mathbb{R}^N \times U\right) : \ \forall \phi \in C_b^1\left(\mathbb{R}^N\right) : \\ \int_{\mathbb{R}^N \times U} \left(\mathcal{U}^u \phi(y) + \phi(x) - \phi(y)\right) \gamma(dy, du) = 0 \end{array} \right\}.$$

Here, $\mathcal{U}^u$ stands for the classical infinitesimal generator (associated to frozen control parameter $u \in U$) given by

$$\mathcal{U}^u \phi(y) = \langle \nabla \phi(y), f(y,u) \rangle + \lambda(y,u) \int_{\mathbb{R}^N} \left(\phi(z) - \phi(y)\right) Q(y,u,dz), \tag{2}$$

for all $u \in U$, $\phi \in C_b^1\left(\mathbb{R}^N\right)$, and all $y \in \mathbb{R}^N$. We recall that $C_b^1\left(\mathbb{R}^N\right)$ stands for the class of bounded, differentiable functions $\phi : \mathbb{R}^N \longrightarrow \mathbb{R}$ for which the first order derivatives are continuous and bounded. With respect to this closed, convex set Θ, one introduces the linear (primal and dual) value functions

$$\Lambda_g(x) := \inf_{\gamma \in \Theta(x)} \int_{\mathbb{R}^N \times U} g(y) \gamma(dy, du), \tag{3}$$

and

$$\Lambda_g^*(x) = \sup \left\{ \begin{array}{c} \mu \in \mathbb{R} : \ \exists \varphi \in C_b^1\left(\mathbb{R}^N\right) \text{ such that } \forall (y,u) \in \mathbb{R}^N \times U, \\ \mu \leq \mathcal{U}^u \varphi(y) + g(y) + \left(\varphi(x) - \varphi(y)\right) \end{array} \right\}, \tag{4}$$

for all $x \in \mathbb{R}^N$. The link between the initial value function and the linear values is given by[14] Theorem 7.

Theorem 3.1 ([14] Theorem 7). *For every $x \in \mathbb{R}^N$, the equality*

$$v_g(x) = \Lambda_g(x) = \Lambda_g^*(x) \tag{5}$$

holds true.

Remark 3.1. As a by-product, the set $\Theta(x)$ can be seen as the closed convex hull of occupation measures associated to the trajectories of the process. In particular, if the assumption (I) holds true, then, for every $x \in \mathbb{K}$, any measure $\gamma \in \Theta(x)$ has its support in $\mathbb{K} \times U$. Moreover, under the assumption (I), by introducing

$$\Lambda^*_{g,\mathbb{K}}(x) := \sup \left\{ \begin{array}{l} \mu \in \mathbb{R} : \exists \varphi \in C^1_b\left(\mathbb{R}^N\right) \text{ such that } \forall (y,u) \in \mathbb{K} \times U, \\ \mu \leq \mathcal{U}^u \varphi(y) + g(y) + (\varphi(x) - \varphi(y)) \end{array} \right\},$$

one gets

$$\Lambda_g(x) \geq \Lambda^*_{g,\mathbb{K}}(x) \geq \Lambda^*_g(x).$$

Hence, $\Lambda^*_{g,\mathbb{K}}(x) = \Lambda^*_g(x)$.

A careful look at the proof of the[14] Theorem 7 and[13] Theorem 3.6 yields the following Krylov-type result.

Lemma 3.1. *There exists a decreasing function (only depending on the continuity moduli of Q and g but not on the functions themselves) η : $\mathbb{R}_+ \longrightarrow \mathbb{R}_+$ that satisfies $\lim_{\delta \to 0} \eta(\delta) = 0$ and a family of C^1_b functions V^δ such that*

$$V^\delta_g(x) - g(x) + H\left(x, \nabla V^\delta_g(x), V^\delta_g\right) \leq 0, \tag{6}$$

for all $x \in \mathbb{R}^N$, where the Hamiltonian H is given by

$$H(x,p,\psi) = \sup_{u \in U} \left\{ -\langle f(x,u), p\rangle - \lambda(x,u) \int_{\mathbb{R}^N} (\psi(z) - \psi(x))\, Q(x,u,dz) \right\} \tag{7}$$

and such that

$$\sup_{x \in \mathbb{R}^N} \left| V^\delta_g(x) - v_g(x) \right| \leq C\eta(C\delta), \tag{8}$$

for all $\delta > 0$. The constant C can be chosen as the maximum between the Lipschitz norms of f and λ.

The function η corresponding to a continuity modulus can be chosen at best of Lipschitz-continuity type (i.e. $\eta(x) = O(x)$ as $x \to 0$).

3.2. *Support Conditions for Optimality*

Due to Theorem 3.1 (the dual formulation Λ^*) and to Remark 3.1, in order to compute v_g for arguments in $\mathbb{K}$, one tries to maximize, over test functions $\varphi \in C^1_b\left(\mathbb{R}^N\right)$, the infimum

$$\inf_{(y,u) \in \mathbb{K} \times U} \mathcal{U}^u \varphi(y) + g(y) + (\varphi(x) - \varphi(y)).$$

Thus, it is natural to introduce the set

$$D_g(x) := \left\{ \begin{array}{c} (\mu, \varphi) \in \mathbb{R} \times C_b^1\left(\mathbb{R}^N\right) \text{ s.t.} \\ mu = \inf_{(y,u) \in \mathbb{K} \times U} \mathcal{U}^u \varphi(y) + g(y) + (\varphi(x) - \varphi(y)) \end{array} \right\}, \tag{9}$$

for all $x \in \mathbb{K}$. The set $D_g(x)$ is well defined and non-empty (consequence of the assumptions on the characteristic triplet and cost function g). One can alternatively write the value function(s) as

$$v_g(x) = \sup\{\mu, (\mu, \varphi) \in D_g(x)\}, \tag{10}$$

for all $x \in \mathbb{K}$. We introduce the following.

Definition 3.1. Whenever $x \in \mathbb{K}$, the couple $(\overline{\mu}, \overline{\varphi}) \in D_g(x)$ is an optimal pair if it satisfies $v_g(x) = \overline{\mu}$. If such pair exists, we denote by

$$\Omega_{g,(\overline{\mu},\overline{\varphi})}(x) := \{(y,u) \in \mathbb{K} \times U, \text{s.t. } \overline{\mu} = \mathcal{U}^u \overline{\varphi}(y) + g(y) + (\overline{\varphi}(x) - \overline{\varphi}(y))\}. \tag{11}$$

As one easily understands, optimal measures have their support in such sets (should they exist). The problem is that optimal pairs do not always exist. A simple case when they do exist is when the value function itself belongs to $C_b^1\left(\mathbb{R}^N\right)$ (one just takes $\overline{\varphi} = v_g$ and uses the subsolution property for the associated HJB equation (see[14] Theorem 18) to infer

$$v_g(x) \leq \mathcal{U}^u v_g(y) + g(y) + (v_g(x) - v_g(y)),$$

for all $(y,u) \in \mathbb{K} \times U$. In this case compactness of $\mathbb{K} \times U$ implies that $\Omega_{g,(\overline{\mu},\overline{\varphi})}(x)$ is non empty.)

Proposition 3.1. *Let $x \in \mathbb{K}$ and assume that $(\overline{\mu}, \overline{\varphi}) \in D_g(x)$ is an optimal pair. Then, $\gamma \in \Theta(x)$ is optimal for $v_g(x)$ if and only if $\gamma\left(\Omega_{g,(\overline{\mu},\overline{\varphi})}(x)\right) = 1$.*

Proof. To prove the necessity of the support condition, let us consider $\gamma \in \Theta(x)$ which is optimal. It follows, by optimality and the definition of $\Theta(x)$, that

$$\begin{aligned} v_g(x) &= \int_{\mathbb{K}\times\mathbb{U}} g(y)\,\gamma(dydu) \\ &= \int_{\mathbb{K}\times\mathbb{U}} \left(\mathcal{U}^u \overline{\varphi}(y) + g(y) + (\overline{\varphi}(x) - \overline{\varphi}(y))\right)\gamma(dydu). \end{aligned}$$

Then, the definition of $D_g(x)$ implies that $v_g(x) \geq \overline{\mu}$ and the equality holds true if and only if

$$\mathcal{U}^u \overline{\varphi}(y) + g(y) + (\overline{\varphi}(x) - \overline{\varphi}(y)) = \overline{\mu},$$

γ-almost surely (or, equivalently, $\gamma\left(\Omega_{g,(\overline{\mu},\overline{\varphi})}(x)\right)=1$). But the optimality of the pair $(\overline{\mu},\overline{\varphi})$ can only happen if $v_g(x)=\overline{\mu}$ and, thus, the conclusion follows.

For sufficiency, one writes

$$\mathcal{U}^u\overline{\varphi}(y)+g(y)+(\overline{\varphi}(x)-\overline{\varphi}(y))=\overline{\mu}=v_g(x),$$

γ-almost surely and integrates on $\mathbb{K}\times\mathbb{U}$ w.r.t. γ to get optimality. □

In general, one simply picks the test functions V_g^δ and μ^δ such that $\left(\mu^\delta,V_g^\delta\right)\in D_g(x)$. If among these functions an optimal pair does not exist (thus implying the existence of a strictly increasing sequence denoted, by abuse of notation, $(\mu^n)_{n\geq 1}$ such that $\sup_n\mu^n=v_g(x)$), one introduces an approximate set

$$\begin{aligned}\Omega_g^n(x):=\{&(y,u)\in\mathbb{K}\times U,\text{s.t. } v_g(x)+\eta\left((v_g(x)-\mu^n)^{\frac{1}{2}}\right)\\&\geq\mathcal{U}^u\overline{\varphi}(y)+g(y)+(\overline{\varphi}(x)-\overline{\varphi}(y))\}\end{aligned}\tag{12}$$

and the usual (inner ant outer) limit sets

$$\Omega_g^{in}(x):=\liminf_{n\to\infty}\Omega_g^n(x)=\underset{n\geq 1}{\cup}\ \underset{n\geq n}{\cap}\ \Omega_g^k(x),$$

$$\Omega_g^{out}(x):=\limsup_{n\to\infty}\Omega_g^n(x)=\underset{n\geq 1}{\cap}\ \underset{k\geq n}{\cup}\ \Omega_g^k(x),$$

$$\Omega_g^{out,cl}(x):=\underset{n\geq 1}{\cap}cl\left(\underset{k\geq n}{\cup}\Omega_g^k(x)\right),$$

where cl is the usual Kuratowski closure operator. Again by abuse of notation, we let V_g^n be such that $\left(\mu^n,V_g^n\right)\in D_g(x)$. The support of optimal measures is characterized as follows.

Proposition 3.2.

(i) For every $x\in\mathbb{K}$, if $\gamma\in\Theta(x)$ is an optimal measure, then its support is contained in $\Omega_g^{out}(x)\subset\Omega_g^{out,cl}(x)$. In particular, when the limit of the sets exists (i.e. $\Omega_g^{in}(x)=\Omega_g^{out}(x)$), one has

$$\sup_{n\geq 1}\gamma\left(\underset{k\geq n}{\cap}\Omega_g^k(x)\right)=1.$$

(ii) As a (partial) converse, for $x\in\mathbb{K}$, if $\gamma\in\Theta(x)$ is such that $\max\limits_{n\geq 1}\gamma\left(\underset{k\geq n}{\cap}\Omega_g^k(x)\right)=1$, then γ is optimal.

Proof. (i) When $\gamma \in \Theta(x)$ is optimal, one has

$$\begin{aligned} v_g(x) &= \int_{\mathbb{K}\times\mathbb{U}} g(y)\gamma(dydu) \\ &= \int_{\mathbb{K}\times\mathbb{U}} \left(\mathcal{U}^u V_g^k(y) + g(y) + \left(V_g^k(x) - V_g^k(y)\right)\right)\gamma(dydu) \\ &\geq \int_{\mathbb{K}\times\mathbb{U}} \left(\mu^k 1_{\Omega_g^k(x)} + \left[v_g(x) + \eta\left(\left(v_g(x) - \mu^k\right)^{\frac{1}{2}}\right)\right] 1_{(\mathbb{K}\times\mathbb{U})\setminus\Omega_g^k(x)}\right)\gamma(dydu) \end{aligned}$$

for all $k \geq 1$. It follows that

$$\begin{aligned} &\left(\eta\left(\left(v_g(x) - \mu^k\right)^{\frac{1}{2}}\right) + 1\right)\gamma\left(\Omega_g^k(x)\right) \\ &\geq \left(\frac{v_g(x) - \mu^k}{\eta\left(\left(v_g(x) - \mu^k\right)^{\frac{1}{2}}\right)} + 1\right)\gamma\left(\Omega_g^k(x)\right) \geq 1. \end{aligned}$$

One passes to $\limsup_{k\to\infty}$ to get $\gamma\left(\Omega_g^{out}(x)\right) = 1$ and the proof is complete.

(ii) Whenever $\gamma\left(\bigcap_{k\geq n_0} \Omega_g^k(x)\right) = 1$, it follows that $\gamma\left(\Omega_g^k(x)\right) = 1$, for all $k \geq n_0$. Then, one simply writes down

$$\begin{aligned} v_g(x) &\leq \int_{\mathbb{K}\times\mathbb{U}} g(y)\gamma(dydu) \\ &= \int_{\mathbb{K}\times\mathbb{U}} \left(\mathcal{U}^u V_g^k(y) + g(y) + \left(V_g^k(x) - V_g^k(y)\right)\right)\gamma(dydu) \\ &\leq v_g(x) + \eta\left(\left(v_g(x) - \mu^k\right)^{\frac{1}{2}}\right), \end{aligned}$$

for all $k \geq n_0$. Passing to the limit as $k \to \infty$, one infers that $\gamma \in \Theta(x)$ is optimal. □

4. Two-Scale Systems. Asymptotic Optimality

4.1. *A PDMP Description of Scales*

We consider a piecewise deterministic model set on $\mathbb{R}^{N+M}$ in which the second component (say regime) is faster than the first one and so is the frequency of its jumps. (To have a simple example, think of this second component as a discrete one rapidly toggling between the elements of the Euclidian basis of $\mathbb{R}^M$ and specifying different regime evolutions for some proteins). At any point, jumps can occur by:

- either changing the regime but, to express in terms of the previous example, keep continuous protein level;

- or, within the same regime, a slower (and much more rare) jump (say transcription) occurs and it is seen at protein level.

This amounts to considering a PDMP governed by a controlled characteristic triple $(f^\varepsilon, \lambda^\varepsilon, Q^\varepsilon)$ such that

$$f^\varepsilon = \left(f_1, \tfrac{1}{\varepsilon} f_2\right), \; \lambda^\varepsilon(y,u) = \lambda_1(y,u) + \tfrac{1}{\varepsilon}\lambda_2(y,u)$$
$$Q^\varepsilon\left((y_1,y_2), u, dz_1 dz_2\right) = \tfrac{\lambda_1(y,u)}{\lambda^\varepsilon(y,u)} Q_1(y_2)(y_1,u,dz_1)$$
$$+ \tfrac{\lambda_2(y,u)}{\varepsilon\lambda^\varepsilon(y,u)} Q_2(y_1)(y_2,u,dz_2).$$

Here, $f_1 : \mathbb{R}^{N+M} \times U \longrightarrow \mathbb{R}^N$, $f_2 : \mathbb{R}^{N+M} \times U \longrightarrow \mathbb{R}^M$ are bounded, uniformly continuous and Lipschitz-continuous in space, uniformly w.r.t. the control (i.e. satisfy the same assumptions as in the classical context). Similar regularity is assumed for $\lambda_{1,2}$ and Q. Moreover, we assume the existence of invariant sets $\mathbb{K}_1$ and $\mathbb{K}_2$ for $(f_1, \lambda_1(\cdot, y_2), Q_1(y_2))$, for all $y_2 \in \mathbb{K}_2$ respectively $(f_2, \lambda_2(y_1, \cdot), Q_2(y_1))$, for all $y_1 \in \mathbb{K}_1$. We will also use the notation $Q_1(y_1, y_2, u, dz_1)$ for $Q_1(y_2)(y_1, u, dz_1)$.

This kind of singular-perturbation problem is particularly challenging because, in many cases, one fails to identify the limit dynamics. This is the case even for simpler systems without jumps. Thus, a fortiori, one has difficulties in characterizing the asymptotic optimal controls. However, in particular cases, the value functions associated to the perturbed problem uniformly converge to the same value as the one obtained for the limit system. Moreover, the speed of convergence can be made explicit.

We do not aim in this short note at giving explicit conditions to have this type of convergence. This makes the object of an ongoing research exploiting non-expansive properties for temperate virus inspired by[8]. The conditions are intended in the spirit of[9] Corollaries 3.4 and 4.3 (for Brownian setting). For already existing conditions of convergence, the interested reader is invited to consult the paper[6] or[3] and references therein.

4.2. *Asymptotics of Linearized Formulations*

Inspired by the considerations on classical discounted control problems presented in the first part, we introduce the perturbed set of constraints

$$\Theta^\varepsilon(x) = \left\{ \begin{array}{c} \gamma \in \mathcal{P}\left(\mathbb{R}^{N+M} \times U\right) : \; \forall \phi \in C_b^1\left(\mathbb{R}^{N+M}\right), \\ \int_{\mathbb{R}^N \times U} \left(\mathcal{U}^{u;\varepsilon}\phi(y) + \phi(x) - \phi(y)\right)\gamma(dy,du) = 0 \end{array} \right\},$$

where $\mathcal{U}^{u.\varepsilon}$ stands for the infinitesimal generator associated to $(f^\varepsilon, \lambda^\varepsilon, Q^\varepsilon)$ and given by

$$\begin{aligned}\mathcal{U}^{u.\varepsilon}\phi(y_1,y_2) &= \langle \nabla_{y_1}\phi(y), f_1(y,u)\rangle + \tfrac{1}{\varepsilon}\langle \nabla_{y_2}\phi(y), f_2(y,u)\rangle \\ &+\lambda_1(y,u)\int_{\mathbb{R}^N}(\phi(z,y_2)-\phi(y_1,y_2))\,Q_1(y,u,dz) \\ &+\tfrac{1}{\varepsilon}\lambda_2(y,u)\int_{\mathbb{R}^M}(\phi(y_1,z)-\phi(y_1,y_2))\,Q_2(y,u,dz)\end{aligned} \tag{13}$$

for all $u \in U$, $\phi \in C_b^1\left(\mathbb{R}^{N+M}\right)$, and all $y = (y_1, y_2) \in \mathbb{R}^{N+M}$. We consider a cost functional g having the same regularity as in the first part, but defined on $\mathbb{R}^{N+M}$ and set

$$v_g^\varepsilon(x) := \inf_u \mathbb{E}\left[\int_0^\infty e^{-t} g\left(X_t^{x,u;\varepsilon}\right) dt\right], \text{ for all } x \in \mathbb{R}^{N+M},$$

where $X_t^{x,u;\varepsilon}$ is the PDMP admitting the local characteristics $(f^\varepsilon, \lambda^\varepsilon, Q^\varepsilon)$ and constructed as described before. Moreover, we set

$$\Lambda_g^\varepsilon(x) := \inf_{\gamma\in\Theta^\varepsilon(x)} \int_{\mathbb{R}^{N+M}\times U} g(y)\,\gamma(dy,du) \tag{14}$$

and

$$\Lambda_g^{\varepsilon*}(x) := \sup\left\{\begin{array}{c}\mu\in\mathbb{R}:\ \exists\varphi\in C_b^1\left(\mathbb{R}^{N+M}\right) \text{ such that,}\\ \forall (y,u)\in\mathbb{R}^{N+M}\times U,\\ \mu \le \mathcal{U}^{u;\varepsilon}\varphi(y) + g(y) + (\varphi(x)-\varphi(y))\end{array}\right\}, \tag{15}$$

for all $x \in \mathbb{R}^{N+M}$. The arguments on classical control problems yield that the value function $v_g^\varepsilon(\cdot)$ associated to a regular cost g can, alternatively, be given by

$$v_g^\varepsilon(x) = \Lambda_g^\varepsilon(x) = \Lambda_g^{\varepsilon*}(x), \text{ for all } x \in \mathbb{R}^{N+M}.$$

It is natural, at this point, to set

$$\Theta\left(y_1^0,y_2^0\right) := \liminf_{\varepsilon\to 0}\Theta^\varepsilon\left(y_1^0,y_2^0\right), \text{ for } \left(y_1^0,y_2^0\right)\in\mathbb{R}^{N+M}. \tag{16}$$

(This is the classical lower limit for sets with the usual weak* convergence of measures.) We begin with some simple considerations on the structure of this set of constraints.

Proposition 4.1. *The following inclusion holds true*

$$\Theta\left(y_1^0,y_2^0\right)\subset$$
$$\left\{\begin{array}{c}\gamma\in\mathcal{P}\left(\mathbb{R}^M\times\mathbb{R}^N\times U\right) \ s.t.\ \forall\psi\in C_b^1\left(\mathbb{R}^N\right) \ and\ \forall\phi\in C_b^1\left(\mathbb{R}^{N+M}\right),\\ \int\limits_{\mathbb{R}^{N+M}\times U}\left[\begin{array}{c}\mathcal{U}^{u.1,y_2}\psi(y_1)+g(y_1,y_2)+\\ \left(-\psi(y_1)+\psi\left(y_1^0\right)\right)\end{array}\right]\gamma(dy_1dy_2du)=0 \ and\\ \int\limits_{\mathbb{R}^{N+M}\times U}\left[\begin{array}{c}\mathcal{U}^{u,2}\phi(y_1,y_2)+g(y_1,y_2)+\\ \left(-\phi(y_1,y_2)+\phi\left(y_1^0,y_2^0\right)\right)\end{array}\right]\gamma(dy_1dy_2du)=0\end{array}\right\},$$

where

$$\mathcal{U}^{u,1,y_2}\psi(y_1) := \langle \nabla\psi(y_1), f_1(y_1,y_2,u)\rangle$$
$$+\lambda_1(y_1,y_2,u)\int_{\mathbb{R}^N}(\psi(z)-\psi(y_1))\,Q_1(y_1,y_2,u,dz)$$

and

$$\mathcal{U}^{u,2}\phi(y_1,y_2) := \langle \nabla_{y_2}\phi(y_1,y_2), f_2(y_1,y_2,u)\rangle$$
$$+\lambda_2(y_1,y_2,u)\int_{\mathbb{R}^M}(\phi(y_1,z)-\phi(y_1,y_2))\,Q_1(y_1,y_2,u,dz),$$

for all $\phi \in C^1(\mathbb{R}^{N+M})$, $\psi \in C^1(\mathbb{R}^N)$ *and all* $(y_1,y_2) \in \mathbb{R}^N \times \mathbb{R}^M$, $u \in U$.

Proof. The proof is standard and will be omitted. □

We define the linearized problem with respect to the new set of constraints

$$\Lambda_g(y_1^0,y_2^0) := \inf_{\gamma\in\Theta(y_1^0,y_2^0)} \int_{\mathbb{R}^{N+M}\times U} g(z)\,\gamma(dzdu).$$

4.3. *First Main Result: An Asymptotic Formulation*

Throughout the remaining of the paper, unless stated otherwise, we assume that there exists a continuity modulus ω only depending on the bound and continuity modulus of the cost function g and the bound and continuous moduli of the characteristics of the PDMP and some function $\xi \in C_b(\mathbb{R}^N)$ (which may depend on g and (f,λ,Q)) such that

$$\sup_{(y_1,y_2)\in\mathbb{K}=\mathbb{K}_1\times\mathbb{K}_2} \left|v_g^{\varepsilon}(y_1,y_2)-\xi(y_1)\right| \le \omega(\varepsilon). \tag{17}$$

In the Brownian setting, for further details and conditions implying this kind of convergence, the reader is referred to [9]; see also [15].

If $\delta > 0$ and $v_g^{\varepsilon,\delta}$ is the value function associated with the "shaken" problem (i.e. in which $\varphi \in \{g,f,\lambda\}$ are replaced with $\varphi^\delta(x,(u,v)) := \varphi(x+\delta v,u)$, $v \in \mathbb{R}^{N+M}$, $|v| \le 1$ respectively Q is replaced by $Q^\delta(x,(u,v),dy) = \widetilde{Q}(x+v,u,dy+v)$), under analogous assumptions, the inequality (17) holds true for some ξ^δ replacing ξ. Since ω only depends on the continuity moduli, the right-hand member is a generic ω independent of $\delta > 0$. In particular,

$$\left|v_g^{\varepsilon,\delta}(y_1,y_2)-v_g^{\varepsilon,\delta}(y_1,y_2')\right| \le 2\omega(\varepsilon),$$

for all $y_1 \in \mathbb{K}_1$ and all $y_2, y_2' \in \mathbb{K}_2$. Now, let us consider $(\psi_\delta)_\delta$ to be a sequence of standard mollifiers $\psi_\delta(x) := \frac{1}{\delta^{M+N}}\psi\left(\frac{x}{\delta}\right)$, $x \in \mathbb{R}^{N+M}$, $\delta > 0$, where $\psi \in C^{\infty}\left(\mathbb{R}^{M+N}\right)$ is a positive function such that

$$Supp(\psi) \subset \overline{B}(0,1) \text{ and } \int_{\mathbb{R}^{N+M}} \psi(x)dx = 1.$$

Then, the convoluted functions $V_g^{\varepsilon,\delta} := v_g^{\varepsilon,\delta} * \psi_\delta$ satisfy (with the modulus η given in Lemma 3.1 and due to the inequality (17)):

$$\begin{aligned}
&\left|V_g^{\varepsilon,\delta}(y_1,y_2) - v_g^{\varepsilon}(y_1,y_2)\right| \leq c_0\left(1+\tfrac{1}{\varepsilon}\right)\eta\left(c_0\left(1+\tfrac{1}{\varepsilon}\right)\delta\right),\\
&\left|V_g^{\varepsilon,\delta}(y_1,y_2) - V_g^{\varepsilon,\delta}(y_1,y_2')\right| \leq 2c_0\left(1+\tfrac{1}{\varepsilon}\right)\eta\left(c_0\left(1+\tfrac{1}{\varepsilon}\right)\delta\right) \qquad (18)\\
&+\left|v_g^{\varepsilon}(y_1,y_2) - \xi(y_1)\right| \leq 2c_0\left(1+\tfrac{1}{\varepsilon}\right)\eta\left(c_0\left(1+\tfrac{1}{\varepsilon}\right)\delta\right) + \omega(\varepsilon),
\end{aligned}$$

where c_0 is independent of δ and ε. Finally, by recalling that $\nabla V_g^{\varepsilon,\delta} = \frac{1}{\delta}v_g^{\varepsilon,\delta} * \nabla\psi_\delta$, one gets

$$\left|\nabla V_g^{\varepsilon,\delta}(y_1,y_2) - \nabla V_g^{\varepsilon,\delta}(y_1,y_2')\right| \leq \frac{1}{\delta}2\omega(\varepsilon) \text{ and } \left\|\nabla V_g^{\varepsilon,\delta}\right\|_\infty \leq \frac{c_0}{\delta}, \qquad (19)$$

for some constant $c_0 > 0$ independent of δ and ε. As consequence, one gets the following representation theorem.

Theorem 4.1. *Let us assume that the value functions for the perturbed systems converge uniformly to a function $\xi \in C_b\left(\mathbb{R}^N\right)$ (in the sense of (17)). Then the limit value function has the following representation (for all $y_2^0 \in \mathbb{K}_2$)*

$$\xi\left(y_1^0\right) = \Lambda_g\left(y_1^0, y_2^0\right),$$

for all $y_1^0 \in \mathbb{K}_1$. This common value is given by

$$\Lambda_g^*\left(y_1^0\right) := \sup\left\{\begin{array}{l}
\eta \in \mathbb{R} : \exists\alpha \in C\left(\mathbb{R}_+;\mathbb{R}_+\right),\ \lim\limits_{\varepsilon\to 0}\alpha(\varepsilon) = 0 \ s.t.\\
\forall\varepsilon > 0,\ \exists\varphi \in C_b^1\left(\mathbb{R}^{N+M}\right) \ s.t.\\
\sup\limits_{y,y'\in\mathbb{R}^M}\left\|\varphi(\cdot,y) - \varphi(\cdot,y')\right\|_\infty \leq \alpha(\varepsilon) \text{ and } s.t.\\
\forall(y,u) \in \mathbb{R}^{N+M} \times U, \eta \leq \mathcal{U}^{u;\varepsilon}\varphi(y_1,y_2) + g(y_1,y_2) +\\
\left\|-\varphi(y_1,\cdot)\right\|_\infty + \left\|\varphi\left(y_1^0,\cdot\right)\right\|_\infty
\end{array}\right\}, \qquad (20)$$

for all $y_1^0 \in \mathbb{K}_1$.

Proof. The first assertion is obvious by invoking the equality between the classical value v_g^ε and its linearized formulation Λ_g^ε and the construction of the asymptotic constraints set (16). We only need to prove that $\Lambda_g^*\left(y_1^0\right)$ gives the same object which is less obvious from simply passing to the limit in the dual formulations $\Lambda_g^{\varepsilon *}$ in (15). To this purpose, let us fix, for the time being, $\gamma \in \Theta\left(y_1^0, y_2^0\right)$ as the limit of some sequence $\gamma^\varepsilon \in \Theta^\varepsilon\left(y_1^0, y_2^0\right)$. Moreover, let us take

$$\begin{aligned}
&\eta \in \mathbb{R} : \exists \alpha \in C\left(\mathbb{R}_+;\mathbb{R}_+\right),\ \lim_{\varepsilon\to 0}\alpha\left(\varepsilon\right) = 0 \text{ s.t. } \forall\varepsilon > 0,\ \exists\varphi \in C_b^1\left(\mathbb{R}^{N+M}\right) \text{ s.t.}\\
&\sup_{y,y'\in\mathbb{R}^M}\left\|\varphi\left(\cdot,y\right) - \varphi\left(\cdot,y'\right)\right\|_\infty \le \alpha\left(\varepsilon\right) \text{ and s.t. } \forall\left(y,u\right) \in \mathbb{R}^M \times U,\\
&\eta \le \mathcal{U}^{u;\varepsilon}\varphi\left(y_1,y_2\right) + g(y_1,y_2) + \left\|-\varphi\left(y_1,\cdot\right)\right\|_\infty + \left\|\varphi\left(y_1^0,\cdot\right)\right\|_\infty
\end{aligned}$$

It is clear that

$$\eta \le \mathcal{U}^{u;\varepsilon}\varphi\left(y_1,y_2\right) + g(y_1,y_2) + \left(\varphi\left(y_1^0,y_2^0\right) - \varphi\left(y_1,y_2\right)\right) + 2\alpha\left(\varepsilon\right)$$

and, by integrating w.r.t. γ^ε, passing to the limit as $\varepsilon \to 0$ and recalling that both η and γ are arbitrary, one infers $\Lambda_g^*\left(y_1^0\right) \le \Lambda_g\left(y_1^0, y_2^0\right)$.

To prove the remaining (converse) inequality, we consider the family of functions $V_g^{\varepsilon,\delta}$ regular subsolutions of the Hamilton Jacobi integrodifferential equation associated to $v_g^{\varepsilon,\delta}$ i.e. satisfying

$$\mathcal{U}^{u;\varepsilon}V_g^{\varepsilon,\delta}\left(y_1,y_2\right) + g(y_1,y_2) - V_g^{\varepsilon,\delta}\left(y_1,y_2\right) \ge 0.$$

Then, one easily gets using (18) and (17)

$$\xi\left(y_1^0\right) - \alpha\left(\delta,\varepsilon\right) \le \mathcal{U}^{u;\varepsilon}V_g^{\varepsilon,\delta}\left(y_1,y_2\right) + g(y_1,y_2) + \left\|-V_g^{\varepsilon,\delta}\left(y_1,\cdot\right)\right\|_\infty + \left\|V_g^{\varepsilon,\delta}\left(y_1^0,\cdot\right)\right\|_\infty$$

and

$$\left|V_g^{\varepsilon,\delta}\left(y_1,y_2\right) - V_g^{\varepsilon,\delta}\left(y_1,y_2'\right)\right| \le \alpha\left(\delta,\varepsilon\right)$$

where

$$\alpha\left(\delta,\varepsilon\right) := 2c_0\left(1 + \frac{1}{\varepsilon}\right)\eta\left(c_0\left(1 + \frac{1}{\varepsilon}\right)\delta\right) + \omega\left(\varepsilon\right).$$

The reader is invited to note that $\liminf_{\varepsilon\to 0}\liminf_{\delta\to 0}\alpha\left(\delta,\varepsilon\right) = \liminf_{\varepsilon\to 0}\omega\left(\varepsilon\right) = 0$ to deduce that $\xi\left(y_1^0\right) \le \Lambda_g^*\left(y_1^0\right)$ which concludes the proof of our theorem. □

4.4. *Second Main Result: Support for Asymptotic Optimality*

In most of the interesting cases, the function η can be computed explicitly (e.g. if Q has some Lipschitz continuity property). For fixed $\varepsilon > 0$, we let

$$\delta^\varepsilon := \varepsilon^2 \times \sup\left\{\delta > 0 : \eta\left(\delta\right) \le \varepsilon^3\right\}. \tag{21}$$

Motivated by the classical case, we let, by abuse of notation, $\xi\left(y_1^0, y_2^0\right) = \xi\left(y_1^0\right)$ and we introduce the one-approximating and two-approximating sets by setting

$$\Omega_g^{1,n}(x) := \left\{ \begin{array}{c} (y,u) \in \mathbb{K} \times U, \text{s.t.} \\ \xi(x) + \eta\left(\left(\xi(x) - \mu^n\right)^{\frac{1}{2}}\right) \geq \mathcal{U}^{u;n^{-1}} V_g^{n^{-1},\delta^{n^{-1}}}(y) + \\ g(y) + \left(V_g^{n^{-1},\delta^{n^{-1}}}(x) - V_g^{n^{-1},\delta^{n^{-1}}}(y)\right) \end{array} \right\}, \quad (22)$$

where $\left(\mu^n, V_g^{n^{-1},\delta^{n^{-1}}}\right) \in D_g^{n^{-1}}(x)$ (defined as in (9) but for the infinitesimal operator $\mathcal{U}^{u;n^{-1}}$), respectively $\Omega_g^{2,n}(x)$ in which $\xi(x)$ is replaced by $v_g^{n^{-1}}$ and the usual (inner ant outer) limit sets

$$\Omega_g^{2,in}(x) := \liminf_{n\to\infty} \Omega_g^{2,n}(x), \ \Omega_g^{2,out,cl}(x) := \bigcap_{n\geq 1} cl\left(\bigcup_{k\geq n} \Omega_g^{2,k}(x)\right).$$

Remark 4.1. A glance at the proof of the main result shows that the value function ξ is the common limit to both $v_g^{n^{-1}}(x)$ and such μ^n.

We get the following result stating that optimal measures can be obtained as limit of optimal measures for the perturbed problems and, then, the support is included in the outer limit. Conversely, if one takes admissible measures whose support is reasonably close to $\Omega_g^{1,n}(x)$, any limit of such sequence provides an optimal measure.

Theorem 4.2. *Let us fix some* $\left(y_1^0, y_2^0\right) \in \mathbb{K}_1 \times \mathbb{K}_2$.

i. If $\gamma_n \in \Theta^{\frac{1}{n}}\left(y_1^0, y_2^0\right)$ *is optimal, then any limit* γ *is optimal and* $\gamma\left(\Omega_g^{2,out,cl}\right) = 1$.

ii. If $\gamma_n \in \Theta^{\frac{1}{n}}\left(y_1^0, y_2^0\right)$ *is such that* $\gamma_n\left(\left(\Omega_g^{1,n}(x)\right)^c\right) = o\left(\frac{\delta^{n^{-1}}}{n}\right)$, *then any limit of* γ_n *is optimal.*

Proof. For the first assertion, since $\gamma_n \in \Theta^{\frac{1}{n}}\left(y_1^0, y_2^0\right)$ is optimal, one has

$$\begin{aligned} v_g^{n^{-1}}\left(y_1^0, y_2^0\right) &= \int_{\mathbb{R}^{N+M}\times U} \left[\mathcal{U}^{u;n^{-1}} V_g^{n^{-1},\delta^{n^{-1}}}(y) + g(y)\right. \\ &+ \left.\left(V_g^{n^{-1},\delta^{n^{-1}}}(x) - V_g^{n^{-1},\delta^{n^{-1}}}(y)\right)\right] \gamma_n(dydu) \geq \mu^n \gamma_n\left(\Omega_g^{2,n}(x)\right) \\ &+ \left[v_g^{n^{-1}}(x) + \eta\left(\left(v_g^{n^{-1}}(x) - \mu^n\right)^{\frac{1}{2}}\right)\right] \gamma_n\left(\left(\Omega_g^{2,n}(x)\right)^c\right). \end{aligned}$$

As in the classical case, one infers that

$$\gamma_n\left(\Omega_g^{2,n}(x)\right) \geq \frac{1}{1 + \frac{v_g^{n^{-1}}(x) - \mu^n}{\eta\left(\left(v_g^{n^{-1}}(x) - \mu^n\right)^{\frac{1}{2}}\right)}}.$$

Therefore, for every $n \geq n_0 \geq 1$, one has

$$\gamma_n\left(cl\left(\underset{k\geq n_0}{\cup}\Omega_g^{2,k}(x)\right)\right) \geq \frac{1}{1+\frac{v_g^{n^{-1}}(x)-\mu^n}{\eta\left(\left(v_g^{n^{-1}}(x)-\mu^n\right)^{\frac{1}{2}}\right)}}$$

and the conclusion follows by letting $n \to \infty$ (and recalling Remark 4.1).

For the second assertion, one begins with noting that, due to (19), the following inequatity holds $\mathcal{U}^{u;n^{-1}}V_g^{n^{-1},\delta^{n^{-1}}}(y) \leq \frac{c_0(1+n)}{\delta^{n^{-1}}}$. Consequently,

$$\begin{aligned}
\xi\left(y_1^0\right) \leq \lim_{n\to\infty}\int_{\mathbb{R}^{N+M}\times U}\Big[\mathcal{U}^{u;n^{-1}}V_g^{n^{-1},\delta^{n^{-1}}}(y)+g(y)\\
+\left(V_g^{n^{-1},\delta^{n^{-1}}}(x)-V_g^{n^{-1},\delta^{n^{-1}}}(y)\right)\Big]\gamma_n(dydu) \leq \lim_{n\to\infty}\Big[\xi\left(y_1^0\right)\\
+\eta\left(\left(\xi\left(y_1^0\right)-\mu^n\right)^{\frac{1}{2}}\right)+\left(\frac{c_0(1+n)}{\delta^{n^{-1}}}+3\left\|g\right\|_\infty\right)\gamma_n\left(\left(\Omega_g^{1,n}(x)\right)^c\right)\Big]
\end{aligned}$$

and the conclusion follows. □

Acknowledgement

The work of the first author has been partially supported by the French National Research Agency project PIECE, number ANR-12-JS01-0006.

References

1. N. V. Krylov, On the rate of convergence of finite-difference approximations for Bellman's equations with variable coefficients, *Probab. Theory Related Fields* **117**, 1 (2000).
2. V. Gaitsgory, Suboptimization of singularly perturbed control systems, *SIAM J. Control and Optim.* **30**, 1228 (1992).
3. G. Yin and Q. Zhang, Control of dynamic systems under the influence of singularly perturbed markov chains, *Journal of Mathematical Analysis and Applications* **216**, 343 (1997).
4. E. Fridman, Exact slow-fast decomposition of a class of non-linear singularly perturbed optimal control problems via invariant manifolds, *International Journal of Control* **72**, 1609 (1999).
5. D. Naidu, Singular perturbations and time scales in control theory and applications: an overview, *Dynamics of Continuous, Discrete and Impulsive Systems, Series B* , 233 (2002).
6. O. L. V. Costa and F. Dufour, Singular perturbation for the discounted continuous control of piecewise deterministic markov processes, *Applied Mathematics & Optimization* **63**, 357 (2010).

7. O. Serea, Characterization of the optimal trajectories for the averaged dynamics associated to singularly perturbed control systems, *accepted in Journal of Differential Equations* (2014).
8. D. Goreac, Asymptotic control for a class of piecewise deterministic markov processes associated to temperate viruses, *SIAM Journal on Control and Optimization* **53**, 1860 (2015).
9. V. Borkar and V. Gaitsgory, Averaging of singularly perturbed controlled stochastic differential equations, *Appl. Math. Optimization* **56**, 169 (2007).
10. M. H. A. Davis, Piecewise-deterministic Markov-processes - A general-class of non-diffusion stochastic-models, *Journal of the Royal Statistical Society Series B-Methodological* **46**, 353 (1984).
11. M. H. A. Davis, Control of Piecewise-deterministic processes via discrete-time dynamic-programming, *Lect. Notes Control Inf. Sci.* **78**, 140 (1986).
12. M. H. A. Davis, *Markov models and optimization*, Monographs on Statistics and Applied Probability, Vol. 49 (Chapman & Hall, London, 1993).
13. D. Goreac, Viability, Invariance and Reachability for Controlled Piecewise Deterministic Markov Processes Associated to Gene Networks, *ESAIM-Control Optimisation and Calculus of Variations* **18**, p. 401 (APR 2012).
14. D. Goreac and O.-S. Serea, Linearization Techniques for Controlled Piecewise Deterministic Markov Processes; Application to Zubov's Method, *Applied Mathematics and Optimization* **66**, 209 (2012).
15. V. Borkar and V. Gaitsgory, On existence of limit occupational measures set of a controlled stochastic differential equation, *SIAM J. Control Optim.* **44**, 1436 (2005).

Inverse problems for control theory

Mohammed Al Horani

Department of Mathematics,
The University of Jordan (Sabbatical Year),
Amman, 11942, Jordan
and Department of Mathematics, Hail University
Hail, Saudi Arabia
E-mail: horani@ju.edu.jo

Angelo Favini

Dipartimento di Matematica, Universita di Bologna,
Bologna, 40126, Italy
E-mail: favini@dm.unibo.it

We are concerned with some inverse problems for abstract differential equations. Their motivation comes from relevant equations of interest in the control theory. Various concrete examples are described.

Keywords: Inverse problem; Hilbert space; Linear differential problem; First-Order differential equations; Second-Order differential equations.

1. Introduction

Given a Hilbert space Y, many abstract differential problems of great interest in control theory take the form

$$\frac{dy}{dt} = Ay(t) + f(t)B\,z\,, \qquad 0 \le t \le \tau\,,$$
$$y(0) = y_0 \in D(A)\,,$$

where A is a generator of an analytic semigroup in Y, the operator B acts linearly from another Hilbert space U into $[D(A^*)]'$, the dual space of $D(A^*)$ with respect to the $Y-$norm, i.e.,

$$\|y\|_{[D(A^*)]'} := \|A^{-1}y\|_Y,$$

and z is a fixed element in U. It is indeed assumed, without loss of generality, that $A^{-1} \in \mathcal{L}(Y)$. Moreover, it happens that, in many important situations, for some fixed $\gamma \in (0,1)$, $(-A)^{-\gamma}B \in \mathcal{L}(U,Y)$, see the monograph[1].

One of the main aims of this paper is to show that under suitable regularity assumptions on $g \in C([0,\tau];\mathbb{C})$, $y_0 \in D(A)$, the inverse problem consisting of finding a pair $(y,f) \in C([0,\tau];Y) \times C([0,\tau];\mathbb{C})$ such that

$$A^{-1}\frac{dy}{dt} = y(t) + f(t)A^{-1}Bz, \quad 0 \le t \le \tau, \tag{1}$$

$$y(0) = y_0, \tag{2}$$

$$\Phi[A^{-1}y(t)] = g(t), \quad 0 \le t \le \tau, \tag{3}$$

where $\Phi \in Y^*$, admits a unique strict solution provided that $\Phi[A^{-1}Bz] \ne 0$. Notice that (1) replaces equivalently the first evolution equation introduced above. We observe that our main results follow from general identification theorems in[6].

In Section 2 we recall some preliminaries. In Section 3 we handle some related first-order equations. In Section 4 we face the case of second-order equations. In Section 5 we furnish some applications to partial differential equations of interest in control theory. We refer to[2] for more details, examples and applications, in particular, to second-order in time equations. See also[3–5].

2. Some Lemmas

Here we recall a general identification result recently described in the paper[6], cfr. Theorem 5.1. Take two closed linear operators L, M, in the Banach space X such that $0 \in \rho(L)$, $D(L) \subseteq D(M)$ and

$$\|M(\lambda M - L)^{-1}\|_{\mathcal{L}(X)} \le c(1+|\lambda|)^{-\beta}, \tag{4}$$

for all $\lambda \in \sum_\alpha := \{z \in \mathbb{C}; \mathrm{Re} z \ge -c(1+|\mathrm{Im} z|)^\alpha, \quad c > 0, \ 0 < \beta \le \alpha \le 1\}$.

We have

Lemma 2.1. *Let L, M be two closed linear operators on X satisfying (4). Assume that $2\alpha + \beta > 2$, $3 - 2\alpha - \beta < \theta < 1$. Let $y_0 \in D(L)$ and suppose $y_1 \in D(L)$ such that*

$$My_1 = Ly_0 + z\big[g'(0) - \Phi[Ly_0] - \Phi[h(0)]\big]/\Phi[z] + h(0),$$

with z a fixed element in $(X, D(A))_{\theta,\infty}$, $A = LM^{-1}$, $D(A) = M(D(L))$, $\Phi \in X^*$, $\Phi[My_0] = g(0)$, $g \in C^2([0,\tau];\mathbb{C})$, $h \in C^1([0,\tau];X)$, $h' \in B\big([0,\tau];(X,D(A))_{\theta,\infty}\big)$. Then the identification problem

$$My'(t) = Ly(t) + f(t)z + h(t), \quad 0 \le t \le \tau, \tag{5}$$

$$y(0) = y_0 \tag{6}$$

$$\Phi[My(t)] = g(t), \quad 0 \le t \le \tau, \tag{7}$$

admits a unique strict solution $(y, f) \in C^1([0,\tau]; D(L)) \times C^1([0,\tau];\mathbb{C})$ *such that*

$$(My')' - f'(.)z - h'(.) \in C^{(2\alpha+\beta-3+\theta)/\alpha}([0,\tau];X) \cap B\left([0,\tau]; X_A^{(2\alpha+\beta-3+\theta)/\alpha}\right).$$

Here $(X, D(A))_{\theta,\infty}$ denotes the well known real interpolation space between X and $D(A)$(endowed with the graph norm) and

$$X_A^\theta = \left\{ u \in X;\ \sup_{t>0} t^\theta \left\| A^0 \left(t - A\right)^{-1} u \right\|_X < \infty \right\},$$

where A^0 is a symbol defined by

$$A^0 \left(t - A\right)^{-1} = -I + t(t - A)^{-1},$$

so that $A^0(t-A)^{-1}$ coincides with $A(t-A)^{-1}$ when A is single-valued. For more details, we refer to the monograph[7] by Favini and Yagi, pp. 48-51.

A weaker formulation of problem (5)-(7) reads

$$\frac{d}{dt}\big(My(t)\big) = Ly(t) + f(t)z + h(t), \quad 0 \le t \le \tau, \tag{8}$$

$$(My)(0) = My_0 \tag{9}$$

$$\Phi[My(t)] = g(t), \quad 0 \le t \le \tau, \tag{10}$$

which is described by Favini and Tanabe in[6], Theorem 6.3. We write it in the form of lemma as follows.

Lemma 2.2. *Suppose* $2\alpha + \beta + \theta > 3$, $2\alpha + \beta > 2$, $y_0 \in D(L)$, $Ly_0 \in (X, D(A))_{\theta,\infty}$, $A = LM^{-1}$, $z \in (X, D(A))_{\theta,\infty}$, $h \in C([0,\tau]; X) \cap B\big([0,\tau]; (X, D(A))_{\theta,\infty}\big)$, $g \in C^1([0,\tau];\mathbb{C})$, $\Phi \in X^*$, $\Phi[My_0] = g(0)$, $\Phi[z] \neq 0$. *Then problem (8)-(10) admits a unique solution* (y, f) *such that*

$$My \in C^1([0,\tau]; X), \quad f \in C([0,\tau];\mathbb{C}),$$

$$Ly \in C^{(2\alpha+\beta-3+\theta)/\alpha}\big([0,\tau]; X\big) \cap B\left([0,\tau]; X_A^{(2\alpha+\beta-3+\theta)/\alpha}\right).$$

Lemma 2.1 and Lemma 2.2 concern in some sense a maximal space regularity of solutions. We could consider a maximal time regularity by using very recent results given by Favaron and Favini in[8] and Favaron, Favini and Tanabe in[9].

3. First-Order Case

In order to solve problem (1)-(3), we shall assume that $-A$ admits bounded imaginary powers. This hypothesis guarantees that

$$(Y, D(A))_{\theta,2} = [Y, D(A)]_\theta = D\big((-A)^\theta\big) \subseteq (Y, D(A))_{\theta,\infty},$$

where $[Y, D(A)]_\theta$ denotes the complex interpolation space of index θ between Y and $D(A)$, see[10]. Recall also that A is assumed to generate an analytic semigroup in Y. Then Lemma 2.1 yields easily the result as follows.

Theorem 3.1. *Suppose* $(-A)^{\theta-1}B \in \mathcal{L}(U,Y)$, $\Phi[A^{-1}y_0] = g(0)$, $A^{-1}Bz \in (Y, D(A))_{\theta,\infty}$, $0 < \theta < 1$, $y_0 + A^{-1}Bz\big(g'(0) - \Phi[y_0] - \Phi[A^{-1}h(0)]\big)/\Phi[A^{-1}Bz] \in D(A)$, $g \in C^2([0,\tau];\mathbb{C})$, $h \in C^1([0,\tau];Y)$. *Then problem (1)-(3) admits a unique solution* $(y, f) \in C^1([0,\tau];Y) \times C^1([0,\tau];\mathbb{C})$ *such that* $\big(A^{-1}y'\big)' - f'(.)A^{-1}Bz - A^{-1}h' \in C^\theta([0,\tau];Y) \cap B\big([0,\tau];(Y, D(A))_{\theta,\infty}\big)$.

Proof. It is an immediate consequence of Lemma 2.1 taking into account $M = A^{-1}$, $L = I$. □

Notice that the strong assumption is related to

$$y_0 + A^{-1}Bz\big(g'(0) - \Phi[y_0] - \Phi[A^{-1}h(0)]\big)/\Phi[A^{-1}Bz].$$

Observe also that $A^{-1}Bz = -(-A)^{-\theta}(-A)^{\theta-1}Bz$ already belongs to $D\big((-A)^\theta\big)$.

Of course, if we confine ourselves to "generalized" solutions, i.e., solutions to

$$\frac{d}{dt}\big(A^{-1}y(t)\big) = y(t) + f(t)A^{-1}Bz + A^{-1}h(t), \quad 0 \le t \le \tau, \tag{11}$$

$$\big(A^{-1}y\big)(0) = A^{-1}y_0, \tag{12}$$

$$\Phi[A^{-1}y(t)] = g(t), \quad 0 \le t \le \tau, \tag{13}$$

then Lemma 2.2 applies immediately.

Theorem 3.2. *Let* A *be the generator of an analytic semigroup in the Hilbert space* Y, $-A$ *having bounded imaginary powers. Assume* $(-A)^{\theta-1}B \in \mathcal{L}(U,Y)$, $0 < \theta < 1$, $\Phi \in Y^*$, $\Phi[A^{-1}y_0] = g(0)$, $g \in C^1([0,\tau];\mathbb{C})$, $A^{-1}Bz \in (Y, D(A))_{\theta,\infty}$, $\Phi[A^{-1}Bz] \neq 0$, $y_0 \in D(A)$, $h \in C([0,\tau];Y)$. *Then problem (11)-(13) admits a unique strict solution* (y, f) *such that* $A^{-1}y \in C^1([0,\tau];Y)$, $f \in C([0,\tau];\mathbb{C})$, $y \in C^\theta([0,\tau];Y) \cap B\big([0,\tau];(Y, D(A))_{\theta,\infty}\big)$.

Remark 3.1. The problem above is immediately reduced to a classical inverse problem by means of the change of variable $A^{-1}y = x$, so that it becomes

$$\begin{aligned} &x'(t) = Ax(t) + f(t)A^{-1}Bz + A^{-1}h(t), \quad 0 \le t \le \tau, \\ &x(0) = A^{-1}y_0, \\ &\Phi\left[x(t)\right] = g(t) \quad 0 \le t \le \tau. \end{aligned}$$

Remark 3.2. Consider the Cauchy problem

$$\text{(P)} \qquad \begin{cases} y'(t) = Ay(t) + f(t)Bz, & t \in [0,\tau], \\ \qquad\qquad y(0) = y_0 \end{cases}$$

under the assumption that $(-A)^{\theta-1}B \in \mathcal{L}(U, Y)$. Its solution, if existing, is necessarily given by

$$\begin{aligned} y(t) &= e^{tA}y_0 + \int_0^t e^{(t-s)A}Bzf(s)\,ds \\ &= e^{tA}y_0 + \int_0^t (-A)^{1-\theta}e^{(t-s)A}(-A)^{\theta-1}Bzf(s)\,ds. \end{aligned}$$

Taking into account the well known results on the complex interpolation spaces

$$\left[D\big((-A)^{\alpha}\big), D\big((-A)^{\beta}\big)\right]_{\theta}, \quad 0 < \theta < 1, \quad \alpha,\, \beta > 0,$$

cfr.[10], it is easy to verify that

$$\|(-A)^{1-\theta}e^{tA}\|_{\mathcal{L}(Y)} \le c\,t^{\theta-1},$$

and thus the above integral is well defined, provided that f is continuous. However, under suitable regularity conditions on the data (1), $(-A)^{\theta-1}y$ satisfies

$$\frac{d}{dt}\big((-A)^{\theta-1}y(t)\big) = A(-A)^{\theta-1}y(t) + f(t)(-A)^{\theta-1}Bz,$$

i.e., the equation obtained by applying $(-A)^{\theta-1}$ to both sides of the equation

$$\frac{dy}{dt} = Ay(t) + f(t)B\,z.$$

This is a motivation to handle the original inverse problem in the new one of finding the pair $(y, f) \in C([0,\tau]; Y) \times C([0,\tau]; \mathbb{C})$ such that

$$\begin{aligned}
&\frac{d}{dt}\left((-A)^{\theta-1}y(t)\right) = A(-A)^{\theta-1}y(t) + f(t)(-A)^{\theta-1}Bz, \quad 0 \le t \le \tau,\\
&(-A)^{\theta-1}y(0) = (-A)^{\theta-1}y_0,\\
&\Phi\left[(-A)^{\theta-1}y(t)\right] = g(t), \quad 0 \le t \le \tau.
\end{aligned}$$

Such a system, after the change of variable $(-A)^{\theta-1}y(t) = x(t)$, is reduced to the regular inverse problem

$$x'(t) = Ax(t) + f(t)(-A)^{\theta-1}Bz, \quad 0 \le t \le \tau, \tag{14}$$

$$x(0) = (-A)^{\theta-1}y_0, \tag{15}$$

$$\Phi\left[x(t)\right] = g(t), \quad 0 \le t \le \tau. \tag{16}$$

By using[11], if $(-A)^{\theta-1}Bz \in (Y, D(A))_{\delta,\infty}$, $\delta \in (0,1)$, (in particular, $(-A)^{\delta+\theta-1}Bz \in Y$), $(-A)(-A)^{\theta-1}y_0 = (-A)^{\theta}y_0 \in D\big((-A)^{\delta}\big)$, $\Phi[(-A)^{\theta-1}Bz] \neq 0$, $g \in C^{1+\delta}([0,\tau]; \mathbb{C})$, $\Phi[(-A)^{\theta-1}y_0] = g(0)$, then problem (14)-(16) admits a unique solution (x, f) such that x', $Ax \in C^{\delta}([0,\tau]; Y)$, $f \in C^{\delta}([0,\tau]; \mathbb{C})$, $x' \in B\big([0,\tau]; (Y, D(A))_{\delta,\infty}\big)$.

Of course, these properties translate into $\big((-A)^{\theta-1}y\big)'$, $(-A)^{\theta}y \in C^{\delta}([0,\tau]; Y)$, $\big((-A)^{\theta-1}y\big)' \in B\big([0,\tau]; (Y, D(A))_{\delta,\infty}\big)$.

Remark 3.3. As already observed, the problem

$$\begin{aligned}
&\frac{dy}{dt} = Ay(t) + f(t)Bz + h(t), \quad t \in [0,\tau],\\
&y(0) = y_0
\end{aligned}$$

admits an integral solution

$$\begin{aligned}
y(t) &= e^{tA}y_0 + \int_0^t e^{(t-s)A}Bzf(s)\,ds + \int_0^t e^{(t-s)A}h(s)\,ds\\
&= e^{tA}y_0 + \int_0^t e^{(t-s)A}(-A)^{1-\theta}(-A)^{\theta-1}Bzf(s)\,ds + \int_0^t e^{(t-s)A}h(s)\,ds\\
&= e^{tA}y_0 + \int_0^t (-A)^{1-\theta}e^{(t-s)A}(-A)^{\theta-1}Bzf(s)\,ds + \int_0^t e^{(t-s)A}h(s)\,ds.
\end{aligned}$$

Apply formally A to both sides of the above equation. We get

$$Ay(t) = e^{tA}Ay_0 - \int_0^t (-A)^{2-\theta}e^{(t-s)A}(-A)^{\theta-1}Bzf(s)\,ds + \int_0^t Ae^{(t-s)A}h(s)\,ds.$$

Using the estimate in[8], Proposition 12, with $\alpha = \beta = 1$

$$\|(-A)^{1+\delta}e^{tA}\|_{\mathcal{L}((Y,D(A))_{\gamma,\infty},Y)} \le ct^{\gamma-\delta-1}, \quad t > 0,$$

we conclude that

$$\|(-A)^{2-\theta}e^{(t-s)A}(-A)^{\theta-1}Bzf(s)\|_Y \le c(t-s)^{\gamma-2+\theta}\, \|(-A)^{\theta-1}Bz\|_{(Y,D(A))_{\gamma,\infty}}$$

provided that $f \in C([0,\tau];\mathbb{C})$.

If $1 < \gamma + \theta$, the integral

$$\int_0^t (-A)^{2-\theta}e^{(t-s)A}(-A)^{\theta-1}Bzf(s)\,ds$$

converges provided that $(-A)^{\theta-1}Bz \in (Y, D(A))_{\gamma,\infty}$ and this is a request of large space regularity for $(-A)^{\theta-1}Bz$. Notice that we could require $(-A)^{\theta-1}Bz \in D\left((-A)^{\gamma}\right), \ \gamma > 1-\theta$, but this is a very restrictive assumption.

The integral term

$$\int_0^t Ae^{(t-s)A}h(s)ds$$

is well defined if $h \in B\big([0,\tau];(Y,D(A))_{\theta,\infty}\big)$ and it does not deserve any problem.

4. Second-Order Problem

The second-order equation of concern is

$$y'' = B_1y' + Ay + f(t)Bz, \quad 0 \le t \le \tau,$$

with suitable initial conditions on y and y'. We can transform such an equation into a system in a convenient product space $W \times Y$, where $D(A) \subseteq W \subseteq Y$

$$\frac{d}{dt}\begin{bmatrix} y \\ x \end{bmatrix} = \begin{bmatrix} 0 & I \\ A & B_1 \end{bmatrix}\begin{bmatrix} y \\ x \end{bmatrix} + f(t)\begin{bmatrix} 0 \\ Bz \end{bmatrix}.$$

It is assumed here that

$$\mathcal{a} = \begin{bmatrix} 0 & I \\ A & B_1 \end{bmatrix}, \quad D(\mathcal{a}) = D(A) \times D(B_1)$$

generates an analytic semigroup in $W \times Y$, so that all previous results apply. Observe that formally

$$\mathcal{a}^{-1} = \begin{bmatrix} -A^{-1}B_1 & A^{-1} \\ I & 0 \end{bmatrix}.$$

Moreover, suppose $A^{-1}B_1$ extends to an operator from Y to Y or A and B_1 commute in the resolvent sense. Then we can consider

$$\begin{bmatrix} -A^{-1}B_1 & A^{-1} \\ I & 0 \end{bmatrix} \frac{d}{dt} \begin{bmatrix} y \\ x \end{bmatrix} = \begin{bmatrix} y \\ x \end{bmatrix} + f(t) \begin{bmatrix} A^{-1}Bz \\ 0 \end{bmatrix}, \quad 0 \le t \le \tau,$$

$y(0) = y_0, \quad x(0) = y_1,$

$$\Psi\left(\begin{bmatrix} -A^{-1}B_1 y + A^{-1}x \\ y \end{bmatrix}\right) = \Phi\left[-A^{-1}B_1 y + A^{-1}x\right] + \Phi\left[A^{-1}B_1 y\right]$$
$$= \Phi\left[A^{-1}x\right] = g'(t).$$

Here $\Phi \in Y^*$ is given.

Notice that the weaker problem

$$\frac{d}{dt}\begin{bmatrix} -A^{-1}B_1 & A^{-1} \\ I & 0 \end{bmatrix} \begin{bmatrix} y \\ x \end{bmatrix} = \begin{bmatrix} y \\ x \end{bmatrix} + f(t) \begin{bmatrix} A^{-1}Bz \\ 0 \end{bmatrix}, \quad 0 \le t \le \tau,$$

after the change of variable, $\xi = -A^{-1}B_1 y + A^{-1}x, \quad \eta = y$, transforms into

$$\frac{d}{dt}\begin{bmatrix} \xi \\ \eta \end{bmatrix} = \begin{bmatrix} 0 & I \\ A & B_1 \end{bmatrix} \begin{bmatrix} \xi \\ \eta \end{bmatrix} + f(t) \begin{bmatrix} A^{-1}Bz \\ 0 \end{bmatrix},$$

$$\Psi\big((\xi,\eta)\big) = \Phi[\xi] + \Phi[A^{-1}B_1\eta] = g'(t), \quad 0 \le t \le \tau,$$

which is immediately solvable by Lemma 2.1 provided that $\Phi[A^{-1}Bz] \neq 0$.

As it is easily seen, this approach applies to a number of systems from control problems, see Lasiecka-Triggiani[1].

5. Examples and Applications

Here we give a number of applications, confining for sake of brevity to a few of them, see AL Horani-Favini[2].

Example 5.1 (Heat Equation with Dirichlet Boundary Conditions). Consider the initial boundary value problem in a bounded domain Ω in $\mathbb{R}^n$, $n \ge 1$, with a smooth boundary $\partial\Omega$

$$\frac{\partial y}{\partial t} = \Delta y + c^2 y, \quad \text{in } (0,\tau) \times \Omega,$$
$$y(0,\cdot) = y_0 \quad \text{in } \Omega,$$
$$y|_\Sigma = f(t)u(\cdot) \quad \text{in } (0,\tau) \times \partial\Omega \equiv \Sigma.$$

We want to determine (y, f) from the additional information

$$\Phi[y(t,.)] = \int_\Omega \eta(x) y(t,x)\,dx = g(t), \quad 0 \le t \le \tau,$$

where $\eta \in L^2(\Omega)$ is known together with $g \in C([0,\tau];\mathbb{C})$. It is seen in[1], p. 181, that the differential problem can be written in the abstract form

$$\frac{\partial y}{\partial t} = Ay(t) + f(t)Bu, \quad 0 \le t \le \tau, \tag{17}$$

$$y(0,\cdot) = y_0, \tag{18}$$

where A is the Laplacian in Ω with $D(A) = H^2(\Omega) \cap H_0^1(\Omega)$ and

$$Bu = -ADu, \quad B : L^2(\partial\Omega) \to [D(A)]', \tag{19}$$

D being the Dirichlet map defined by

$$h = Dg \ \text{ if and only if } (\Delta + c^2)h = 0 \text{ in } \Omega, \ \text{ and } h|_{\partial\Omega} = g.$$

A is the isomorphic extension $L^2(\Omega) \to [D(A)]'$ of the operator A introduced previously. It is known that A generates an analytic semigroup in $L^2(\Omega) = Y$. Moreover,

$$(-A+\omega)^{-\gamma}B \in \mathcal{L}\big(L^2(\partial\Omega), L^2(\Omega)\big)$$

provided that $\gamma = 3/4 + \epsilon, \ \forall \epsilon > 0$. Therefore, our results apply provided that the additional information reads

$$\Phi[(-A)^{-1}y(t)] = g(t),$$

so that the basic assumption reads

$$\Phi[(-A)^{-1}Bu] \ne 0.$$

Correspondingly, we can handle a heat equation with Neumann boundary conditions, see[1], Section 3.3, pp. 194-204.

Example 5.2 (The Strongly Damped Wave Equation). Consider, for $\rho > 0$, the wave equation with a strong degree damping

$$\begin{aligned}
&\frac{\partial^2 w}{\partial t^2} - \Delta w - \rho\Delta\frac{\partial w}{\partial t} = f(t)\delta(x-x^0)u, \quad \text{in } (0,\tau)\times\Omega,\\
&w(0,\cdot) = w_0, \quad \frac{\partial w}{\partial t}(0,\cdot) = w_1 \quad \text{in } \Omega,\\
&w|_\Sigma \equiv 0 \quad \text{in } (0,\tau]\times\partial\Omega,\\
&\int_\Omega h(x)\Delta^{-1}w(t,x)\,dx = g(t), \quad 0 \le t \le \tau,
\end{aligned}$$

where $x^0 \in \Omega, \quad h \in L^2(\Omega)$.

Such a problem can be written in the abstract form

$$\frac{\partial^2 w}{\partial t^2} + \mathcal{A}w + \rho\mathcal{A}\frac{\partial w}{\partial t} = f(t)\delta(x-x^0)u,$$

where

$$\mathcal{A} = -\Delta, \quad D(\mathcal{A}) = H^2(\Omega) \cap H_0^1(\Omega).$$

By choosing

$$Y = D(\mathcal{A}^{1/2}) \times L^2(\Omega), \quad U = \mathbb{R}, \quad A = \begin{bmatrix} 0 & I \\ -\mathcal{A} & -\rho\mathcal{A} \end{bmatrix},$$

one recognizes that A generates an analytic semigroup in Y, but only for $n = 1$ we have $(-A)^{-1/2}Bu \in Y$.

However, if the ambient space is changed to $L^2(\Omega) \times L^2(\Omega)$, then, see[1], p. 218, $(-A)^{-\theta}B \in \mathcal{L}(U; Y)$ for $\theta = 3/4 + \epsilon, \ \epsilon > 0$.

References

1. I. Lasiecka and R. Triggiani, *Control theory for partial differential equations: continuous and approximation theories. I. Abstract parabolic systems*, 1st. edn. (Cambridge University Press, Cambridge, 2000).
2. M. Al Horani, A. Favini, and H. Tanabe, *Inverse Problems for Parabolic Differential Equations and Control Theory*, to appear
3. M. Al Horani and A. Favini, *An Identification Problem for First-Order Degenerate Differential Equations*, J. Optim. Theory. Appl. **130**, 41 (2006).
4. M. Al Horani and A. Favini, *Degenerate First-Order Inverse Problems In Banach Spaces*, Nonlinear Analysis, Theory, Methods, and Applications, **75**, 68 (2012).
5. M. Al Horani and A. Favini, *Degenerate First-Order Identification Problems in Banach Spaces*, in Differential Equations Inverse and Direct Problems, eds. A. Favini, A. Lorenzi, Pure and Applied Mathematics, Vol. 251 (Taylor and Francis Group, Boca Raton, 2006), pp. 1–15.
6. A. Favini and H. Tanabe, *Degenerate Differential Equations and Inverse Problems*, in Prod. Partial Differential Equations, eds. A. Yagi, Y. Yamamoto, (Osaka, Japan, 2013), pp. 89-100.
7. A. Favini, A. Yagi, *Degenerate Differential Equations in Banach Spaces*, 1st edn. (Marcel Dekker. Inc. New York, 1999).
8. A. Favaron and A. Favini, *On the Behaviour of Singular Semigroups in Intermediate and Interpolation Spaces and Its Applications to Maximal Regularity for Degenerate Integro-Differential Evolution Equations*, Abstract and Applied Analysis, 2013, 1 (2013).

9. A. Favaron, A. Favini and H. Tanabe, *Perturbation Methods For Inverse Problems in Degenerate Differential Equations*, to appear.
10. H. Triebel, *Interpolation Theory, Function Spaces, Differential Operators*, 2nd. edn. (North-Holland, Amesterdam, 1978).
11. M. Al Horani and A. Favini, *Perturbation Method for First and Complete Second Order Differential Equations*, J. Optim. Theory. Appl., 166, 949 (2015).

On the ill-posedness of active scalar equations with odd singular kernels

Igor Kukavica

Department of Mathematics, University of Southern California,
Los Angeles, CA 90089, USA
E-mail: kukavica@usc.edu

Vlad Vicol

Department of Mathematics, Princeton University, Princeton, NJ 08544, USA
E-mail: vvicol@math.princeton.edu

Fei Wang

Department of Mathematics, University of Southern California,
Los Angeles, CA 90089, USA
E-mail: wang828@usc.edu

We consider active scalar equations with constitutive laws that are *odd* and *very singular*, in the sense that the velocity field loses more than one derivative with respect to the active scalar. We provide an example of such a constitutive law for which the equation is ill-posed: Either Sobolev solutions do not exist, from Gevrey-class datum, or the solution map fails to be Lipschitz continuous in the topology of a Sobolev space, with respect to Gevrey class perturbations in the initial datum.

Keywords: Active scalar equations, ill-posedness, stability, Gevrey-class.

1. Introduction

In this paper we address the well-posedness of the Cauchy problem for active scalar equations

$$\partial_t \theta + u \cdot \nabla \theta = 0 \tag{1}$$

$$\nabla \cdot u = 0 \tag{2}$$

$$\theta(x,0) = \theta_0(x) \tag{3}$$

posed on $\mathbb{T}^2 \times [0,\infty) = [-\pi,\pi]^2 \times [0,\infty)$ with a certain constitutive law for the incompressible drift

$$u = T\theta,$$

to be specified precisely below (cf. (9) below). The datum θ_0 and the solution θ are taken to have zero mean on $\mathbb{T}^2$.

The study of active scalar equations of type (1)–(2) is motivated by several important fluid models: When $T = \nabla^\perp(-\Delta)^{-1}$, the system becomes the vorticity formulation of the 2D Euler equations; the case $T = \nabla^\perp(-\Delta)^{-1/2}$ corresponds to the surface quasi-geostrophic equation[1]; the constitutive law $T = \partial_1\nabla^\perp(-\Delta)^{-1}$ models flow in an incompressible porous medium with Darcy's law[2]; while $T = \nabla^\perp B$, where B is a scalar bounded Fourier multiplier, appears in magneto-geostrophic dynamics[3]. The well-posedness theory of active scalar equations (either inviscid or with fractional dissipation) has attracted considerable attention in the last two decades. We refer the readers to[1–23] and the references therein.

Recently, these equations have been considered with velocity fields determined by a *singular constitutive law* (i.e., the map $T\colon \theta \mapsto u$ is unbounded) which is also *odd* (by which we mean the corresponding Fourier multiplier is an odd function of frequency). In particular, in[24] the authors considered the velocity field given by

$$u = \mathcal{R}^\perp \Lambda^\beta \theta, \tag{4}$$

where $\beta \in (0,1]$, $\Lambda = \sqrt{-\Delta}$, $\mathcal{R}^\perp = \nabla^\perp \Lambda^{-1}$, and showed the local existence and uniqueness of solutions for (1)–(3) with (4), in the Sobolev space H^4. The two key ingredients in their proof were an estimate for the commutator $\|[\partial_i \Lambda^s, g] f\|_{L^2}$ and the identity

$$\int f A f\, g = -\frac{1}{2}\int f[A, g] f \tag{5}$$

which holds for smooth f, g, where $A = \partial_i \Lambda^\beta$ *is an odd operator*. This result was later sharpened in[12] where the local existence was established in H^σ with $\sigma > 2 + \beta$, using an improved commutator estimate.

In contrast when the constitutive law is *even and singular* there are some ill-posedness results available for these equations. In[25,26], the authors established ill-posedness for even singular constitutive laws, i.e., for operators T that have an even and unbounded Fourier multiplier symbol. The main ingredients in these works were a linear ill-posedness result (severe linear instability) in the spirit of[27,28], and a classical linear implies nonlinear ill-posedness argument[29,30].

On the other hand, we are not aware of *any ill-posedness result for (1)–(3) with an odd constitutive law.* This is due to the special cancelation property (5), which technically reduces the order of the constitutive law by one. The main purpose of this paper is to establish an ill-posedness result for (1)–(3) with an odd constitutive law

$$u = T\theta = \mathcal{R}^{\perp}\Lambda^{\beta}M\theta \tag{6}$$

where $1 < \beta \leq 2$ and M is a zero-order, scalar, even Fourier multiplier operator with symbol m. That is, $\widehat{M\theta}(k) = m(k)\hat{\theta}(k)$, where the specific Fourier multiplier symbol m we consider is given by

$$m(k) = k_1^2|k|^{-2}\phi(k) \tag{7}$$

for all $k \in \mathbb{Z}_*^2$, where we define

$$\phi(k) = \begin{cases} -1, & k_2 = \pm 1, \\ 1, & \text{otherwise.} \end{cases} \tag{8}$$

For simplicity of notation, we denote by $\phi(i\nabla)$ the Fourier multiplier operator with symbol $\phi(k)$. We may then rewrite (6)–(8) concisely as

$$u = -\mathcal{R}^{\perp}\mathcal{R}_1^2\phi(i\nabla)\Lambda^{\beta}\theta \tag{9}$$

where we recall that $\beta \in (1, 2]$ throughout this paper.

Our main result (Theorem 2.2 below) is to prove that the active scalar equation (1)–(3) with constitutive law (9) is ill-posed in any Gevrey space G^s with $s > (4-\beta)/((\beta-1)(3-\beta))$. Here, by ill-posedness we mean that:

- either, given an initial datum θ_0 in G^s, there is no local in time Sobolev solution $\theta \in L_t^{\infty}H_x^{\beta/2+1+\delta}$, where $\delta > 0$ is arbitrary;
- or, the solution map $\theta_0 \to \theta$ is not Lipschitz continuous in $H^{\beta/2+1+\delta}$, with respect to G^s perturbations in the initial datum (see Definition 2.2 below).

We note that the condition $\beta > 1$ is strictly necessary for our result to hold. Indeed, for $\beta \in (0, 1]$, by the results in[12,24], we know that for the system (1)–(3) and (9) we have local existence and uniqueness in the Sobolev space H^{σ} with $\sigma > 2 + \beta$. Additionally, using the identity (5) and the aforementioned commutator estimates, one may show that for $\beta \in (0, 1]$ the equations are locally Lipschitz $(H^{\beta/2+1}, H^{\beta+3})$ well-posed, in the sense of Definition 2.2 below. Thus, the requirement $\beta \in (1, 2]$ in this paper is necessary and sufficient.

The proof of our main result is in the spirit of the earlier Lipschitz ill-posedness works[25,26] with the main difference being that as opposed to these works the eigenfunctions are constructed as sums of sines and cosines; this interplay leads to an eigenvalue problem involving a continued fraction with all positive signs (cf. (29) below). Such continued fractions require a different treatment than[25,26] due to non-monotonicity of the corresponding approximation sequence. Finally, we need a way to bound the Sobolev and Gevrey norms ensuring that the L^2 norm of the solutions grow arbitrarily fast in any short period of time.

The paper is organized as follows. In Section 2, we linearize the active scalar equation around a steady solution and state our main linear and non-linear results. In Section 3, we give the proof of the ill-posedness in Sobolev spaces and Gevrey classes for the linear equation. Finally, in Section 4, we conclude the ill-posedness for the non-linear problem in the corresponding Sobolev and Gevrey spaces using a perturbation argument.

2. The linearized problem and main results

Consider a scalar function of the form $\bar{\theta}(x,t) = \bar{\theta}(x_2)$, where $\bar{\theta}(x_2)$ is a real function in $\mathbb{T}$ with zero mean. According to (9) we have that $\bar{u} = T\bar{\theta} = 0$, in view of the differentiation with respect to x_1 inherent in T. Thus, any such $\bar{\theta}$ is a steady state of (1). We choose a particular function

$$\bar{\theta}(x_2) = \cos(x_2) \tag{10}$$

and consider the linearization of the nonlinear term in (1) around it. Denote the corresponding linear operator by

$$\begin{aligned} L\theta &= -\bar{u}\cdot\nabla\theta - u\cdot\nabla\bar{\theta} = \mathcal{R}_1^3\phi(i\nabla)\Lambda^\beta\theta\,\partial_2\bar{\theta} \\ &= -\sin(x_2)\,\mathcal{R}_1^3\phi(i\nabla)\Lambda^\beta\theta. \end{aligned} \tag{11}$$

We shall use the method of continued fractions (see e.g.[25–28]) to prove that the operator L has a sequence of eigenvalues whose positive real parts diverge to infinity (cf. Theorem 2.3 below).

In order to state our main results, we recall here the definition of the Gevrey-space G^s (cf.[31]).

Definition 2.1 (Gevrey-space). *A function $\theta \in C^\infty(\mathbb{R}^2)$ belongs to the Gevrey class G^s, where $s \geq 1$, if there exists a positive constant $\tau > 0$, called the Gevrey-class radius, such that the G^s_τ-norm is finite, i.e.,*

$$\|\theta\|^2_{G^s_\tau} = \sum_{k\in\mathbb{Z}^2_*} |\hat{\theta}(k)|^2|k|^4 e^{2\tau|k|^{1/s}} < \infty. \tag{12}$$

With this definition, $G^s_\tau = \{\theta \in C^\infty(\mathbb{T}^2)\colon \|\theta\|_{G^s_\tau} < \infty\}$ is an algebra for $s \geq 1, \tau > 0$, and we have $G^s = \bigcup_{\tau>0} G^s_\tau$.

Next we recall the definition of the Lipschitz well-posedness (cf.[26,32]).

Definition 2.2 (Locally Lipschitz (X, Y) well-posedness). *Let $Y \subseteq X \subseteq H^{\beta/2+1+\delta}$ be Sobolev spaces, where $\delta > 0$ is arbitrary. We say that the Cauchy problem for the active scalar equation (1)–(3) with (6) is* locally Lipschitz (X, Y) well-posed, *if there exist continuous functions $T, K\colon [0,\infty)^2 \to (0,\infty)$, so that for every pair of initial data $\theta^{(1)}(0,\cdot), \theta^{(2)}(0,\cdot) \in Y$ there exist unique solutions $\theta^{(1)}, \theta^{(2)} \in L^\infty(0,T;X)$ of the initial value problem associated to (1)–(3) with (6), such that*

$$\|\theta^{(1)}(t,\cdot) - \theta^{(2)}(t,\cdot)\|_X \leq K\|\theta^{(1)}(0,\cdot) - \theta^{(2)}(0,\cdot)\|_Y \tag{13}$$

for every $t \in [0,T]$, where $T = T(\|\theta^{(1)}(0,\cdot)\|_Y, \|\theta^{(2)}(0,\cdot)\|_Y)$ and $K = K(\|\theta^{(1)}(0,\cdot)\|_Y, \|\theta^{(2)}(0,\cdot)\|_Y)$.

The role of the space $H^{\beta/2+1+\delta}$ in the above definition is to ensure that the ranges of the linear operator L defined in (11), and that of the nonlinear operator in (1), defined as

$$N[\theta] = T\theta \cdot \nabla\theta,$$

lie in L^2. Here, recall that $u = T\theta$ is given by (9). More precisely, by the Sobolev embedding we have

$$\|L\theta\|_{L^2} \leq C\|\theta\|_{H^\beta} \leq C\|\theta\|_{H^{\beta/2+1+\delta}} \tag{14}$$

and

$$\|N[\theta]\|_{L^2} = \|T\theta \cdot \nabla\theta\|_{L^2} \leq C\|\theta\|^2_{H^{\beta/2+1+\delta}} \tag{15}$$

for a sufficiently large constant $C > 0$. Note that we can set $\delta = 0$ if $\beta < 2$. The role of the two-spaces X and Y in Definition 2.2 is to allow the solution to lose regularity, as it is expected due to the derivative losses of order β (from T) and 1 (from ∇) present in the nonlinearity.

The first main result in this paper asserts that singular odd active scalar equations are locally Lipschitz ill-posed in Sobolev spaces.

Theorem 2.1. *Assume $\beta \in (1, 2]$, let $\delta > 0$, and define $r = \beta/2 + 1 + \delta$. Then the system (1)–(3) with (6) is locally Lipschitz (H^r, H^s) ill-posed for any $s > r$.*

In fact, in view of the bound (18), one may obtain a stronger ill-posedness result: The solution is not well-posed in a class of Gevrey spaces.

Theorem 2.2. *Let $\beta \in (1, 2]$, $r = \beta/2 + 1 + \delta$ for some $\delta > 0$, and let $s > (4-\beta)/(\beta-1)(3-\beta)$. Then the system (1)–(3) with (6) is locally Lipschitz (H^r, G^s_τ) ill-posed for $\tau > 0$.*

Both theorems above hold with $\delta = 0$ in the case $\beta < 2$. The main ingredient in the proofs of Theorems 2.1 and 2.2 is the following bound on the eigenvalues of the linearized operator L defined in (11).

Theorem 2.3. *The linearized operator L defined in (11) has a sequence of entire real-analytic eigenfunctions $\{\theta_k\}_{k\geq 1}$ with corresponding eigenvalues $\{\lambda_k\}_{k\geq 1}$, such that*

$$\lambda_k \geq C_0^{-1} k^{\beta-1} \tag{16}$$

for all $k \geq 1$, where $C_0 \geq 10$ is a universal constant. Moreover, we may normalize θ_k such that either one of the following statements holds:

(a) Given $s \geq 0$

$$\|\theta_k\|_{H^s} = 1 \quad \textit{and} \quad \|\theta_k\|_{L^2} \geq C_s^{-1} k^{-s(4-\beta)/(3-\beta)} \tag{17}$$

for all $k \geq 1$, where $C_s \geq 1$ is a constant that depends only on s;

(b) given $s \geq 1$ and $\tau > 0$

$$\|\theta_k\|_{G_\tau^s} = 1 \quad \textit{and} \quad \|\theta_k\|_{L^2} \geq C_{s,\tau}^{-1} \exp\left(-C_{s,\tau} k^{(4-\beta)/s(3-\beta)}\right) \tag{18}$$

for all $k \geq 1$, where $C_{s,\tau} \geq 1$ is a constant that depends only on s, τ.

A standard perturbation argument (see e.g.[29,30]) then implies the ill-posedness (1)–(3) with the constitutive law (9).

3. Proof of linear ill-posedness

Proof of Theorem 2.3. Fix an integer $k \geq 1$. We seek an eigenvalue-eigenfunction pair $(\lambda, \theta) = (\lambda_k, \theta_k)$ for L, with θ oscillating at frequency k with respect to x_1, i.e., we are looking for θ of the form

$$\begin{aligned}\theta(x_1, x_2) = &- c_1 \sin(kx_1)\cos(x_2) + \sin(kx_1) \sum_{\text{odd } n\geq 3} c_n \cos(nx_2) \\ &+ \cos(kx_1) \sum_{\text{even } n\geq 2} c_n \sin(nx_2)\end{aligned} \tag{19}$$

where the real coefficients $\{c_n\}_{n\geq 1}$ are to be determined. Using

$$\begin{aligned}&\phi(i\nabla)\theta(x_1, x_2) \\ &\quad = \sin(kx_1) \sum_{\text{odd } n\geq 1} c_n \cos(nx_2) + \cos(kx_1) \sum_{\text{even } n\geq 2} c_n \sin(nx_2)\end{aligned}$$

we get

$$\begin{aligned}
L\theta = &\, k^3 \cos(kx_1) \sum_{\text{odd } n\geq 1} (n^2+k^2)^{(\beta-3)/2} c_n \cos(nx_2)\sin(x_2) \\
&- k^3 \sin(kx_1) \sum_{\text{even } n\geq 2} (n^2+k^2)^{(\beta-3)/2} c_n \sin(nx_2)\sin(x_2) \\
= &\cos(kx_1) \sum_{\text{odd } n\geq 1} k^3 (n^2+k^2)^{(\beta-3)/2} c_n \\
&\qquad\qquad \times \frac{\sin((n+1)x_2) - \sin((n-1)x_2)}{2} \\
&+ \sin(kx_1) \sum_{\text{even } n\geq 2} k^3 (n^2+k^2)^{(\beta-3)/2} c_n \\
&\qquad\qquad \times \frac{\cos((n+1)x_2) - \cos((n-1)x_2)}{2}.
\end{aligned} \tag{20}$$

We introduce the positive coefficients

$$p_n = p_{n,k} = 2(n^2+k^2)^{(3-\beta)/2} k^{-3} \tag{21}$$

omitting the k dependence for ease of notation. Then for all $n \geq 1$ we have

$$p_n \geq 2n^{3-\beta} k^{-3}, \tag{22}$$

where by assumption $3-\beta \geq 1$. With (21) we simplify (20) as

$$\begin{aligned}
L\theta = &\cos(kx_1) \sum_{\text{even } n\geq 2} \left(\frac{c_{n-1}}{p_{n-1}} - \frac{c_{n+1}}{p_{n+1}} \right) \sin(nx_2) - \frac{c_2}{p_2} \sin(kx_1)\cos(x_2) \\
&+ \sin(kx_1) \sum_{\text{even } n\geq 3} \left(\frac{c_{n-1}}{p_{n-1}} - \frac{c_{n+1}}{p_{n+1}} \right) \cos(nx_2).
\end{aligned} \tag{23}$$

The equation $\lambda\theta = L\theta$ thus becomes

$$\begin{aligned}
&- \lambda c_1 \sin(kx_1)\cos(x_2) + \sin(kx_1) \sum_{\text{odd } n\geq 3} \lambda c_n \cos(nx_2) \\
&\quad + \cos(kx_1) \sum_{\text{even } n\geq 2} \lambda c_n \sin(nx_2) \\
&= \cos(kx_1) \sum_{\text{even } n\geq 2} \left(\frac{c_{n-1}}{p_{n-1}} - \frac{c_{n+1}}{p_{n+1}} \right) \sin(nx_2) - \frac{c_2}{p_2} \sin(kx_1)\cos(x_2) \\
&\quad + \sin(kx_1) \sum_{\text{even } n\geq 3} \left(\frac{c_{n-1}}{p_{n-1}} - \frac{c_{n+1}}{p_{n+1}} \right) \cos(nx_2).
\end{aligned} \tag{24}$$

From (24) we obtain the recurrence relationship

$$\frac{c_{n-1}}{p_{n-1}} - \frac{c_{n+1}}{p_{n+1}} = \lambda c_n, \qquad n \geq 2, \tag{25}$$

$$\frac{c_2}{p_2} = \lambda c_1, \qquad n = 1. \tag{26}$$

Denoting

$$\eta_n = \left(\frac{c_n}{p_n}\right)\left(\frac{c_{n-1}}{p_{n-1}}\right)^{-1}, \qquad n \geq 2$$

the recurrence relation (25)–(26) becomes

$$\frac{1}{\eta_n} - \eta_{n+1} = \lambda p_n, \qquad n \geq 2 \tag{27}$$

$$\eta_2 = \lambda p_1. \tag{28}$$

A real number λ yields a solution of (27)–(28) if and only if it solves the continued fraction equation

$$\lambda p_1 = \cfrac{1}{\lambda p_2 + \cfrac{1}{\lambda p_3 + \cdots}}. \tag{29}$$

Note that here the coefficients $\{p_n\}_{n\geq 1}$ are given by (21), and that λ is the only unknown.

Next we show the existence of a solution $\lambda > 0$ of (29). For this purpose, we need to prove first that the continued fraction on the right side of (29) *converges* (note that since the associated sequence is not monotone, this is not immediately clear), and then establish that it *defines a continuous function of* λ *on* $(0, \infty)$.

For $n \in \{2, 3, \cdots\}$ and $m \in \{0, 1, 2, 3, \cdots\}$, define

$$F_{n,m}(\lambda) = \cfrac{1}{\lambda p_n + \cfrac{1}{\cdots + \frac{1}{\lambda p_{n+m}}}}.$$

It is clear that $F_{n,m}(\lambda) > 0$ for $\lambda \in (0, \infty)$. From the identity

$$F_{n,m}(\lambda) = \frac{1}{\lambda p_n + F_{n+1,m-1}(\lambda)}$$

valid for $n \in \{2,3,4,\cdots\}$ and $m \in \{1,2,3,\cdots\}$, we deduce

$$F_{n,m+1}(\lambda) - F_{n,m}(\lambda) = \frac{1}{\lambda p_n + F_{n+1,m}(\lambda)} - \frac{1}{\lambda p_n + F_{n+1,m-1}(\lambda)} = \frac{F_{n+1,m-1}(\lambda) - F_{n+1,m}(\lambda)}{(\lambda p_n + F_{n+1,m}(\lambda))(\lambda p_n + F_{n+1,m-1}(\lambda))}$$

from where we obtain

$$|F_{n,m+1}(\lambda) - F_{n,m}(\lambda)| \le \frac{1}{(\lambda p_n)^2}|F_{n+1,m}(\lambda) - F_{n+1,m-1}(\lambda)|.$$

Using induction on m and the bound

$$|F_{l,1}(\lambda) - F_{l,0}(\lambda)| = \left|\frac{1}{\lambda p_l + \frac{1}{\lambda p_{l+1}}} - \frac{1}{\lambda p_l}\right| \le \frac{1}{\lambda^3 p_l^2 p_{l+1}}$$

which is valid for all $l \ge 2$, we obtain

$$|F_{n,m+1}(\lambda) - F_{n,m}(\lambda)| \le \frac{1}{(\lambda p_n)^2}\frac{1}{(\lambda p_{n+1})^2}\cdots\frac{1}{(\lambda p_{n+m-1})^2}\frac{1}{\lambda^3 p_{n+m}^2 p_{n+m+1}} = \frac{1}{\lambda^{2m+3} p_{n+m+1}(p_n p_{n+1}\cdots p_{n+m})^2} \le \frac{k^{6m+9}}{2^{2m+3}\lambda^{2m+3}}\left(\frac{(n-1)!^2}{(n+m)!(n+m+1)!}\right)^{3-\beta}$$

where used (22) in the last inequality. Hence,

$$|F_{2,m+l}(\lambda) - F_{2,m}(\lambda)| \le \sum_{j=0}^{l-1} \frac{k^{6m+6j+9}}{2^{2m+2j+3}\lambda^{2m+2j+3}}\left(\frac{1}{(2+m+j)!(3+m+j)!}\right)^{3-\beta}.$$

Since $3-\beta \ge 1$, the sequence $F_{2,m}(\lambda)$ is uniformly Cauchy on every compact subset of $(0,\infty)$. Therefore, as $m \to \infty$ the sequence $F_{2,m}(\lambda)$ converges uniformly on compact subsets to a continuous function

$$F_2(\lambda) = \cfrac{1}{\lambda p_2 + \cfrac{1}{\lambda p_3 + \cdots}}.$$

Since

$$F_2(\lambda) \le \frac{1}{\lambda p_2} \tag{30}$$

we moreover have $\lim_{\lambda\to\infty} F_2(\lambda) = 0$. The function

$$G(\lambda) = F_2(\lambda) - \lambda p_1 \tag{31}$$

is continuous on $(0,\infty)$ and satisfies

$$\lim_{\lambda\to\infty} G(\lambda) = -\infty. \tag{32}$$

On the other hand, we have the lower bound

$$G(\lambda) = F_2(\lambda) - \lambda p_1 \geq \frac{1}{\lambda p_2 + \frac{1}{\lambda p_3}} - \lambda p_1 = \frac{1 - \lambda^2 p_1 p_2 - \frac{p_1}{p_3}}{\lambda p_2 + \frac{1}{\lambda p_3}}. \tag{33}$$

Since $p_3 > p_1$, it follows that there exists $\lambda_0 > 0$, such that

$$G(\lambda_0) > 0. \tag{34}$$

From (32), (34), and the intermediate value theorem, we conclude that there exists $\lambda = \lambda_k \in (\lambda_0, \infty)$ such that

$$G(\lambda_k) = 0 \tag{35}$$

providing a solution of (29). Note that (33) and (35) imply

$$0 \geq 1 - \lambda_k^2 p_1 p_2 - \frac{p_1}{p_3} \tag{36}$$

providing a lower bound

$$\lambda_k \geq \frac{\sqrt{1 - p_1/p_3}}{\sqrt{p_1 p_2}}.$$

Recalling (21), we obtain

$$\lambda_k \geq \frac{k^3 \sqrt{1 - (1 - 8/(9+k^2))^{(3-\beta)/2}}}{2(4+k^2)^{(3-\beta)/2}}$$

for all $k \geq 1$. An explicit computation shows that

$$\lim_{k\to\infty} \frac{1}{k^{\beta-1}} \left(\frac{k^3 \sqrt{1 - (1 - 8/(9+k^2))^{(3-\beta)/2}}}{2(4+k^2)^{(3-\beta)/2}} \right) = \sqrt{3-\beta} \geq 1$$

for any $\beta \in (1,2]$. Using $(1-x)^a \geq 1-x$ for $0 \leq a, x \leq 1$, we further obtain

$$\lambda_k \geq C^{-1} k^{\beta-1} \tag{37}$$

for any $k \geq 1$ and any $\beta \in (1,2]$, where $C \geq 1$ is a universal constant.

In particular, $\lambda_k \to \infty$ as $k \to \infty$ at least as fast as the power law $k^{\beta-1}$.

Having found a root $\lambda = \lambda_k$ of (29), we next need to estimate the size of the coefficients c_n defining θ in (19), and verify that they decay sufficiently fast so that θ is real-analytic. The function

$$F_n(\lambda) = \cfrac{1}{\lambda p_n + \cfrac{1}{\lambda p_{n+1} + \cdots}},$$

obeys the recurrence relationship

$$F_n(\lambda) = \frac{1}{\lambda p_n + F_{n+1}(\lambda)},$$

or equivalently

$$F_{n+1}(\lambda) = -\lambda p_n + \frac{1}{F_n(\lambda)}. \tag{38}$$

Since also

$$F_2(\lambda_k) = \lambda_k p_1, \tag{39}$$

by the definition of λ_k, we get, comparing the recurrence relations (27)–(28) for η_n and (38)–(39) for $F_n(\lambda_k)$, that

$$F_n(\lambda_k) = \eta_n.$$

Using $c_n = (p_n/p_{n-1})c_{n-1}\eta_n$ for $n \geq 2$ and

$$F_n(\lambda_k) \leq \frac{1}{\lambda_k p_n}, \qquad n \geq 2$$

we get

$$\begin{aligned} c_n &= \eta_n \cdots \eta_2 \frac{p_n\, c_1}{p_1} = F_n(\lambda_k) \cdots F_2(\lambda_k) \frac{p_n\, c_1}{p_1} \\ &\leq \frac{1}{\lambda_k p_n} \frac{1}{\lambda_k p_{n-1}} \cdots \frac{1}{\lambda_k p_2} \frac{p_n\, c_1}{p_1} = \frac{c_1}{\lambda_k^{n-1} p_{n-1} \cdots p_1} \end{aligned}$$

showing rapid convergence of c_n to 0. Without loss of generality we choose

$$c_1 = 1.$$

Then, by (22) and (37) we have

$$c_n \leq C^n \left(\frac{1}{k^{\beta-1}}\right)^{n-1} (k^3)^{n-1} \frac{1}{((n-1)!)^{3-\beta}} \leq \frac{C^n \left(k^{4-\beta}\right)^{n-1}}{((n-1)!)^{3-\beta}} \tag{40}$$

from where

$$c_n \leq C^n \left(k^{4-\beta}\right)^n (n!)^{\beta-3}, \qquad n \geq 1. \tag{41}$$

Furthermore, using $n! \geq (n/C)^n$ for a sufficiently large C, we obtain

$$c_n \leq \exp\left(n \log\left(Ck^{4-\beta} n^{\beta-3}\right)\right) \tag{42}$$

for all $\beta \in (1,2]$, $n \geq 1$, and $k \geq 1$. In particular, for any $\tau > 0$, we see that for

$$n \geq n(\tau, k, \beta) := Ce^{4\tau} k^{(4-\beta)/(3-\beta)}, \tag{43}$$

where C is sufficiently large, it holds that

$$c_n \leq \exp\left(-2\tau(n^2+k^2)^{1/2}\right).$$

Using (19), we see that for $\ell = (\ell_1, \ell_2) \in \mathbb{Z}^2$, we have

$$|\hat{\theta}_k(\ell)|^2 = \begin{cases} c_n^2, & \ell_1 = \pm k, \ell_2 = \pm n, \\ 0, & |\ell_1| \neq k. \end{cases}$$

Therefore,

$$\begin{aligned} \|\theta_k\|^2_{G^1_\tau} &= \sum_{n \geq 1} c_n^2 (k^2+n^2)^2 \exp\left(2\tau(k^2+n^2)^{1/2}\right) \\ &\leq \sum_{n=1}^{n(\tau,k,\beta)} (k^2+n^2)^2 \exp\left(2\tau(k^2+n^2)^{1/2} + 2n \log\left(\frac{50k^{4-\beta}}{n^{3-\beta}}\right)\right) \\ &\quad + \sum_{n > n(\tau,k,\beta)} (n^2+k^2)^2 \exp\left(-2\tau(n^2+k^2)^{1/2}\right) < \infty \end{aligned}$$

which shows that the function θ_k whose Fourier series n-th coefficients is c_n, is in fact entire real-analytic.

Next, we provide a proof of (17). For $s \geq 0$, recalling (43) we estimate

$$\begin{aligned} \|\theta_k\|^2_{H^s} &= \sum_{n \geq 1} (n^2+k^2)^s c_n^2 \\ &\leq 2^s n(1,k,\beta)^{2s} \sum_{n=1}^{n(1,k,\beta)} c_n^2 + \sum_{n > n(1,k,\beta)} (n^2+k^2)^s \exp\left(-2(n^2+k^2)^{1/2}\right) \\ &\leq C_s k^{2s(4-\beta)/(3-\beta)} \|\theta_k\|^2_{L^2} + C_s \exp\left(-k^{(4-\beta)/(3-\beta)}\right) \\ &\leq 2C_s k^{2s(4-\beta)/(3-\beta)} \|\theta_k\|^2_{L^2} \end{aligned} \tag{44}$$

where $C_s > 0$ is a sufficiently large constant that depends only on s, and is independent of $\beta \in (1,2]$; we have also used $\|\theta_k\|^2_{L^2} \geq 1$, which follows from $c_1 = 1$. Since the equation $L\theta_k = \lambda_k \theta_k$ is linear, we may renormalize

θ_k to have the unit H^s norm. Then, upon multiplying both sides of (44) with this constant, we obtain (17).

The proof of (18) is similar. Given $s \geq 1$ and $\tau > 0$ we have

$$\begin{aligned}
\|\theta_k\|^2_{G^s_\tau} &= \sum_{n\geq 1} c_n^2 (k^2+n^2)^2 \exp\left(2\tau(k^2+n^2)^{1/2s}\right) \\
&\leq n(\tau,k,\beta)^4 \exp\left(4\tau(n(\tau,k,\beta))^{1/s}\right) \sum_{n=1}^{n(\tau,k,\beta)} c_n^2 \\
&\quad + \sum_{n>n(\tau,k,\beta)} (k^2+n^2)^2 \exp\left(2\tau(k^2+n^2)^{1/(2s)} - 2\tau(n^2+k^2)^{1/2}\right) \\
&\leq 2C_{s,\tau} \exp\left(C_{s,\tau} k^{(4-\beta)/s(3-\beta)}\right) \|\theta_k\|^2_{L^2} \qquad (45)
\end{aligned}$$

where we have also used that $c_1 = 1$.

In order to conclude the proof we renormalize θ_k to have unit G^s_τ norm, and then multiply both sides of (45) with this constant we obtain (18). $\square$

4. Proof of nonlinear Lipschitz ill-posedness

Proof for Theorem 2.1. The proof of the above theorem follows directly from Theorem 2.3 and a classical perturbative argument (see, e.g.[25,26,29,30,32]), so we only give here a sketch of these details.

Fix $\theta_0^{(1)}(x_2) = \theta^{(1)}(x_2,t) = \cos(x_2)$ a steady state of the system (1)–(3) with (6). Clearly $\|\theta_0^{(1)}\|_{H^s} = 1$. For $\epsilon \in (0,1]$, let $\theta_0^{(2,\epsilon)}(x) = \theta_0^{(1)}(x_2) + \epsilon\psi_0(x)$ where $\psi_0(x)$ is a smooth function such that $\|\psi_0(x)\|_{H^s} = 1$, to be determined below. We have $\|\theta_0^{(2,\epsilon)}\|_{H^s} \leq 2$ for all $\epsilon \in (0,1]$.

If the system (1)–(3) with (6) would be locally Lipschitz well posed, then there would exist $T_0, K_0 > 0$ (uniform in ϵ), such that the family of solutions $\theta^{(2,\epsilon)}(x,t) \in L^\infty(0,T;H^r)$ of (1)–(3) with (6) with the initial datum $\theta_0^{(2,\epsilon)}(x)$ obey

$$\sup_{t\in[0,T_0]} \|\theta^{(2,\epsilon)}(t) - \theta_0^{(1)}\|_{H^r} \leq K_0 \|\theta_0^{(2,\epsilon)} - \theta_0^{(1)}\|_{H^s} = K_0\epsilon \qquad (46)$$

for all $\epsilon \in (0,1]$. We define

$$\psi_\epsilon(x,t) = \epsilon^{-1}\left(\theta^{(2,\epsilon)}(x,t) - \theta_0^{(1)}\right),$$

to be the ϵ-perturbation of $\theta^{(2,\epsilon)}$ about $\theta_0^{(1)}$. In view of (46) we have that ψ_ϵ is uniformly bounded in $L^\infty(0,T;H^r)$, with the bound

$$\sup_{t\in[0,T_0]} \|\psi_\epsilon(t)\|_{H^r} \leq K_0. \qquad (47)$$

The equation obeyed by ψ_ϵ is

$$\partial_t \psi_\epsilon = L\psi_\epsilon - \epsilon N[\psi_\epsilon]$$

so that by (14)–(15) we have that $\partial_t \psi_\epsilon$ is uniformly bounded in $L^\infty(0,T;L^2)$. Thus, Aubin–Lions lemma $\psi_\epsilon \to \psi$ in $L^2(0,T;L^2)$, where $\psi \in L^\infty(0,T;H^r)$ with a bound inherited from (47), solves the equation

$$\partial_t \psi = L\psi, \qquad \psi(0) = \psi_0.$$

In order to conclude the proof, let $\psi_0(x) = \theta_k(x)$, where θ_k is an eigenfunction of L as constructed in Theorem 2.3, with eigenvalue λ_k, and we choose k large enough such that

$$\frac{\exp\left(T_0 k^{\beta-1}/(2C_0)\right)}{C_s k^{s(4-\beta)/(3-\beta)}} \geq 2K_0$$

where C_0 is the constant from (16), and C_s is the constant from (17). Then using (17) we obtain that

$$\|\psi(T_0/2)\|_{L^2} = \exp(\lambda_k T_0/2)\|\psi_0\|_{L^2} \geq 2K_0$$

which contradicts (47) since $r > 0$, thereby concluding the proof. □

Proof of Theorem 2.2. The proof is the same as the proof of Theorem 2.1, except that the eigenfunction $\psi_0 = \theta_k$ is normalized to have a unit G^s_τ norm, and we pick k large enough so that

$$\frac{\exp(T_0 k^{\beta-1}/(2C_0))}{C_{s,\tau}\exp(C_{s,\tau} k^{(4-\beta)/(s(3-\beta))})} \geq 2K_0,$$

where C_0 is the constant from (16), and $C_{s,\tau}$ is the constant from (18). Finding such a values of k uses the restriction on s. Using (18) we then obtain that $\|\psi(T_0/2)\|_{L^2} \geq 2K_0$ which yields a contradiction. □

Acknowledgments

IK and FW were supported in part by the NSF grant DMS-1311943, while VV was supported in part by the NSF grant DMS-1514771 and by an Alfred P. Sloan Research Fellowship.

References

1. P. Constantin, A. Majda and E. Tabak, Formation of strong fronts in the 2-D quasigeostrophic thermal active scalar, *Nonlinearity* **7**, 1495 (1994).

2. D. Córdoba, F. Gancedo and R. Orive, Analytical behavior of two-dimensional incompressible flow in porous media, *J. Math. Phys.* **48**, 065206, 19 (2007).
3. S. Friedlander and V. Vicol, Global well-posedness for an advection-diffusion equation arising in magneto-geostrophic dynamics, *Ann. Inst. H. Poincaré Anal. Non Linéaire* **28**, 283 (2011).
4. L. Caffarelli and A. Vasseur, Drift diffusion equations with fractional diffusion and the quasi-geostrophic equation, *Ann. of Math. (2)* **171**, 1903 (2010).
5. P. Constantin, D. Cordoba and J. Wu, On the critical dissipative quasi-geostrophic equation, *Indiana Univ. Math. J.* **50**, 97 (2001).
6. P. Constantin, G. Iyer and J. Wu, Global regularity for a modified critical dissipative quasi-geostrophic equation, *Indiana Univ. Math. J.* **57**, 2681 (2008).
7. P. Constantin, A. Tarfulea and V. Vicol, Long time dynamics of forced critical SQG, *Comm. Math. Phys.* **335**, 93 (2015).
8. P. Constantin and J. Wu, Behavior of solutions of 2D quasi-geostrophic equations, *SIAM J. Math. Anal.* **30**, 937 (1999).
9. A. Córdoba and D. Córdoba, A maximum principle applied to quasi-geostrophic equations, *Comm. Math. Phys.* **249**, 511 (2004).
10. M. Dabkowski, A. Kiselev, L. Silvestre and V. Vicol, Global well-posedness of slightly supercritical active scalar equations, *Analysis & PDE* **7**, 43 (2014).
11. H. Dong, Dissipative quasi-geostrophic equations in critical Sobolev spaces: smoothing effect and global well-posedness, *Discrete Contin. Dyn. Syst.* **26**, 1197 (2010).
12. W. Hu, I. Kukavica and M. Ziane, Sur l'existence locale pour une équation de scalaires actifs, *C. R. Acad. Sci. Paris Sér. I Math.* **353**, 241 (2015).
13. N. Ju, Dissipative 2D quasi-geostrophic equation: local well-posedness, global regularity and similarity solutions, *Indiana Univ. Math. J.* **56**, 187 (2007).
14. A. Kiselev, Nonlocal maximum principles for active scalars, *Advances in Mathematics* **227**, 1806 (2011).
15. A. Kiselev and F. Nazarov, A variation on a theme of Caffarelli and Vasseur, *Zap. Nauchn. Sem. S.-Peterburg. Otdel. Mat. Inst. Steklov. (POMI)* **370**, 58 (2009).
16. A. Kiselev, F. Nazarov and A. Volberg, Global well-posedness for the critical 2D dissipative quasi-geostrophic equation, *Invent. Math.* **167**,

445 (2007).
17. I. Kukavica and F. Wang, Weighted decay for the surface quasi-geostrophic equation, *Commun. Math. Sci.* (2015).
18. H. Miura, Dissipative quasi-geostrophic equation for large initial data in the critical Sobolev space, *Comm. Math. Phys.* **267**, 141 (2006).
19. S. Resnick, *Dynamical problems in non-linear advective partial differential equations* (ProQuest LLC, Ann Arbor, MI, 1995), Thesis (Ph.D.)–The University of Chicago.
20. M. Schonbek and T. Schonbek, Asymptotic behavior to dissipative quasi-geostrophic flows, *SIAM J. Math. Anal.* **35**, 357 (2003).
21. L. Silvestre, Eventual regularization for the slightly supercritical quasi-geostrophic equation, *Ann. Inst. H. Poincaré Anal. Non Linéaire* **27**, 693 (2010).
22. J. Wu, Dissipative quasi-geostrophic equations with L^p data, *Electron. J. Differential Equations* , No. 56, 13 (2001).
23. J. Wu, Existence and uniqueness results for the 2-D dissipative quasi-geostrophic equation, *Nonlinear Anal.* **67**, 3013 (2007).
24. D. Chae, P. Constantin, D. Córdoba, F. Gancedo and J. Wu, Generalized surface quasi-geostrophic equations with singular velocities, *Comm. Pure Appl. Math.* **65**, 1037 (2012).
25. S. Friedlander, F. Gancedo, W. Sun and V. Vicol, On a singular incompressible porous media equation, *J. Math. Phys.* **53**, 1 (2012).
26. S. Friedlander and V. Vicol, On the ill/well-posedness and nonlinear instability of the magneto-geostrophic equations, *Nonlinearity* **24**, 3019 (2011).
27. S. Friedlander, W. Strauss and M. Vishik, Nonlinear instability in an ideal fluid, *Ann. Inst. H. Poincaré Anal. Non Linéaire* **14**, 187 (1997).
28. L. Mešalkin and J. Sinaĭ, Investigation of the stability of a stationary solution of a system of equations for the plane movement of an incompressible viscous liquid, *J. Appl. Math. Mech.* **25**, 1700 (1961).
29. M. Renardy, Ill-posedness of the hydrostatic Euler and Navier-Stokes equations, *Arch. Ration. Mech. Anal.* **194**, 877 (2009).
30. T. Tao, *Nonlinear dispersive equations*, CBMS Regional Conference Series in Mathematics, Vol. 106 (Conference Board of the Mathematical Sciences, Washington, DC, 2006), Local and global analysis.
31. C. Levermore and M. Oliver, Analyticity of solutions for a generalized Euler equation, *J. Differential Equations* **133**, 321 (1997).
32. Y. Guo and T. Nguyen, A note on Prandtl boundary layers, *Comm. Pure Appl. Math.* **64**, 1416 (2011).

Equilibrium in an individual - societal SIR vaccination model in presence of discounting and finite vaccination capacity

Laetitia Laguzet*, Gabriel Turinici*,** and Ghozlane Yahiaoui*
*CEREMADE, Université Paris Dauphine, Paris 75016, France
**Institut Universitaire de France,
Tel.: +33-1-44054858
E-mail: laguzet@ceremade.dauphine.fr, Gabriel.Turinici@dauphine.fr
ghozlane.yahiaoui@dauphine.eu

This paper analyzes individual optimal vaccination strategies for a Susceptible-Infected-Recovered (SIR) model. The individuals vaccinate according to a criterion taking into account the risk of infection, the possible side effects of the vaccine and the overall epidemic course. We treat the situation when the vaccination capacity is limited and the individual discounts the future at some rate $\mathcal{D} > 0$. Since the course of the epidemic is determined by the sum of individual decisions a natural question is whether an equilibrium between the individual decisions and the epidemic evolution exists. We show that the answer is positive.

Keywords: optimal vaccination, SIR model; mean field games.

1. Introduction

The vaccination, if available, is often invoked in order to limit the propagation of epidemics. A "benevolent planner" approach can be used to find the vaccination program that will strike the optimal balance between vaccination costs and benefits, expressed in terms of money, medical status, QALY/ DALY (see Refs. 1–3). Refs. 4–9 focused on describing the optimal vaccination policy as function of the vaccine cost, infection cost and epidemic dynamics.

However, failures in vaccination campaigns were observed due to worries related to the cost, risk, inefficiency or side-effects of the vaccine. The individual perception and choices will therefore depart from the "benevolent planner" approach and each individual will take his own best vaccination decision according to his subjective perceived risks and costs. As such other parameters should be taken into account and included in the model such as the individual point of view. Individual decisions, whether vaccination is accepted or not may influence the propagation of an epidemic. Several works analyzed this effect, see Refs. 10–20. We refer the reader to Ref. 21 for a recent literature review.

In this context, we use a SIR model and techniques from the "Mean Field Games" theory to show that an equilibrium exists and determines completely the individual optimal vaccination strategy. This work extends the analysis in Ref. 21 to the situation when the discount factor $\mathcal{D}$ is strictly positive but the vaccination capacity u_{max} is limited. We prove the existence (but not the uniqueness) of an equilibrium.

The work is close in spirit to Refs. 22,23 that show existence and uniqueness of the *Mean Field Games* equilibrium for a framework involving a finite number of possible states, but the hypothesis used there do not apply here (in particular the super-linearity of the cost with respect to the controls).

2. Model and nomenclature

We use a SIR model (see Refs. 8,24 for a general description and Ref. 21 for this particular model) described by the following equations:

$$\begin{cases} S'(t) = -\beta S(t)I(t) - u(t), & S(0) = S_0, \\ I'(t) = \Big(\beta S(t) - \gamma\Big)I(t), & I(0) = I_0, \\ \varphi_I'(t) = \Big(1 - \varphi_I(t)\Big)\beta I(t), & \varphi_I(0) = 0. \end{cases} \tag{1}$$

Note that $S(\cdot)$, $I(\cdot)$, $\varphi_I(\cdot)$ depend on S_0, I_0 and $u(t)$; however to simplify notations this dependence will not be stated explicitly. The signification of the parameters is as follows:

(1) the parameter $\beta > 0$ is the transmission rate of the disease and the parameter $\gamma > 0$ is the recovery rate;
(2) $S(t) \geq 0$, $I(t) \geq 0$ and $\int_0^t \gamma I(s)ds \geq 0$ are the proportions of individuals in the Susceptible, Infected and Recovered classes, respectively.
(3) $u : [0, \infty[\to [0, u_{max}]$ is the vaccination rate; it is a measurable function; $u_{max} < \infty$ is the maximal vaccination capacity. Therefore $\int_0^t u(s)ds$ is the proportion of people that vaccinated by the time t. Since $S(t)$ cannot be negative, this imposes some constraints on the set of admissible vaccination rates $u(t)$ (no more vaccination once $S(t) = 0$). The admissible vaccinations rates (at the societal level) are:

$$\mathcal{U} = \Big\{u : [0, \infty[\to [0, u_{max}] \Big| u \text{ measurable}, S(t) \geq 0 \ \forall t \geq 0\Big\}. \tag{2}$$

We also introduce the projection to $\mathcal{U}$. For any $u : [0, \infty[\to [0, u_{max}]$ let t_u be the largest value t such that $u \cdot \mathbb{1}_{[0,t]}(\cdot) \in \mathcal{U}$. Then:

$$\mathcal{R}(u) := u \cdot \mathbb{1}_{[0,t_u]}(\cdot). \tag{3}$$

By convention for $t_u = \infty$ we take $\mathcal{R}(u) := u$ (in this case $u \in \mathcal{U}$).

(4) $\varphi_I(t)$ is the cumulative probability of infection in $[0, t]$ (in absence of vaccination). Its main use is to formulate the individual criterion for vaccination, see below equation (6).

We suppose that β, γ, $S_0 > 0$ and $I_0 > 0$ (with $S_0 + I_0 \leq 1$) are known and fixed from now on.

The first two equations of the system (1) represent the societal dynamics as illustrated in Figure 1; the individual dynamics is modeled by a continuous time Markov chain with individual jumping between the Susceptible, Infected, Recovered and Vaccinated classes (see Ref. 21 for the details of the Markov process).

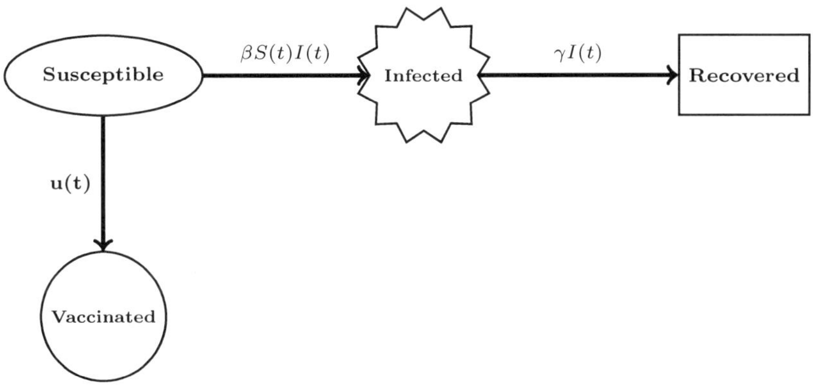

Fig. 1. Societal overall dynamics.

Moving from the Susceptible to the Vaccinated class depends only on the will of the individual. Moreover, his choice will not influence, on a global scale, the societal propagation dynamics. We suppose that each individual is aware of the propagation dynamics at the societal level (1). Therefore, the individual can choose between taking the risk of being vaccinated or taking the risk of contracting the disease. He makes a choice only according to his well-being.

A coherence check of the model requires that the individual decisions sum up to give the overall vaccination dynamics (thus epidemic evolution). A natural question is whether such an equilibrium exists. We give a positive answer in a particular case that has not yet been considered in the literature.

3. Optimal individual strategies

The individual strategy is described by $\varphi_V(t)$ which is the probability of the individual to vaccinate in $[0,t]$. By convention $1-\varphi_V(\infty)$ is the probability to never vaccinate. Thus φ_V is a cumulative distribution function (CDF) on $[0,\infty[\cup\{\infty\}$.

The requirement $u(t) \leq u_{max}$ imposes some constraints on the CDF φ_V, see Ref. 21 for a detailed discussion. Let

$$Z_{S(\cdot)} = \min_{\tau \geq 0}\{\tau | S(\tau) = 0\}. \tag{4}$$

By convention if $S(\cdot)$ is always strictly positive we set $Z_{S(\cdot)} = \infty$. Two situations arise:

- On $[0, Z_{S(\cdot)}[$ we can work, by density, with functions satisfying:

$$\varphi_V'(t) = \Big(1 - \varphi_V(t)\Big)\lambda(t), \varphi_V(0) = 0, \tag{5}$$

  where $\lambda(\cdot)$ is a positive measurable function. The individual can choose his instantaneous vaccination rate $\lambda(\cdot)$; the bound u_{max} on the overall societal vaccination rate $u(t)$ induces a bound on $\lambda(t)$ of the form $\lambda(t) \leq \frac{u_{max}}{S(t)}$.
- On the other side, $S(\cdot)\Big|_{[Z_{S(\cdot)},\infty[} \equiv 0$ therefore on $[Z_{S(\cdot)}, \infty[$ there are no susceptibles left to challenge the capacity of the vaccination centers and thus an additional individual arriving into such a population can vaccinate at any instant (s)he desires; consequently the function φ_V is only required, as any CDF function, to be increasing, right-continuous and with $\varphi_V(\infty) \leq 1$.

In order to choose his best vaccination strategy the individual defines a cost, denoted $J_{indi}(\varphi_V; u)$:

$$J_{indi}(\varphi_V; u) = r_I \Phi_I(\infty) + \int_0^\infty \Big[r_I(\Phi_I(t) - \Phi_I(\infty)) \\ + r_V e^{-\mathcal{D}t}(1 - \varphi_I(t))\Big] d\varphi_V(t). \tag{6}$$

Here $\mathcal{D} > 0$ is the discount factor, and $\Phi_I(t)$ is a function describing the risk of infection weighted according to the discount factor:

$$\Phi_I(t) = \int_0^t e^{-\mathcal{D}s}\varphi_I'(s)ds = \int_0^t e^{-\mathcal{D}s}(1-\varphi_I(s))\beta I(s)ds,\ \Phi_I(0) = 0. \quad (7)$$

The parameter r_V is the cost of the vaccination (measured in money or as medical side-effects) and r_I the cost of being infected.

Set:

$$g(t) = r_I(\Phi_I(t) - \Phi_I(\infty)) + r_V e^{-\mathcal{D}t}(1-\varphi_I(t)). \quad (8)$$

In particular $g(t) \in [-r_I, r_I + r_V]$, $\forall t \geq 0$.

Thus, the individual wants to minimize $J_{indi}(\varphi_V; u) = r_I\Phi_I(\infty) + \int_0^\infty g(t)d\varphi_V(t)$, which, after elimination of the term $r_I\Phi_I(\infty)$ (independent of φ_V), amounts to looking for a minimum of

$$\int_0^\infty g(t)d\varphi_V(t) = \int_0^{Z_{S(\cdot)}} g(t)(1-\varphi_V(t))\lambda(t)dt + \int_{Z_{S(\cdot)}}^\infty g(t)d\varphi_V(t). \quad (9)$$

As in Ref. 21 the minimum of the second quantity is readily seen to be $-(1-\varphi_V(Z_{S(\cdot)}^-)) \cdot \left(\inf_{t\geq Z_{S(\cdot)}} g(t)\right)_-$ (for any real number x, we denote $x_- = \max\{-x, 0\}$).

3.1. *The HJB equation associated to the individual optimal solutions*

Lemma 3.1. *Let $u \in \mathcal{U}$. Then*

(1) $I(t)$ vanishes to 0 at ∞ uniformly with respect to $u \in \mathcal{U}$ (but depending on β, γ, S_0, I_0);

(2) there exists $T_{max} > 0$ depending only on β, γ, S_0, I_0 such that the function $g(t)$ is positive on $[T_{max}, \infty[$. In particular the optimal $d\varphi_V$ is null on $[T_{max}, \infty[$.

Proof. Let us compute the derivative of $g(t)$:

$$\begin{aligned} g'(t) &= r_I e^{-\mathcal{D}t}(1-\varphi_I(t))\beta I(t) - \mathcal{D}r_V e^{-\mathcal{D}t}(1-\varphi_I(t)) \\ &\quad - r_V e^{-\mathcal{D}t}(1-\varphi_I(t))\beta I(t) \\ &= e^{-\mathcal{D}t}(1-\varphi_I(t))\Big[(r_I - r_V)\beta I(t) - r_V\mathcal{D}\Big]. \end{aligned} \quad (10)$$

The second equation in (1) shows that $I(t)$ can be either decreasing (if $S_0 \leq \gamma/\beta$) or increasing while $S(t) \geq \gamma/\beta$ and then decreasing. In particular $I(t)$ can be larger than $\frac{r_V \mathcal{D}}{(r_I - r_V)\beta}$ at most on some bounded interval (depending on S_0, I_0, u). Thus in general $g(t)$ has the form in Figures 2 and 3: decreasing, then increasing then decreasing again. Moreover $\lim_{t\to\infty} g(t) = 0$.

The optimal $d\varphi_V$ is null on any interval $[T, \infty[$ such that $g(t)$ is decreasing (thus positive) on the whole interval. Therefore it is enough to find T_{max} such that $I(t) \leq \frac{r_V \mathcal{D}}{(r_I - r_V)\beta}$ for $t \geq T_{max}$. To this end we will prove that $I(t)$ vanishes at infinity uniformly with respect to u (but depending on S_0, I_0, γ and β).

Let $C \leq I_0$ and $T_C = \frac{\ln(\frac{I_0}{C}) + \frac{S_0}{C}}{\gamma} > 0$, chosen such that $I_0 e^{-\gamma t + \frac{S_0}{C}} \leq C$, $\forall t \geq T_C$. Suppose, by contradiction, that there exists $t^* \geq T_C$ with $I(t^*) > C$ (note that t^* may depend on u). Thus, as I is continuous and $\lim_{t\to\infty} I(t) = 0$, we find $t^{**} \geq t^* \geq T_C$ such that $I(t^{**}) = C$. But, since $C \leq I_0$, we also have that $I(t^{**}) = C \leq I_0$ implies $I(t) \geq C, \forall t \leq t^{**}$ and thus $I(t) \geq C$, $\forall t \leq T_C$.

On the other side, $S'(t) \leq -\beta S(t) I(t)$ thus $S(t) \leq S_0 e^{-\beta t C}$, $\forall t \leq T_C$. Moreover $I'(t) = \Big(\beta S(t)I(t) - \gamma I(t)\Big)$ implies : $I(t) \leq I_0 e^{-\gamma t + S_0 \beta \int_0^t e^{-\beta C \tau} d\tau} < I_0 e^{-\gamma t + \frac{S_0}{C}} \leq C$, which is absurd.

Thus for any C small enough ($C \leq I_0$), there exists $T_C > 0$ independent of u such that $I(t) \leq C$, $\forall t \geq T_C$ which means that I vanishes at infinity independently of u. This gives the first point of the result.

The second point follows from the behavior of the function $g(t)$ and the definition of the individual cost. □

Remark 3.1. The point 1 of Lemma 3.1 can also be proved in the following way: for any $\varepsilon < I_0$ let t_ε be the unique point such that $I(t_\varepsilon) = \varepsilon$ (unique because $I(t)$ can only decrease once it is below I_0). Then for any $t \leq t_\varepsilon$ we have $I(t) \geq I(t_\varepsilon) = \varepsilon$. But since

$$1 \geq \int_0^\infty \gamma I(t) dt \geq \int_0^{t_\varepsilon} \gamma I(t) dt \geq \gamma I(t_\varepsilon) t_\varepsilon, \tag{11}$$

one obtains $t_\varepsilon \leq \frac{1}{\gamma\varepsilon}$ thus any $t \geq \frac{1}{\gamma\varepsilon}$ has $I(t) \leq I(t_\varepsilon) = \varepsilon$.

Note that on $[Z_{S(\cdot)}, \infty[$ the variable $I(t)$ can only decrease. Thus, considering the arguments of Lemma 3.1 the function $g(t)$ can be increasing and then decreasing or only decreasing. In both situations, the minimum value, if it is negative, is attained at point $Z_{S(\cdot)}$ (otherwise it is null). Thus:

$$-\Big(\inf_{t \geq Z_{S(\cdot)}} g(t)\Big)_- = -\Big(g(Z_{S(\cdot)})\Big)_-. \tag{12}$$

The individual minimization problem can therefore be rephrased as minimizing

$$\int_0^\infty g(t)d\varphi_V(t) = \int_0^{Z_{S(\cdot)}} g(t)(1-\varphi_V(t))\lambda(t)dt-(1-\varphi_V(Z^-_{S(\cdot)})\cdot\Big(g(Z_{S(\cdot)})\Big)_- . \tag{13}$$

For any $t \geq 0$ we introduce the notation

$$V(t) :=\inf\Big\{\int_t^\infty g(\tau)d\varphi_V(\tau)\Big|\varphi_V \text{ increasing, right continuous with left limits, } \varphi_V(t) = 0, \varphi_V(\infty) \leq 1, \varphi'_V(\tau) = (1-\varphi_V(\tau))\lambda(\tau), \forall\tau \in [t, Z_{S(\cdot)}[\text{ with } 0 \leq \lambda(\tau) \leq u_{max}/S(\tau)\Big\}. \tag{14}$$

Previous arguments show that for $t \in [0, Z_{S(\cdot)}[$

$$V(t) := \inf\Big\{\int_t^{Z_{S(\cdot)}} g(\tau)(1-\varphi_V(\tau))\lambda(\tau)d\tau - (1-\varphi_V(Z^-_{S(\cdot)}))\cdot\Big(g(Z_{S(\cdot)})\Big)_-\Big| \varphi_V(t) = 0, \varphi_V(\infty) \leq 1, \varphi'_V(\tau) = (1-\varphi_V(\tau))\lambda(\tau), \forall\tau \in [t, Z_{S(\cdot)}[\text{ with } 0 \leq \lambda(\tau) \leq u_{max}/S(\tau)\Big\}. \tag{15}$$

Standard manipulations allow to see that V is continuous on $[0, Z_{S(\cdot)}[\cup]Z_{S(\cdot)}, \infty[$.

Moreover the limit $\lim_{t\nearrow Z_{S(\cdot)}} V(t)$ equals $-\Big(g(Z_{S(\cdot)})\Big)_-$ and can be attained with a φ_V that is continuous. To see this, consider the relevant case $Z_{S(\cdot)} < \infty$ and treat first the situation when $g(Z_{S(\cdot)}) \geq 0$. Then in the limit $t \nearrow Z_{S(\cdot)}$ the best possible value for $V(t)$ is null, attained with $\varphi_V = 0$. When $g(Z_{S(\cdot)}) < 0$, since $g(\cdot)$ is continuous and the total mass of φ_V cannot be larger than one, in the limit $t \nearrow Z_{S(\cdot)}$ the best possible value for $V(t)$ is $g(Z_{S(\cdot)})$, attained for the law obtained setting $\lambda(\tau) = u_{max}/S(\tau)$. To see that in this case the mass is one it is enough to prove that $\varphi_V(Z^-_{S(\cdot)}) = 1$, or equivalently $\int_t^{Z_{S(\cdot)}} u_{max}/S(\tau)d\tau = \infty$; the equation of $S'(t)$ integrated between $Z_{S(\cdot)} - \epsilon$ and $Z_{S(\cdot)}$ gives the upper bound $S(Z_{S(\cdot)} - \epsilon) \leq \epsilon u_{max}$ and the conclusion follows.

Thus setting $V(Z_{S(\cdot)}) = -\Big(g(Z_{S(\cdot)})\Big)_-$ we obtain that $V(\cdot)$ is continuous on $[0, Z_{S(\cdot)}]$ and is a viscosity solution of the following Hamilton Jacobi Bellman equation (see Ref. 25):

$$-\Big[V'(t) + \min_{\lambda(t)\in[0,u_{max}/S(t)]} (g(t) - V(t))\lambda(t)\Big] = 0,\ t < Z_{S(\cdot)} \tag{16}$$

$$V(Z_{S(\cdot)}) = -\Big(g(Z_{S(\cdot)})\Big)_- . \tag{17}$$

On the other hand, for $t \geq Z_{S(\cdot)}$: $V(t) = -\Big(g(t)\Big)_-$ and also $V(t) = 0$ for any $t \geq T_{max}$ which means that on $[Z_{S(\cdot)}, T_{max}[$ (if it is non-void) the value function V is solution of the HJB equation

$$- \min\{V'(t), g(t) - V(t)\} = 0, \; Z_{S(\cdot)} \leq t < T_{max} \tag{18}$$

$$V(T_{max}) = 0. \tag{19}$$

Combining equations(16)-(17) and (18)-(19) we obtain that V is the continuous viscosity solution of:

$$\forall t \leq T_{max} : \; - \min\{V'(t), V'(t)S(t) + u_{max}(g(t) - V(t))\} = 0, \; V(T_{max}) = 0. \tag{20}$$

Note that, for any given $g(\cdot)$, $S(\cdot)$, the problem (20) can be written

$$-F_{g(\cdot),S(\cdot)}(t, V, V') = 0, \; V(T_{max}) = 0. \tag{21}$$

where $F_{g(\cdot),S(\cdot)} : [0, T_{max}] \times \mathbb{R} \times \mathbb{R} \to \mathbb{R}$ is defined as

$$F_{g(\cdot),S(\cdot)}(t, v, p) = \min\{p, pS(t) + u_{max}(g(t) - v)\}. \tag{22}$$

In particular $F_{g(\cdot),S(\cdot)}$ is continuous.

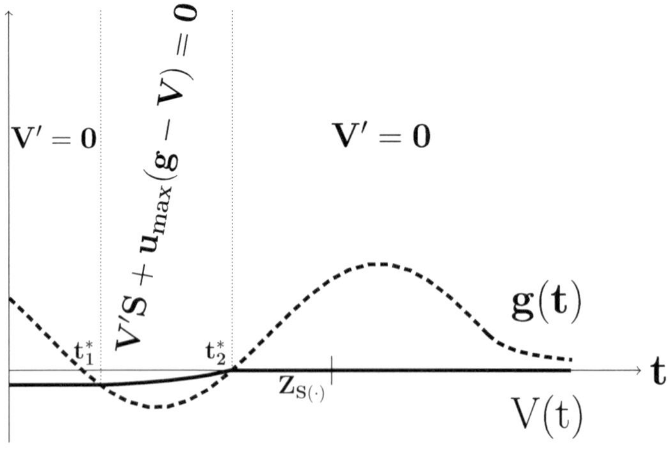

Fig. 2. Representation of the functions $g(t)$ (dashed line) and $V(t)$ (solid line) for the situation $g(t) \geq 0$ for all $t \geq Z_{S(\cdot)}$. In this case the equation of V can either be $V' = 0$ or $V'S + u_{max}(g - V) = 0$.

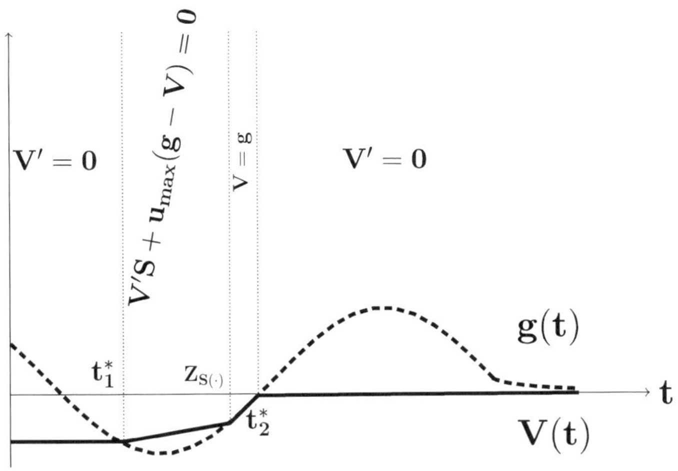

Fig. 3. Representation of the functions $g(t)$ (dashed line) and $V(t)$ (solid line) for the situation $g(Z_{S(\cdot)}) < 0$. With respect to the Figure 2 a new region appears, for $t \in [Z_{S(\cdot)}, t_2^*]$ where $V = g$.

3.2. *Structure of individual optimal solutions*

The considerations of the previous section allow to prove:

Lemma 3.2. *Let u be an arbitrary control in $\mathcal{U}$ and denote by $S(\cdot)$ and $g(\cdot)$ the corresponding solutions of (1). Then the value function $V(\cdot)$ defined in (14) is the unique solution of the HJB equation (20). Moreover $V(0)$ is attained by at least one optimal strategy φ_V having the following form:*

(1) If $\inf_{t\geq 0} g(t) = 0$ then $V(0) = 0$ and $\varphi_V = 0$;
(2) If $\inf_{t\geq 0} g(t) < 0$, let $t_u \leq T_{max}$ be such that $g(t_u) = \inf_{t\geq 0} g(t)$. If $t_u \geq Z_{S(\cdot)}$ then $d\varphi_V = \delta_{t_u}$ and $V(0) = g(t_u)$;
(3) If $t_u < Z_{S(\cdot)}$ there exist t_u^1, t_u^2 with $0 \leq t_u^1 \leq t_u < t_u^2 \leq \min\{Z_{S(\cdot)}, T_{max}\}$ such that:

$$\begin{cases} d\varphi_V(t) = (1 - \varphi_V)\frac{u_{max}}{S(t)}dt, \ t \in [t_u^1, t_u^2[\\ d\varphi_V(t) = 0, \ t \notin [t_u^1, t_u^2[. \end{cases} \tag{23}$$

In this case $\inf_{t\geq 0} g(t) < V(0) < 0$.

Proof. The considerations in Section 3.1 ensure that the value function is solution of (20). To prove the uniqueness, we apply the theorem 2.4 page 26 from Ref. 26 (see also chapter II.3 from Ref. 27 and the equation

2.20 section II.2 from Ref. 28). To this end consider $W(t) = e^t V(t)$, thus $V'(t) = e^{-t}(W'(t) - W(t))$. We replace $V(t)$ and $V'(t)$ in (20) and obtain that $W(t)$ is solution of the following problem:

$$\begin{aligned} &\forall t \leq T_{max} : \\ &-\min\{W'(t) - W(t), (W'(t) - W(t))S(t) + u_{max}(e^t g(t) - W(t))\} = 0, \\ &W(T_{max}) = 0. \end{aligned} \tag{24}$$

The uniqueness of the solution of (24) is obtained under assumptions of strict monotonicity of $F_W(t, v, p) = -\min\{p-v, (p-v)S(t)+u_{max}(e^t g(t)-v)\}$ with respect to its second argument. In practice one has to prove

$$\forall t \geq 0, \forall p \in \mathbb{R}, \forall v_1 \geq v_2 : F_W(t, v_1, p) - F_W(t, v_2, p) \geq C_m(v_1 - v_2), \quad \text{with } C_m > 0. \tag{25}$$

In our case, F_W satisfies (25) with $C_m = \min(1, u_{max}) > 0$.

To prove the second part of the result, we propose a solution candidate $\mathcal{V}$. First of all, $\mathcal{V}(t) := -(g(t))_-$ on $[\min\{Z_{S(\cdot)}, T_{max}\}, \infty[$. In this case $d\varphi_V$ is either null or a Dirac mass (depending on whether g is positive or negative).

For $t < \min\{Z_{S(\cdot)}, T_{max}\}$ two alternatives appear:

- When $Z_{S(\cdot)} < \infty$ and $g(Z_{S(\cdot)}) < 0$ then $S(t)\mathcal{V}'(t)+u_{max}(g(t)-\mathcal{V}(t)) = 0$ on some interval $[t_1^*, Z_{S(\cdot)}[$ with t_1^* the first (solving backwards from $Z_{S(\cdot)}$) such that $\mathcal{V}(t_1^*) = g(t_1^*)$. In this case $d\varphi_V = (1 - \varphi_V)u_{max}/S(t)$ on some interval and null elsewhere. On $[0, t_1^*]$, the function $\mathcal{V}(t)$ is constant.
- For other situations, denote

$$t_2^* = \inf\{t \geq 0 | g(\tau) \geq 0, \forall \tau \geq t\}. \tag{26}$$

  Then $S(t)\mathcal{V}'(t) + u_{max}(g(t) - \mathcal{V}(t)) = 0$ on some interval $[t_1^*, t_2^*[$ with t_1^* the first (solving backwards from t_2^*) such that $\mathcal{V}(t_1^*) = g(t_1^*)$. In this case $d\varphi_V = (1 - \varphi_V)u_{max}/S(t)$ on some interval and null elsewhere. On $[0, t_1^*]$, the function $\mathcal{V}(t)$ is constant.

Elementary computations allow to prove that $\mathcal{V}$ is a solution of (20), therefore the unique solution, therefore $\mathcal{V} = V$. In particular, this implies the alternatives 1-3. See Figures 2 and 3 for an illustration. □

3.3. *Equilibrium*

An equilibrium between the individual vaccination and the societal vaccination satisfies $u(t) = S(t)\lambda(t)$ (see Ref. 21). The Lemma 3.2 suggests that an equilibrium is necessarily of the form $u(t) = u_{max}\mathbb{1}_{[T_1^*,T_2^*]}(t)$ with $T_1^*, T_2^* \le T_{max}$.

To prove the existence of an equilibrium we introduce $K = \{(t_1,t_2) \in [0,T_{max}]^2 | t_1 \le t_2\}$ and the function $\mathcal{F} : K \to 2^K$ (2^K are the sub-ensembles of K) defined as follows. For any $x = (T_1,T_2) \in K$ consider V^x the solution of the HJB equation (20) corresponding to $u^x(t) = \mathcal{R}(u_{max}\mathbb{1}_{[T_1,T_2]}(\cdot))$. If, with the notations of the Lemma 3.2, the alternative 3 arises then set $\mathcal{F}(x) = \{(t^1_{u^x}, t^2_{u^x})\}$; otherwise set $\mathcal{F}(x) = \{(t,t)|0 \le t \le T_{max}\}$.

Theorem 3.1. *An equilibrium strategy exists. It is of the form* $u^e = u_{max}\mathbb{1}_{[T_1^e,T_2^e]}(\cdot)$, $(T_1^e,T_2^e) \in K$.

Proof. An equilibrium corresponds to a fixed point of $\mathcal{F}$. Moreover the definition (3) and equation (11) ensure that any point x^* with $x^* \in \mathcal{F}(x^*)$ corresponds necessarily to an admissible control (see also the arguments that allow to prove the continuity of V defined in equation (15)). We use the Kakutani fixed-point theorem and will check its assumptions:

- for all $x \in K$, $\mathcal{F}(x)$ is a non-empty convex subset of K ;
- the graph of the set-valued function $\mathcal{F}$ is closed in $K \times K$.

The first property follows from the definition of $\mathcal{F}$. For the second consider two sequences $(x_n = (T_1^n,T_2^n))_{n\ge 0}, (y_n)_{n\ge 0} \subset K$ such that $y_n \in \mathcal{F}(x_n)$, $x_n \to x = (T_1,T_2)$, $y_n \to y$. We have to prove that $y \in \mathcal{F}(x)$. Denote $u^x = u_{max}\mathbb{1}_{[T_1,T_2]}(\cdot)$, $u^{x_n} = u_{max}\mathbb{1}_{[T_1^n,T_2^n]}(\cdot)$.

We use the following stability properties from Ref. 25 page 176 : let $(V_n)_{n\ge 1}$ satisfy (in the viscosity sense):

$$-F_n(t,V_n(t),V_n'(t)) = 0,\ t \le T_{max},\ V_n(T_{max}) = 0. \tag{27}$$

Suppose that the sequence $(F_n)_{n\ge 1}$ converges uniformly on compact sets to some F and the sequence $(V_n)_{n\ge 1}$ converges to V in $C([0,T_{max}])$. Then V is a viscosity solution of

$$-F(t,V(t),V'(t)) = 0,\ t \le T_{max},\ V(T_{max}) = 0. \tag{28}$$

Here we take, $F = F_{g(\cdot),S(\cdot)}$, S, g, I correspond to the control $\mathcal{R}(u^x)$; $F_n = F_{g_n(\cdot),S_n(\cdot)}$, S_n, g_n, I_n correspond to the control $\mathcal{R}(u^{x_n})$.

We use the a priori estimate (16) of the theorem 2.1 from Ref. 29 which asserts that, denoting $U(t) = \int_0^t \mathcal{R}(u^x)(\tau)d\tau$, $U_n(t) = \int_0^t \mathcal{R}(u^{x_n})(\tau)d\tau$, there exists a constant C such that:

$$|S(t)-S_n(t)| + \int_0^t |S(\tau)-S_n(\tau)|d\tau \le C\Big[|U(t)-U_n(t)| + \int_0^t |U(\tau)-U_n(\tau)|d\tau\Big], \tag{29}$$

for any $t \in [0, T_{max}]$. The same works for $I(\cdot)$. But the convergence of T_1^n to T_1 and T_2^n to T_2 implies the uniform convergence of U_n to U and thus uniform convergence of S_n to S, I_n to I; hence g_n converges to g uniformly. This is enough to ensure the uniform convergence of F_n to F. By Lemma 3.2 to any solution of equations (28) and (27) we can associate an element of K; using the compactness of K, we obtain the conclusion. □

References

1. R. Zeckhauser and D. Shepard, Where now for saving lives?, *Law and Contemporary Problems* **40**, 5 (1976).
2. S. Anand and K. Hanson, Disability-adjusted life years: a critical review, *Journal of Health Economics* **16**, 685 (1997).
3. F. Sassi, Calculating QALYs, comparing QALY and DALY calculations, *Health Policy and Planning* **21**, 402 (2006).
4. H. W. Hethcote and P. Waltman, Optimal vaccination schedules in a deterministic epidemic model, *Mathematical Biosciences* **18**, 365 (December 1973).
5. A. Abakuks, Optimal immunisation policies for epidemics, *Advances in Appl. Probability* **6**, 494 (1974).
6. R. Morton and K. H. Wickwire, On the optimal control of a deterministic epidemic, *Advances in Appl. Probability* **6**, 622 (1974).
7. S. P. Sethi and P. W. Staats, Optimal control of some simple deterministic epidemic models, *J. Oper. Res. Soc.* **29**, 129 (1978).
8. O. Diekmann and J. Heesterbeek, *Mathematical epidemiology of infectious diseases. Model building, analysis and interpretation.* (Wiley Series in Mathematical and Computational Biology. Chichester: Wiley., 1999).
9. L. Laguzet and G. Turinici, Global optimal vaccination in the SIR model: Properties of the value function and application to cost-effectiveness analysis, *Mathematical Biosciences* **263**, 180 (2015).
10. C. T. Bauch, A. P. Galvani and D. J. D. Earn, Group interest versus self-interest in smallpox vaccination policy, *Proceedings of the National Academy of Sciences* **100**, 10564 (2003).
11. P. J. Francis, Optimal tax/subsidy combinations for the flu season, *Journal of Economic Dynamics and Control* **28**, 2037 (2004).
12. B. Buonomo, A. dOnofrio and D. Lacitignola, Global stability of an {SIR} epidemic model with information dependent vaccination, *Mathematical Biosciences* **216**, 9 (2008).

13. A. dOnofrio, P. Manfredi and E. Salinelli, Vaccinating behaviour, information, and the dynamics of SIR vaccine preventable diseases, *Theoretical Population Biology* **71**, 301 (2007).
14. A. P. Galvani, T. C. Reluga and G. B. Chapman, Long-standing influenza vaccination policy is in accord with individual self-interest but not with the utilitarian optimum, *Proceedings of the National Academy of Sciences* **104**, 5692 (2007).
15. T. C. Reluga, C. T. Bauch and A. P. Galvani, Evolving public perceptions and stability in vaccine uptake, *Math. Biosci.* **204**, 185 (2006).
16. M.-G. Cojocaru, Dynamic equilibria of group vaccination strategies in a heterogeneous population, *Journal of Global Optimization* **40**, 51 (2008).
17. F. H. Chen, Modeling the effect of information quality on risk behavior change and the transmission of infectious diseases, *Mathematical Biosciences* **217**, 125 (2009).
18. P. E. M. Fine and J. A. Clarkson, Individual versus public priorities in the determination of optimal vaccination policies, *American Journal of Epidemiology* **124**, 1012 (1986).
19. D. L. Brito, E. Sheshinski and M. D. Intriligator, Externalities and compulsary vaccinations, *Journal of Public Economics* **45**, 69 (1991).
20. S. Bhattacharyya and C. T. Bauch, Wait and see vaccinating behaviour during a pandemic: A game theoretic analysis, *Vaccine* **29**, 5519 (2011).
21. L. Laguzet and G. Turinici, Individual vaccination as Nash equilibrium in a SIR model with application to the 2009-10 Influenza A(H1N1) epidemic in France, *Bulletin of Mathematical Biology* **in press**, p. 30pp (2015), doi:10.1007/s11538-015-0111-7.
22. D. A. Gomes, J. Mohr and R. R. Souza, Continuous time finite state mean field games, *Applied Mathematics & Optimization* **68**, 99 (2013).
23. O. Guant, Existence and uniqueness result for mean field games with congestion effect on graphs, *Applied Mathematics & Optimization* **72**, 291 (2015).
24. R. M. Anderson and R. M. May, *Infectious Diseases of Humans Dynamics and Control* (Oxford University Press, 1992).
25. A. Bressan and B. Piccoli, *Introduction to the mathematical theory of control*, AIMS Series on Applied Mathematics, Vol. 2 (American Institute of Mathematical Sciences (AIMS), Springfield, MO, 2007).
26. G. Barles, *Solutions de viscosité des équations de Hamilton-Jacobi*, Mathématiques & Applications (Berlin) [Mathematics & Applications], Vol. 17 (Springer-Verlag, Paris, 1994).

27. M. Bardi and I. Capuzzo-Dolcetta, *Optimal control and viscosity solutions of Hamilton-Jacobi-Bellman equations*Systems & Control: Foundations & Applications, Systems & Control: Foundations & Applications (Birkhäuser Boston Inc., Boston, MA, 1997), With appendices by Maurizio Falcone and Pierpaolo Soravia.
28. M. G. Crandall and P.-L. Lions, Viscosity solutions of Hamilton-Jacobi equations, *Trans. Amer. Math. Soc.* **277**, 1 (1983).
29. A. Bressan, JR. and F. Rampazzo, Impulsive control systems with commutative vector fields., *J. Optim. Theory Appl.* **71**, 67 (1991).

On some minimization problems in $\mathbf{R}^N$

Mihai Mariş
Institut de Mathématiques de Toulouse UMR 5219, Université Paul Sabatier, 118 route de Narbonne, 31062 Toulouse Cedex 9, France
E-mail: mihai.maris@math.univ-toulouse.fr

For a large class of constrained minimization problems in $\mathbf{R}^N$ invariant under translations, we prove that dichotomy of minimizing sequences may occur only if vanishing also occurs. Then we discuss the dependence of the Lagrange multipliers associated to minimizers with respect to the constraint and we show that continuity cannot be expected. Finally, we show that if we minimize under Pohozaev constraints we may control the Lagrange multipliers.

Keywords: Constrained minimization, Concentration-compactness principle, Lagrange multipliers, symmetry of minimizers.

1. Introduction

In this paper we consider general variational problems of the form

$$(\mathcal{P}_\lambda)\qquad \begin{aligned} &\text{minimize } E(u) := \int_{\mathbf{R}^N} F(u(x), \nabla u(x))\, dx \quad \text{under the constraint} \\ &Q(u) := \int_{\mathbf{R}^N} G(u(x), \nabla u(x))\, dx = \lambda. \end{aligned}$$

Here $u : \mathbf{R}^N \longrightarrow \mathbf{R}^m$ is a function that belongs to some function space $\mathscr{X}$,$\nabla u = (\frac{\partial u}{\partial x_1}, \dots, \frac{\partial u}{\partial x_N})$, and F, G are real-valued functions defined on $\mathbf{R}^m \times \mathbf{R}^{mN}$. We denote by $(s, \xi) = (s, \xi_1, \dots \xi_N)$ vectors in $\mathbf{R}^m \times \mathbf{R}^{mN}$, where $\xi_i = (\xi_i^1, \dots, \xi_i^m) \in \mathbf{R}^m$ for $1 \leqslant i \leqslant N$.

Let

$$E_{\min}(\lambda) = \inf\{E(u) \mid u \in \mathscr{X}, Q(u) = \lambda\}, \tag{1}$$

with the convention $\inf(\emptyset) = \infty$.

A minimizing sequence for $(\mathcal{P}_\lambda)$ is a sequence $(u_n)_{n\geqslant 1} \subset \mathscr{X}$ such that $Q(u_n) \longrightarrow \lambda$ and $E(u_n) \longrightarrow E_{\min}(\lambda)$. Our aim is to show that, under suitable conditions, problem $(\mathcal{P}_\lambda)$ admits minimizers and, moreover, any minimizing sequence has a convergent subsequence (we say that it is *precompact*). The precompactness of minimizing sequences is important: in many cases, special solutions (such as solitary waves, standing waves, stationary solutions) of Hamiltonian systems are obtained as minimizers of problems like $(\mathcal{P}_\lambda)$. If E and Q are both conserved by the flow associated to the evolution equation and if any minimizing sequence is precompact, the set of minimizers is orbitally stable for the evolution equation[3].

The main tool in proving precompactness of minimizing sequences for $(\mathcal{P}_\lambda)$ is the concentration-compactness principle, introduced by P.-L. Lions in his celebrated paper[9]. In its simplest form this principle asserts that given a bounded sequence of nonnegative functions $(f_n)_{n\geqslant 1} \subset L^1(\mathbf{R}^N)$, there exists a subsequence $(f_{n_k})_{k\geqslant 1}$ that either "spreads" over the space (that is, $\sup\limits_{x\in\mathbf{R}^N} \int_{B(x,1)} f_{n_k}(y)\,dy \longrightarrow 0$ as $k \longrightarrow \infty$), a phenomenon called "vanishing", or f_{n_k} can be "split" into two nontrivial parts whose supports are far away from each other (this situation is called "dichotomy"), or most of the mass of f_{n_k} remains on balls of fixed radius (the last case is called "concentration"). In applications, one may use the concentration-compactness principle for $f_n = |F(u_n, \nabla u_n)|$ or $f_n = |G(u_n, \nabla u_n)|$ or $f_n = |F(u_n, \nabla u_n)| + |G(u_n, \nabla u_n)|$. The next step consists in showing that vanishing and dichotomy cannot happen, hence concentration occurs. Finally, one needs to prove that concentration implies convergence.

In general, the most difficult part in using the concentration-compactness principle is to rule out dichotomy. We show that, under very general assumptions, if dichotomy occurs for a minimizing sequence $(u_n)_{n\geqslant 1}$ then vanishing must occur for another minimizing sequence $(w_n)_{n\geqslant 1}$. In other words: if we rule out vanishing, then we rule out dichotomy, too.

We introduce some notation. Given an affine hyperplane $\Pi \subset \mathbf{R}^N$ we denote by Π^+ and Π^- the two half-spaces determined by Π and by S_Π the symmetry with respect to Π. For any function f defined on $\mathbf{R}^N$ we denote

$$f_{\Pi^+}(x) = \begin{cases} f(x) & \text{if } x \in \Pi^+ \cup \Pi, \\ f(S_\Pi(x)) & \text{if } x \in \Pi^-, \end{cases} \qquad \text{respectively}$$
$$f_{\Pi^-}(x) = \begin{cases} f(x) & \text{if } x \in \Pi^- \cup \Pi, \\ f(S_\Pi(x)) & \text{if } x \in \Pi^+. \end{cases} \tag{2}$$

Definition 1.1. We say that a space $\mathscr{Y}$ of functions defined on $\mathbf{R}^N$ has the reflection property with respect to a direction $v \in \mathbf{R}^N \setminus \{0\}$ if for any

$f \in \mathscr{Y}$ and for any affine hyperplane $\Pi \subset \mathbf{R}^N$ orthogonal to v we have $f_{\Pi^+}, f_{\Pi^-} \in \mathscr{Y}$, where f_{Π^+}, f_{Π^-} are as in (5.1).

For instance, the spaces $W^{1,p}(\mathbf{R}^N)$ with $1 \leqslant p \leqslant \infty$ as well as $H^s(\mathbf{R}^N)$ and $\dot{H}^s(\mathbf{R}^N)$ with $-\frac{1}{2} < s < \frac{3}{2}$ have the reflection property with respect to any direction (see[10] for the last two).

We will make use of the following assumptions:

(A1) The function space $\mathscr{X}$ satisfies:

- **(i)** For any function with compact support $u \in \mathscr{X}$ and any $a \in \mathbf{R}^N$ there holds $u(\cdot + a) \in \mathscr{X}$.
- **(ii)** If $u, v \in \mathscr{X}$ have compact and disjoint supports, then $u + v \in \mathscr{X}$.
- **(iii)** If the set $\{u \in \mathscr{X} \mid Q(u) = \lambda\}$ is not empty, there exists a sequence $(u_n)_{n \geqslant 1} \subset \mathscr{X}$ such that each u_n has compact support, $Q(u_n) = \lambda$ and $E(u_n) \longrightarrow E_{min}(\lambda)$.
- **(iv)** $F(u, \nabla u)$ and $G(u, \nabla u)$ belong to $L^1(\mathbf{R}^N)$ for any $u \in \mathscr{X}$.

In most applications assumption **(A1)-(iii)** can be checked by using the density of functions with compact support in usual function spaces.

If $\mathscr{X}$ satisfies **(A1) (i)-(iii)**, the function $E_{\min}$ is subadditive, that is

$$E_{\min}(\lambda_1 + \lambda_2) \leq E_{\min}(\lambda_1) + E_{\min}(\lambda_2). \tag{3}$$

This is a well-known fact and has already been pointed out by P.-L. Lions[9]. We recall here the proof. Fix $\varepsilon > 0$. Consider u and v with compact support such that $Q(u) = \lambda_1$, $E(u) \leq E_{\min}(\lambda_1) + \frac{\varepsilon}{2}$, $Q(v) = \lambda_2$, $E(v) \leq E_{\min}(\lambda_2) + \frac{\varepsilon}{2}$. Let x be such that the supports of u and $v(x + \cdot)$ are disjoint. Then $Q(u + v(x + \cdot)) = \lambda_1 + \lambda_2$ and $E(u + v(x + \cdot)) \leq E_{\min}(\lambda_1) + E_{\min}(\lambda_2) + \varepsilon$. Since ε is arbitrary, (3) follows.

Our next assumption is that the minimization problem $(\mathcal{P}_\lambda)$ is, in some sense, well-posed, that is $E(u)$ controls $Q(u)$. More precisely, we assume

(A2) $\lim\limits_{\lambda \to 0,\ \lambda > 0} E_{\min}(\lambda) = 0 = E_{\min}(0)$.

In applications, assumption **(A2)** is usually checked by using "energy estimates" involving the specific properties of the functions F and G.

Notice that **(A2)** implies that $E_{\min}$ is finite on some interval $(0, \delta)$. If **(A1) (i)-(iii)** and **(A2)** are both satisfied, it is easy to see that $E_{\min}(\lambda) < \infty$ for any $\lambda > 0$. Indeed, given $\lambda > 0$ one may choose an integer n such that $\frac{\lambda}{n} \in (0, \delta)$. Then (3) implies $E_{\min}(\lambda) \leq n E_{\min}(\frac{\lambda}{n}) < \infty$.

(A3) The functions F and G are symmetric with respect to one direction $\mathbf{e} \in \mathbf{R}^N \setminus \{0\}$ and $\mathscr{X}$ has the reflection property with respect to $\mathbf{e}$.

Changing coordinates if necessary we may assume that $\mathbf{e} = (1, 0, \ldots, 0)$ (and we will do so in the sequel). Then the symmetry of F and G means

$$F(s, -\xi_1, \xi_2, \ldots, \xi_N) = F(s, \xi_1, \xi_2, \ldots, \xi_N) \quad \text{and}$$
$$G(s, -\xi_1, \xi_2, \ldots, \xi_N) = G(s, \xi_1, \xi_2, \ldots, \xi_N)$$

for all $(s, \xi) \in \mathbf{R}^m \times \mathbf{R}^{mN}$.

Sometimes we need a stronger assumption, namely:

(A4) The functions F and G have a common two-dimensional symmetry, for instance

$$F(s, \xi_1 \cos\theta + \xi_2 \sin\theta, \xi_1 \sin\theta - \xi_2 \cos\theta, \xi_3, \ldots, \xi_N) = F(s, \xi_1, \xi_2, \xi_3, \ldots, \xi_N),$$
$$G(s, \xi_1 \cos\theta + \xi_2 \sin\theta, \xi_1 \sin\theta - \xi_2 \cos\theta, \xi_3, \ldots, \xi_N) = G(s, \xi_1, \xi_2, \xi_3, \ldots, \xi_N)$$

for all $(s, \xi) \in \mathbf{R}^m \times \mathbf{R}^{mN}$ and all $\theta \in \mathbf{R}$, and $\mathscr{X}$ has the reflection property with respect to any direction $(x_1, x_2, 0, \ldots, 0) \in \mathbf{R}^N$ with $(x_1, x_2) \neq (0, 0)$.

Our main results can be summarized as follows.

Theorem 1.1. *Suppose that* ***(A1), (A2), (A3)*** *hold. Then:*

(i) The function $E_{\min}$ *is concave on* $[0, \infty)$.
(ii) If there exist $\lambda_1, \lambda_2 > 0$ *such that* $E_{\min}(\lambda_1 + \lambda_2)$ *is finite and* $E_{\min}(\lambda_1 + \lambda_2) = E_{\min}(\lambda_1) + E_{\min}(\lambda_2)$, *then* $E_{\min}$ *is linear on* $[0, \lambda_1 + \lambda_2]$.
(iii) If $E_{\min}$ *is linear on* $[0, \lambda]$, *there exists a sequence* $(u_n)_{n \geqslant 1} \subset \mathscr{X}$ *such that* $Q(u_n) = \lambda$, $E(u_n) \longrightarrow E_{\min}(\lambda)$, *each* u_n *has compact support contained in the union of* n *disjoint balls* $B(a_i, R_n)$ *that we may choose such that the distances* $|a_i - a_j|$ *are arbitrarily large, and*

$$\int_{B(a_i, R_n)} F(u_n(x), \nabla u_n(x))\, dx < \tfrac{1}{n} E_{\min}(\lambda) + \tfrac{1}{n^2},$$

$$\int_{B(a_i, R_n)} G(u_n(x), \nabla u_n(x))\, dx = \tfrac{\lambda}{n}.$$

Using the standard terminology associated with the concentration-compactness principle, part (iii) in Theorem 1.1 gives a minimizing sequence for $(\mathcal{P}_\lambda)$ having the "vanishing property", which means that its "energy" and "mass" spread over $\mathbf{R}^N$.

Theorem 1.2. *Assume that* ***(A1), (A2), (A3)*** *are satisfied. Suppose that for any* $\lambda \geqslant 0$, *all minimizers of* $(\mathcal{P}_\lambda)$ *are* C^1 *functions on* $\mathbf{R}^N$.

(i) If $E_{\min}$ *is linear on* $[0, \Lambda]$, *then for any* $\lambda \in (0, \Lambda)$ *the problem* $(\mathcal{P}_\lambda)$ *does not admit minimizers.*
(ii) Assume in addition that ***(A4)*** *is satisfied.*

If $E_{\min}$ is affine on some interval $[a,b]$ with $0 < a < b$, then for any $\lambda \in (a,b)$ the problem $(\mathcal{P}_\lambda)$ does not admit minimizers.

In most applications, under suitable assumptions on the functions F and G, the minimizers of $(\mathcal{P}_\lambda)$ satisfy Euler-Lagrange equations. In many situations the Euler-Lagrange equations are quasilinear elliptic systems. The regularity of solutions of such systems received considerable attention during the last decades and there is a huge literature devoted to this topic. Various sufficient conditions that guarantee the regularity of solutions have been given. We refer the reader to the standard textbooks[4–8,15] and references therein for regularity results.

This paper is organized as follows. In the next section we give the proof of Theorem 1.1. In section 3 we show that results similar to Theorems 1.1 and 1.2 (or even stronger) are true for a class of nonlocal operators. In section 4 we discuss the dependence of Lagrange multipliers associated to $(\mathcal{P}_\lambda)$ with respect to λ and we show why it is interesting to use Pohozaev constraints. The last section is devoted to the symmetry of minimizers.

2. Proof of Theorem 1.1

(i) We already know that $E_{\min}(\lambda) < \infty$ for all $\lambda \geqslant 0$. If $E_{\min}(\lambda_0) = -\infty$, using (3) we infer that $E_{\min}(\lambda) = -\infty$ for all $\lambda > \lambda_0$. Similarly (3) and **(A2)** imply

$$\limsup_{\lambda \to \lambda_0, \lambda > \lambda_0} E_{\min}(\lambda) \leqslant E_{\min}(\lambda_0) \qquad \text{for any } \lambda_0 \geqslant 0.$$

Consider $0 < \lambda_1 < \lambda_2$ such that $E_{\min}(\lambda_1)$ and $E_{\min}(\lambda_2)$ are finite. Then $E_{\min}\left(\frac{\lambda_1+\lambda_2}{2}\right)$ is also finite. Fix $\varepsilon > 0$. Choose $u \in \mathscr{X}$ such that $Q(u) = \frac{\lambda_1+\lambda_2}{2}$ and $E(u) < E_{\min}\left(\frac{\lambda_1+\lambda_2}{2}\right) + \frac{\varepsilon}{2}$. For $t \in \mathbf{R}$ define

$$\begin{aligned}
u_{1,t}(x) &= \begin{cases} u(x) \text{ if } x_1 \leqslant t, \\ u(2t - x_1, x_2, \ldots, x_N) \text{ if } x_1 > t, \end{cases} \\
u_{2,t}(x) &= \begin{cases} u(2t - x_1, x_2, \ldots, x_N) \text{ if } x_1 \leqslant t, \\ u(x) \text{ if } x_1 > t. \end{cases}
\end{aligned} \tag{4}$$

By **(A3)** we have $u_{1,t}, u_{2,t} \in \mathscr{X}$ for all $t \in \mathbf{R}$ and a change of variables gives

$$\begin{aligned} Q(u_{1,t}) &= 2\int_{\{x_1 \leqslant t\}} G(u(x), \nabla u(x))\,dx, \\ E(u_{1,t}) &= 2\int_{\{x_1 \leqslant t\}} F(u(x), \nabla u(x))\,dx, \\ Q(u_{2,t}) &= 2\int_{\{x_1 > t\}} G(u(x), \nabla u(x))\,dx, \\ E(u_{2,t}) &= 2\int_{\{x_1 > t\}} F(u(x), \nabla u(x))\,dx. \end{aligned} \tag{5}$$

Since $G(u, \nabla u) \in L^1(\mathbf{R}^N)$, it is an easy consequence of Lebesgue's dominated convergence theorem that the function $t \longmapsto Q(u_{1,t})$ is continuous on $\mathbf{R}$, tends to 0 as $t \longrightarrow -\infty$ and tends to $2Q(u) = \lambda_1 + \lambda_2$ as $t \longrightarrow \infty$. Hence there exists t_0 such that $Q(u_{1,t_0}) = \lambda_1$. It follows from (5) that $Q(u_{2,t_0}) = \lambda_2$. From the definition of $E_{\min}$ it is clear that $E(u_{1,t_0}) \geqslant E_{\min}(\lambda_1)$ and $E(u_{2,t_0}) \geqslant E_{\min}(\lambda_2)$. Using (5) again we infer that

$$E_{\min}(\lambda_1)+E_{\min}(\lambda_2) \leqslant E(u_{1,t_0})+E(u_{2,t_0}) = 2E(u) \leqslant 2E_{\min}\left(\frac{\lambda_1+\lambda_2}{2}\right)+\varepsilon.$$

Since ε is arbitrary, we get $\frac{1}{2}\left(E_{\min}(\lambda_1) + E_{\min}(\lambda_2)\right) \leqslant E_{\min}\left(\frac{\lambda_1+\lambda_2}{2}\right)$. Then the concavity of $E_{\min}$ follows from the following elementary result.

Lemma 2.1. *Let $I \subset \mathbf{R}$ be an interval. Assume that $f : I \longrightarrow \mathbf{R}$ is a function satisfying:*

(i) $f\left(\frac{a+b}{2}\right) \geqslant \frac{1}{2}(f(a) + f(b))$ *for any $a, b \in I$, and*
(ii) $\limsup\limits_{x \to a, x > a} f(x) \leqslant f(a)$ *whenever a is an interior point of I.*

Then f is concave.

(ii) Consider $\lambda_1, \lambda_2 > 0$ such that $E_{\min}(\lambda_1 + \lambda_2)$ is finite. By the concavity of $E_{\min}$ and the fact that $E_{\min}(0) = 0$ we get $E_{\min}(\lambda_i) \geqslant \frac{\lambda_i}{\lambda_1+\lambda_2} E_{\min}(\lambda_1 + \lambda_2)$ for $i = 1, 2$ and we may have equality in one of these inequalities if and only if $E_{\min}$ is linear on $[0, \lambda_1 + \lambda_2]$. Summing up we find $E_{\min}(\lambda_1) + E_{\min}(\lambda_2) \geqslant E_{\min}(\lambda_1 + \lambda_2)$ and equality may occur if and only if $E_{\min}$ is linear on $[0, \lambda_1 + \lambda_2]$.

(iii) Assume that $E_{\min}$ is linear on $[0, \lambda]$. Fix $n \in \mathbb{N}^*$. By **(A1)-(iii)** there exists a function $v_n \in \mathscr{X}$ such that $Q(v_n) = \frac{\lambda}{n}$,

$$E(v_n) \leq E_{min}\left(\frac{\lambda}{n}\right) + \frac{1}{n^2} = \frac{1}{n}E_{min}(\lambda) + \frac{1}{n^2}$$

and v_n has compact support contained in a ball $B(0, R_n)$. Choose $a_1, \ldots, a_n \in \mathbf{R}^N$ such that $|a_i - a_j| > 2R_n$ for $i \neq j$ (of course that $a_1, \ldots, a_n$ can be chosen as "far away from each other" as we want). It is clear that the supports of $v_n(\cdot + a_j)$ et $v_n(\cdot + a_k)$ are disjoint if $j \neq k$. Let $u_n = v_n(\cdot + a_1) + \cdots + v_n(\cdot + a_n)$. Then $u_n \in \mathscr{X}$ by **(A1)-(ii)**, $Q(u_n) = nQ(v_n) = \lambda$, $E(u_n) = nE(v_n) \leq E_{min}(\lambda) + \frac{1}{n}$ and $(u_n)_{n\geq 1}$ satisfies the conclusion of Theorem 1.1 (iii). □

3. Some nonlocal problems

The previous results can be extended to some minimization problems involving nonlocal functionals (but in each case the constraint has to be *local*, namely $Q(u) = \int_{\mathbf{R}^N} G(u(x), \nabla u(x))\, dx$). In the sequel we denote by $\mathcal{F}$ or $\widehat{\ }$ the Fourier transform. We consider the following functionals:

$$E_1(u) = \int_{\mathbf{R}^N} |\xi|^{2s} |\widehat{u}(\xi)|^2\, d\xi + \int_{\mathbf{R}^N} F(u(x), \nabla u(x))\, dx, \ s \in (0,1), \tag{6}$$

$$E_2(u) = \int_{\mathbf{R}^N} (1 + |\xi|^2)^s |\widehat{u}(\xi)|^2\, d\xi + \int_{\mathbf{R}^N} F(u(x), \nabla u(x))\, dx, \ s \in (0,1), \tag{7}$$

$$E_3(u) = \int_{\mathbf{R}^N} F(u(x), \nabla u(x))\, dx - \int_{\mathbf{R}^N}\int_{\mathbf{R}^N} \frac{H(u(x))H(u(y))}{|x-y|^{N-2}}\, dx\, dy, \tag{8}$$

$$E_4(u) = \int_{\mathbf{R}^3} |\nabla u|^2 dx + \int_{\mathbf{R}^3} F(u(x), \nabla u(x))\, dx - \int_{\mathbf{R}^3} \frac{\xi_3^2}{|\xi|^2} |\widehat{u^2}(\xi)|^2\, d\xi. \tag{9}$$

Functionals E_1 and E_2 arise in problems involving the fractional powers of the Laplacian, $(-\Delta)^s$, respectively fractional powers of the operator $Id - \Delta$, E_3 is the generalized Choquard functional, and E_4 is the Davey-Stewartson functional.

We will consider the generalized Choquard functional in space dimension $N \geqslant 3$. It follows from [10], Lemma 4.5 p. 573 that under suitable conditions on H, the nonlocal term in E_3 can be written as

$$\int_{\mathbf{R}^N}\int_{\mathbf{R}^N} \frac{H(u(x))H(u(y))}{|x-y|^{N-2}}\, dx\, dy = c_N \int_{\mathbf{R}^N} \frac{1}{|\xi|^2} \left|\widehat{H(u)}(\xi)\right|^2 d\xi. \tag{10}$$

We will use the following natural assumption:

(NL) We have $\mathscr{X} \subset \mathcal{S}'(\mathbf{R}^N)$ and for any $u \in \mathscr{X}$ we have

- $|\xi|^s \widehat{u} \in L^2(\mathbf{R}^N)$ when we deal with E_1,

- $(1+|\xi|^2)^{\frac{s}{2}}\widehat{u} \in L^2(\mathbf{R}^N)$ when we deal with E_2,
- $\frac{1}{|\xi|}\widehat{H(u)} \in L^2(\mathbf{R}^N)$ when we deal with E_3,
- $\frac{\xi_3}{|\xi|}\widehat{|u|^2} \in L^2(\mathbf{R}^3)$ when we deal with E_4.

The next variant of definition (1) is more appropriate here:

$$\begin{aligned} E^i_{\min}(\lambda) &= \inf\{a \in \mathbf{R}\cup\{-\infty\} \mid \text{ there exists } (u_n)_{n\geqslant 1} \subset \mathscr{X} \\ &\qquad \text{such that } Q(u_n) \longrightarrow \lambda \text{ and } E_i(u_n) \longrightarrow a\} \\ &= \sup_{\delta>0}\left(\inf\{E(u) \mid u \in \mathscr{X}, Q(u) \in (\lambda-\delta,\lambda+\delta)\ \}\right). \end{aligned} \tag{11}$$

If **(A1)** and **(NL)** are satisfied, proceeding as in the case of local functionals and using the Riemann-Lebesgue Lemma it is easily seen that $E^i_{\min}$ is subadditive, that is (3) holds for $i = 1, 2, 3, 4$. The following results hold.

Theorem 3.1. *Suppose that **(A1)**, **(A2)**, **(A3)**, **(NL)** are satisfied. Then the conclusion of Theorem 1.1 is true with $E^i_{\min}$ instead of $E_{\min}$ for $i = 1, 2, 3, 4$.*

Theorem 3.2. *Assume that **(A1)**, **(A2)**, **(A3)**, **(NL)** are satisfied. Let $0 \leqslant a < b$. If $E^1_{\min}$ or $E^2_{\min}$ is affine on $[a, b]$, then for any $\lambda \in (a, b)$ the problem $(\mathcal{P}_\lambda)$ (where E is replaced by E_1 or by E_2) does not admit minimizers.*

Notice that Theorem 3.2 gives a better nonexistence result for minimizers of E_1 and E_2 than we have for local functionals. In particular, we do not need the C^1 regularity of minimizers and, in the case $a > 0$, we do not need assumption **(A4)**.

As in the case of local functionals, the main step in proving Theorem 3.1 is to establish the concavity of the function

$$E^i_{min}(\lambda) = \inf\{E_i(u) \mid u \in \mathscr{X}, Q(u) = \lambda\}.$$

Assume that **(A1)** holds. Let $a, b > 0$ and suppose that $u \in \mathscr{X}$ satisfies $Q(u) = \frac{a+b}{2}$. After translation we may assume that

$$\int_{\{x_1<0\}} G(u, \nabla u)\, dx = \frac{a}{2} \quad \text{and} \quad \int_{\{x_1>0\}} G(u, \nabla u)\, dx = \frac{b}{2}.$$

Let u_1 and u_2 be the functions obtained from u by mirror symmetry with respect to $\{x_1 = 0\}$, as in (4). Then $Q(u_1) = a$ and $Q(u_2) = b$. We show that $2E_i(u) \geqslant E_i(u_1) + E_i(u_2)$; it is then straightforward to get the desired conclusion. To do this we use some integral identities established in[10].

Denote $Au(x) = \frac{1}{2}(u(x_1, x') - u(-x_1, x'))$, where $x' = (x_2, \ldots, x_N)$. Using identities (3.3) and (3.4) p. 558 and Corollary 3.4 p. 565 in[10] we get for $s \in (-\frac{1}{2}, \frac{3}{2})$:

$$E_1(u_1) + E_1(u_2) - 2E_1(u) = -\frac{16\sin(s\pi)}{\pi^2} N_s^2(Au), \tag{12}$$

where

$$N_s^2(f) = \int_{\mathbf{R}^{N-1}} \int_{|\xi'|}^{\infty} \left(t^2 - |\xi'|^2\right)^s \left| \int_0^\infty \widehat{f}(\xi_1, \xi') \frac{\xi_1}{t^2 + \xi_1^2} \, d\xi_1 \right|^2 dt \, d\xi' \geqslant 0,$$

respectively

$$E_2(u_1) + E_2(u_2) - 2E_2(u) = -\frac{16\sin(s\pi)}{\pi^2} \tilde{N}_s^2(Au), \tag{13}$$

where

$$\tilde{N}_s^2(f) = \int_{\mathbf{R}^{N-1}} \int_{\sqrt{|\xi'|^2+1}}^{\infty} \left(t^2 - |\xi'|^2 - 1\right)^s \left| \int_0^\infty \widehat{f}(\xi_1, \xi') \frac{\xi_1}{t^2 + \xi_1^2} \, d\xi_1 \right|^2 dt \, d\xi'.$$

From identity (4.12) p. 576 in[10] it follows that there is $C_N > 0$ such that

$$E_3(u_1) + E_3(u_2) - 2E_3(u) = -C_N \int_{\mathbf{R}^{N-1}} \frac{1}{|\xi'|} \left| \int_0^\infty \widehat{A(H(u))}(\xi) \frac{\xi_1}{|\xi|^2} d\xi_1 \right|^2 d\xi' \leqslant 0,$$

where A is as above, while identity (4.30) and Lemma 4.10 p. 585 in[10] imply that there is $k > 0$ such that

$$E_4(u_1) + E_4(u_2) - 2E_4(u) = -k \int_{\mathbf{R}^2} \frac{\xi_3^2}{|\xi'|} \left| \int_0^\infty \widehat{A(u^2)}(\xi) \frac{\xi_1}{|\xi|^2} d\xi_1 \right|^2 d\xi' \leqslant 0.$$

The main ingredients in the proof of Theorem 3.2 are identities (12) and (13) and the fact that N_s (resp. $\tilde{N}_s$) is a norm on the space of functions that are in $\dot{H}^s(\mathbf{R}^N)$ (resp. in $H^s(\mathbf{R}^N)$) and are odd with respect to x_1. A result similar to Theorem 1.2 above is true for E_3 and E_4, see[13] for details.

4. Lagrange multipliers and Pohozaev constraints

If u is a solution of the minimization problem $(\mathcal{P}_\lambda)$, E and Q are differentiable at u and $Q'(u) \neq 0$, there exists a Lagrange multiplier $\beta(u)$ such that

$$E'(u) = \beta(u) Q'(u).$$

In many applications the functionals E and Q are not differentiable in the classical sense and the following result is more suitable.

Proposition 4.1. *Assume that u is a minimizer for $(\mathcal{P}_\lambda)$, $E_{\min}$ is concave in a neighborhood of λ and*

(i) for any $\phi \in C_c^\infty(\mathbf{R}^N)$, the function $u+\phi$ belongs to $\mathscr{X}$,

(ii) there exist two distributions $S, T \in \mathcal{D}'(\mathbf{R}^N)$, $S \neq 0$, such that for any $\phi \in C_c^\infty(\mathbf{R}^N)$ there holds as $t \downarrow 0$

$$\frac{E(u+t\phi)-E(u)}{t} \longrightarrow \langle T, \phi\rangle \qquad \text{and} \qquad \frac{Q(u+t\phi)-Q(u)}{t} \longrightarrow \langle S, \phi\rangle.$$

Then there exists $\beta \in [E'_{\min,r}(\lambda), E'_{\min,\ell}(\lambda)]$ such that $T = \beta S$. Here $E'_{\min,r}(\lambda)$ and $E'_{min,\ell}(\lambda)$ denote the right and left derivatives of $E_{\min}$ at λ.

In applications, one usually wants to solve a given PDE, corresponding to a specific value of β. In our framework, an important question is whether it is possible to choose λ in order to get a specific value for β or whether we can reach all β in some interval. The control of Lagrange multipliers is a difficult problem in Calculus of Variations which is still unsolved, except when the considered functionals are homogeneous.

Remark 4.1. If $E_{\min}$ is not linear on $[0, \lambda]$ and all minimizing sequences for $(\mathcal{P}_\lambda)$ have convergent subsequences, we expect to find minimizers u_ℓ, u_r for $(\mathcal{P}_\lambda)$ with Lagrange multipliers $E'_{\min,\ell}(\lambda)$ and $E'_{\min,r}(\lambda)$, respectively. If $0 < \lambda_1 < \lambda_2$, u_1 and u_2 are minimizers for $(\mathcal{P}_{\lambda_1})$ and $(\mathcal{P}_{\lambda_2})$, respectively, and β_1, β_2 are the corresponding Lagrange multipliers, then $\beta_1 \geqslant \beta_2$ (and equality may hold only if $E_{\min}$ is affine on $[\lambda_1, \lambda_2]$). Indeed, $E_{\min}$ is concave on $[0,\infty)$, hence $\beta_1 \geqslant E'_{\min,r}(\lambda_1) \geqslant E'_{\min,\ell}(\lambda_2) \geqslant \beta_2$.

Let $J \subset (0,\infty)$ be an interval. The set of Lagrange multipliers associated to minimizers of $(\mathcal{P}_\lambda)$ when $\lambda \in J$ is an interval if and only if for each $\lambda \in J$ and for each $\beta \in [E_{\min,r}(\lambda), E_{\min,\ell}(\lambda)]$ there exists a minimizer for $(\mathcal{P}_\lambda)$ with Lagrange multiplier β. The next examples show that this is not always the case in applications.

Example 4.2 ([13]) Let $N \geqslant 3$ and let $q \in (2, 2^*)$, where $2^* = \frac{2N}{N-2}$ is the Sobolev exponent associated to 2. Denote

$$I_q(\lambda) = \inf\Big\{ \int_{\mathbf{R}^N} |\nabla u|^2 + |u|^2 \, dx \;\Big|\; u \in H^1(\mathbf{R}^N),\ \int_{\mathbf{R}^N} |u|^q \, dx = \lambda \Big\}.$$

It is easy to see that $I_q(\lambda) = \lambda^{\frac{2}{q}} I_p(1)$ and the problem $(\mathcal{P}_\lambda^1)$ that consists in minimizing $\displaystyle\int_{\mathbf{R}^N} |\nabla u|^2 + |u|^2 dx$ in $H^1(\mathbf{R}^N)$ under the constraint $\displaystyle\int_{\mathbf{R}^N} |u|^q \, dx = \lambda$ admits minimizers for any $\lambda > 0$. Since I_q is differentiable on $(0, \infty)$, by Proposition 4.1 any minimizer of $(\mathcal{P}_\lambda^1)$ has the Lagrange multiplier $\frac{2}{p}\lambda^{\frac{2}{p}-1}I_p(1)$.

Fix $q, r \in (2, 2^*)$, $q \neq r$. Let $G(s) = \begin{cases} s^q \text{ if } s \geq 0, \\ |s|^r \text{ if } s < 0. \end{cases}$ Then $G \in C^2(\mathbf{R})$ and G is convex. Consider the minimization problem

$$(\mathcal{P}_\lambda^2) \qquad \begin{array}{c} \text{minimize } E(u) = \displaystyle\int_{\mathbf{R}^N} |\nabla u|^2 + |u|^2 dx \text{ in } H^1(\mathbf{R}^N) \\ \text{under the constraint } Q(u) = \displaystyle\int_{\mathbf{R}^N} G(u)\,dx = \lambda. \end{array}$$

It is easily seen that $E_{\min}(\lambda) = \inf\{E(u) \mid u \in H^1(\mathbf{R}^N),\ Q(u) = \lambda\} = \min(I_q(\lambda), I_r(\lambda))$ and minimizers exist for any $\lambda > 0$. If u is a minimizer we have either $u \geqslant 0$ or $u \leqslant 0$ a.e. (for otherwise, denoting $\lambda_\pm = Q(u_\pm)$ we would have $\lambda_\pm > 0$, $\lambda_+ + \lambda_- = \lambda$ and $E_{\min}(\lambda) = E(u_+) + E(u_-)$; then Theorem 1.1 would imply that $E_{\min}$ is linear on $[0, \lambda]$, a contradiction).

There exists a unique $\lambda_* > 0$ such that $\lambda_*^{\frac{2}{q}} I_q(1) = \lambda_*^{\frac{2}{r}} I_r(1)$. If u is a solution of $(\mathcal{P}_{\lambda_*})$, then either $u \geqslant 0$, u is a minimizer for $I_q(\lambda_*)$ and the associated Lagrange multiplier is $\frac{2}{q}\lambda_*^{\frac{2}{q}-1} I_p(1)$, or $u \leqslant 0$, u is a minimizer for $I_r(\lambda_*)$ and the Lagrange multiplier associated to u is $\frac{2}{r}\lambda_*^{\frac{2}{r}-1} I_p(1)$. In this case we see that $E'_{\min,r}(\lambda_*) < E'_{\min,\ell}(\lambda_*)$ and the set of all Lagrange multipliers associated to minimizers of $(\mathcal{P}_\lambda^2)$ is precisely $\{E'_{\min,r}(\lambda), E'_{\min,\ell}(\lambda)\}$.

Pohozaev constraints. A fruitful idea is to use Pohozaev constraints to get rid of Lagrange multipliers. For instance, assume that β is fixed and we want to find u satisfying $E'(u) = \beta Q'(u)$. A direct approach is to search for critical points of the functional $E - \beta Q$. If critical points of this functional satisfy a Pohozaev-type identity that can be written as $P(u) = 0$, it might be interesting to find minimizers (or merely critical points) of $E - \beta Q$ in the set $\{u \in \mathscr{X} \mid P(u) = 0\}$. If they exist, they should satisfy an Euler-Lagrange equation of the form $E'(u) - \beta Q'(u) = \gamma P'(u)$, and the Pohozaev identity corresponding to this equation should be of the form $P(u) = \gamma \tilde{P}(u)$, where $\tilde{P}$ is the "Pohozaev functional" corresponding to P. Since $P(u) = 0$, we get $\gamma \tilde{P}(u) = 0$. If a nontrivial function u cannot satisfy simultaneously the identities $P(u) = 0$ and $\tilde{P}(u) = 0$, we see that necessarily $\gamma = 0$, hence u is a critical point of $E - \beta Q$, as desired.

Notice that this method has been successfully used in[12] to solve a variational problem in dimension $N \geqslant 3$. However, exactly the same variational problem could not be solved in a similar way in space dimension $N = 2$ (there are no minimizers under Pohozaev constraints in dimension two).

Instead of developing an abstract theory, we prefer to give a simple example to illustrate the above idea.

Example 4.3 ([13]) Consider the nonlinear vector field equation

$$-\Delta u = g(u) \quad \text{in } \mathbf{R}^N, \qquad |u(x)| \to 0 \text{ as } |x| \to \infty, \tag{14}$$

where $u : \mathbf{R}^N \longrightarrow \mathbf{R}^m$ and $g = (g_1, \dots, g_m)$. We assume that $g(0) = 0$ and there exists $G \in C^1(\mathbf{R}^m \setminus \{0\}, \mathbf{R})$ such that $g = \nabla G$ in $\mathbf{R}^m \setminus \{0\}$. We make the standard growth assumptions on G, namely

$$\limsup_{|s|\to 0} \frac{G(s)}{|s|^{2^*}} \leqslant 0, \qquad \limsup_{|s|\to\infty} \frac{G(s)}{|s|^{2^*}} \leqslant 0, \qquad |g(s)| \leqslant C\left(1 + |s|^{2^*}\right) \text{ in } \mathbf{R}^m$$

for some $C > 0$. Moreover, we assume that there exists $s_0 \in \mathbf{R}^m$ such that $G(s_0) > 0$. These assumptions are classical in the literature and are almost optimal for the existence of a solution to (14), see[1,2,13,14]. At least formally, solutions of (14) are critical points of the functional

$$E(u) = \frac{1}{2}\int_{\mathbf{R}^N} |\nabla u|^2 dx - \int_{\mathbf{R}^N} G(u) dx.$$

It is well-known (see[1] in the scalar case,[2,14] in the vector case) that under the previous assumptions, *least energy* solutions to (14) (that is, solutions which minimize E among all solutions) can be found by minimizing $\int_{\mathbf{R}^N} |\nabla v|^2\, dx$ under the constraint $\int_{\mathbf{R}^N} G(v)\, dx = \lambda$, then choosing an appropriate value for λ. It is also known (see[2]) that any solution $u \in \mathscr{X}$ of (14) satisfies the Pohozaev identity

$$P(u) = 0, \qquad \text{where } P(u) = \frac{N-2}{2}\int_{\mathbf{R}^N} |\nabla u|^2\, dx - N\int_{\mathbf{R}^N} G(u)\, dx. \tag{15}$$

This identity comes from the behavior of E with respect to dilations in $\mathbf{R}^N$. Indeed, denote $u_\sigma(x) = u(\frac{x}{\sigma})$. Then $E(u_\sigma) = \frac{\sigma^{N-2}}{2}\int_{\mathbf{R}^N} |\nabla u|^2\, dx - \sigma^N \int_{\mathbf{R}^N} G(u)\, dx$. If u is a critical point of E, formally we should have $\frac{d}{d\sigma}_{|\sigma=1}(E(u_\sigma)) = 0$ and this is precisely (15). See[2] for a rigorous proof.

Thus any solution $u \in \mathscr{X}$ of (14) belongs to the set

$$\mathcal{C} = \{u \in \dot{H}^1(\mathbf{R}^N) \mid P(u) = 0\}.$$

We may find least energy solutions to (14) by minimizing E in the set $\mathcal{C}$. One can prove that minimizers exist and any minimizer u satisfies the Euler-Lagrange equation $E'(u) = \beta P'(u)$ for some $\beta \in \mathbf{R}$. If $u \in \mathscr{X}$ is any finite energy solution of this Euler-Lagrange equation (not necessarily a minimizer), then u satisfies the Pohozaev identity $\frac{d}{d\sigma}_{|\sigma=1}(E - \beta P)(u_\sigma) = 0$, that is

$$P(u) - \beta\left(\frac{(N-2)^2}{2}\int_{\mathbf{R}^N} |\nabla u|^2\, dx - N^2\int_{\mathbf{R}^N} G(u)\, dx\right) = 0.$$

Since $P(u) = 0$, we get $\beta \int_{\mathbf{R}^N} |\nabla u|^2 \, dx = 0$. If u is not constant we have necessarily $\beta = 0$, that is $E'(u) = 0$, as desired.

5. Symmetry of minimizers

A natural question is: if the functionals E and Q have common symmetries, do the minimizers of $(\mathcal{P}_\lambda)$ inherit the same symmetry properties?

In this section we will consider the symmetric version of $(\mathcal{P}_\lambda)$, namely

$$(\mathcal{S}_\lambda) \qquad \begin{array}{l} \text{minimize } E(u) = \displaystyle\int_{\mathbf{R}^N} F(u(x), |\nabla u(x)|) \, dx \text{ in } \mathscr{X} \\ \text{under the constraint} \quad Q(u) = \displaystyle\int_{\mathbf{R}^N} G(u(x), |\nabla u(x)|) \, dx = \lambda. \end{array}$$

Let us mention the following result, which is a particular case of Theorem 2 p. 314 in[11].

Theorem 5.1. *([11]) Assume that $\lambda \neq 0$, $(\mathcal{S}_\lambda)$ admits minimizers, any minimizer is C^1 in $\mathbf{R}^N$, and $\mathscr{X}$ has the reflection property with respect to any direction. Then, after a translation, any minimizer of $(\mathcal{S}_\lambda)$ is radially symmetric.*

Notice that Theorem 5.1 is valid for vector-valued minimizers without any assumption on the sign of their components. We give below a simple proof in space dimension two. In the general case the proof is similar, but a bit more involved.

Proof of Theorem 5.1 in the case $N = 2$. We start with the following elementary

Lemma 5.1. *Assume that $N = 2$ and the conditions of Theorem 5.1 hold. Let $u \in \mathscr{X}$ be a minimizer for $(\mathcal{S}_\lambda)$. Suppose in addition that for any straight line Π containing $(0,0)$ there holds*

$$\int_{\Pi+} G(u(x), |\nabla u(x)|) \, dx = \int_{\Pi-} G(u(x), |\nabla u(x)|) \, dx = \frac{\lambda}{2},$$

where Π_+ and Π_- are the two half-planes determined by Π. Then u is radially symmetric with respect to $(0,0)$.

Proof. Let Π be any straight line containing the origin. Choose a coordinate system such that $\Pi = Oy$. Let

$$v_1(x,y) = \begin{cases} u(x,y) \text{ if } x \leq 0 \\ u(-x,y) \text{ if } x > 0, \end{cases} \qquad v_2(x,y) = \begin{cases} u(-x,y) \text{ if } x < 0 \\ u(x,y) \text{ if } x \geq 0. \end{cases} \tag{16}$$

Then $v_1, v_2 \in \mathscr{X}$ and $Q(v_1) = Q(v_2) = \lambda$, hence $E(v_1) \geqslant E(u)$ and $E(v_2) \geqslant E(u)$. On the other hand we have $E(v_1) + E(v_2) = 2E(u)$, so that necessarily $E(v_1) = E(v_2) = E(u)$ and v_1, v_2 are also minimizers for $(\mathcal{S}_\lambda)$. We infer that v_1, $v_2 \in C^1(\mathbf{R}^2)$. Taking into account (1), we see that necessarily $\frac{\partial u}{\partial x}(0, y) = 0$ for any $y \in \mathbf{R}$. Hence we have proved that for any line $\Pi \subset \mathbf{R}^2$ containing the origin there holds

$$\frac{\partial u}{\partial n} = 0 \quad \text{ on } \Pi, \text{ where } n \text{ is the normal to } \Pi.$$

Using polar coordinates $x = r\cos\theta$, $y = r\sin\theta$ we get $\dfrac{\partial u}{\partial \theta} = 0$ in $\mathbf{R}^2 \setminus \{(0,0)\}$ and this implies that u does not depend on θ, hence u is radial. $\square$

Now we come back to the proof of Theorem 5.1. Let u be a minimizer for $(\mathcal{S}_\lambda)$. After translation, we may assume that

$$\int_{\{x<0\}} G(u, |\nabla u|)\, dx\, dy = \int_{\{x>0\}} G(u, |\nabla u|)\, dx\, dy = \frac{\lambda}{2}.$$

Let u_1 and u_2 be the two functions obtained from u by symmetry with respect to Oy, as in (1). As above we see that u_1, u_2 are minimizers. By construction, u_1 and u_2 are symmetric with respect to x.

After translation in the y direction, we may assume that

$$\int_{\{y<0\}} G(u_1, |\nabla u_1|)\, dx\, dy = \int_{\{y>0\}} G(u_1, |\nabla u_1|)\, dx\, dy = \frac{\lambda}{2}.$$

Let $u_{1,1}$ and $u_{1,2}$ be the two functions obtained from u_1 by mirror symmetry with respect to Ox. As previously we show that $u_{1,1}$ and $u_{1,2}$ are also minimizers for $(\mathcal{S}_\lambda)$ and satisfy

$$u_{1,i}(-x, -y) = u_{1,i}(-x, y) = u_{1,i}(x, y) \qquad \text{for any } x, y \in \mathbf{R} \text{ and } i = 1, 2.$$

Using Lemma 5.1 we infer that $u_{1,1}$ and $u_{1,2}$ are radial with respect to $(0, 0)$. Since $u_{1,1}(x, 0) = u_1(x, 0) = u_{1,2}(x, 0)$ for all $x \in \mathbf{R}$ we get $u_{1,1} = u_{1,2} = u_1$. Thus we have proved that u_1 is radial.

Similarly there exists $k \in \mathbf{R}$ such that

$$\int_{\{y<k\}} G(u_2, |\nabla u_2|)\, dx\, dy = \int_{\{y>k\}} G(u_2, |\nabla u_2|)\, dx\, dy = \frac{\lambda}{2}.$$

The same argument as above implies that u_2 is radial with respect to $(0, k)$.

If $k \neq 0$, we see that the function of one variable $u(0, \cdot) = u_1(0, \cdot) = u_2(0, \cdot)$ is symmetric both with respect to 0 and with respect to k, consequently this function is $2|k|-$periodic. Then $G(u_1, |\nabla u_1|)$ is a radial function, say $G(u_1, |\nabla u_1|)(x, y) = g(\sqrt{x^2 + y^2})$ for some function

g, and g is $2|k|-$periodic. It is then easy to see that if the integral $\int_{\mathbf{R}^2} G(u_1, |\nabla u_1|)\, dx\, dy$ converges, the value of this integral must be 0, which means that $\lambda = 0$, a contradiction. We conclude that necessarily $k = 0$. Since $u_1(0, \cdot) = u(0, \cdot) = u_2(0, \cdot)$, we infer that $u = u_1 = u_2$ and u is radial with respect to the origin. □

Acknowledgments

I gratefully acknowledge support from the French *Agence Nationale de la Recherche* under grant ANR-14-CE25-0009-01 "MAToS".

References

1. H. Berestycki, P.-L. Lions, *Nonlinear scalar field equations, I. Existence of a ground state,* Arch. Rational Mech. Anal. 82 (1983), pp. 313-345.
2. H. Brézis, E. H. Lieb, *Minimum Action Solutions for Some Vector Field Equations,* Comm. Math. Phys. 96 (1984), pp. 97-113.
3. T. Cazenave, P.-L. Lions, *Orbital stability of standing waves for some nonlinear Schrödinger equations,* Commun. Math. Phys. 85, (4) (1982), pp. 549-561.
4. Y.-Z. Chen, L.-C. Wu, *Second Order Elliptic Equations and Elliptic Systems,* Translations of Mathematical Monographs Vol. 174, AMS, Providence, RI, 1998.
5. M. Giaquinta, *Multiple Integrals in the Calculus of Variations and Nonlinear Elliptic Systems,* Princeton University Press, Princeton, NJ, 1983.
6. M. Giaquinta, *Introduction to the Regularity Theory for Nonlinear Elliptic Systems,* Birkhäuser Verlag, Basel, 1993.
7. D. Gilbarg, N. S. Trudinger, *Elliptic Partial Differential Equations of Second Order,* 2nd ed., Springer-Verlag, New York, 1983.
8. O. A. Ladyzhenskaya, N. N. Ural'tseva, *Linear and quasilinear elliptic equations,* Academic Press, New York, 1968.
9. P.-L. Lions, *The concentration-compactness principle in the Calculus of Variations, The locally compact case, part 1, and part 2,* Ann. Inst. H. Poincaré Anal. Non-linéaire, 1 (1984), pp. 109-145 and 223-283.
10. O. Lopes, M. Mariş, *Symmetry of minimizers for some nonlocal variational problems,* J. Funct. Anal. **254**, No. 2, (2008), pp. 535-592.
11. M. Mariş, *On the symmetry of minimizers,* Arch. Rational Mech. Anal. 192 (2) (2009), pp. 311-330.

12. M. Mariş, *Traveling waves for nonlinear Schrödinger equations with nonzero conditions at infinity,* Ann. of Math. 178 (2013), pp. 107-182.
13. M. Mariş, *On some minimization problems in* $\mathbf{R}^N$*: concentration-compactness revisited, Lagrange multipliers and Pohozaev constraints,* preprint.
14. M. Mariş, T. L. Nguyen, *Least energy solutions for general quasilinear elliptic systems,* preprint.
15. J. Malý, W. P. Ziemer, *Fine Regularity of Solutions of Elliptic Partial Differential Equations,* Mathematical Surveys and Monographs, AMS, 1997.

Recent results on multiple periodic solutions of forced relativistic pendulum-type continuous and discrete systems

Jean Mawhin

Université Catholique de Louvain, IRMP, chemin du cyclotron 2, Louvain-la-Neuve, Belgium
E-mail: jean.mawhin@uclouvain.be

We survey some recent results on the existence of at least $N+1$ geometrically distinct periodic solutions for periodically forced N-dimensional relativistic pendulum differential and difference systems.

Keywords: Relativistic pendulum-type systems; relativistic difference system; periodic solutions; critical point theory.

1. Introduction

In this paper we survey some recent results on the multiplicity of T-periodic solutions for N-dimensional differential systems of type

$$-(\varphi(u'))' = \nabla_u F(t,u) + h(t), \tag{1}$$

N-dimensional differential inclusions of type

$$-(\varphi(u'))' \in \partial F(t,u) + h(t), \tag{2}$$

and for N-dimensional difference systems of type

$$-\Delta(\varphi(\Delta u(n-1)) = \nabla_u F(n, u(n)) + h(n) \quad (n \in \mathbb{Z}), \tag{3}$$

where φ is defined by

$$\varphi(y) = \frac{y}{\sqrt{1-|y|^2}} \qquad (y \in B(1) \subset \mathbb{R}^N). \tag{4}$$

Here and below, $|\cdot|$ stands for the Euclidean norm on $\mathbb{R}^N$ and $B(\sigma) \subset \mathbb{R}^N$ denotes the open ball of center 0 and radius σ.

In the case of Eq. (1) and (2), h satisfies the following assumption:

(H_h) $\quad h : \mathbb{R} \to \mathbb{R}^N$ *is T-periodic,* $h \in L^1(0,T;\mathbb{R}^N)$ *and* $\int_0^T h(t)\,dt = 0$.

In Eq. (2), $\partial F(t,u)$ stands for Clarke's generalized gradient[3] of $F(t,\cdot)$ at $u \in \mathbb{R}^N$. In Eq. (3), $\Delta u(n) = u(n+1) - u(n)$ is the forward difference operator, T is a positive integer and $h : \mathbb{Z} \to \mathbb{R}^N$ satisfies the hypothesis:

(H_h'') $h(n+T) = h(n)$ *for all* $n \in \mathbb{Z}$ *and* $\sum_{k=1}^{T} h(k) = 0$.

In the case of Eq. (1), the potential $F : \mathbb{R} \times \mathbb{R}^N \to \mathbb{R}$ is assumed to satisfy the following hypothesis (which, for the sake of simplicity, we take a little less general than in the original papers) :

(H_F) $F : \mathbb{R} \times \mathbb{R}^N \to \mathbb{R}$ *is* T*-periodic in* t*,* T_j*-periodic in each* u_j *for some* $T_j > 0$ $(j = 1, \ldots, N)$*, continuous,* $F(\cdot, 0) = 0$ *and* $F(t, \cdot) \in C^1(\mathbb{R}^N, \mathbb{R})$ *for each* $t \in \mathbb{R}$.

In the case of Eq. (2), the following more general assumption on F is made.

(H_F') $F : \mathbb{R} \times \mathbb{R}^N \to \mathbb{R}$ *is* T*-periodic in* t*,* T_j*-periodic in each* u_j *for some* $T_j > 0$ $(j = 1, \ldots, N)$*, continuous,* $F(\cdot, 0) = 0$*, and there exists* $\alpha \in L^1(0, T)$ *such that for all* $t \in \mathbb{R}$ *and* $x, y \in \mathbb{R}^N$*, it holds*

$$|F(t, x) - F(t, y)| \le \alpha(t)|x - y|.$$

In the case of Eq. (3), the hypothesis upon F is the following one.

(H_F'') $F : \mathbb{Z} \times \mathbb{R}^N \to \mathbb{R}$ *is* T*-periodic in* t*,* T_j*-periodic in each* u_j *for some* $T_j > 0$ $(j = 1, \ldots, N)$*, continuous,* $F(\cdot, 0) = 0$*, and* $F(n, \cdot) \in C^1(\mathbb{R}^N, \mathbb{R})$ *for each* $n \in \mathbb{Z}$.

By *T-periodic solution* of Eq. (1), we mean a T-periodic function $u \in C^1(\mathbb{R}, \mathbb{R}^N)$ such that

$$|u'|_\infty := \max_{t \in [0,T]} |u'(t)| < 1,$$

$\varphi(u')$ is absolutely continuous on $\mathbb{R}$ and Eq. (1) is satisfied for a.e. $t \in \mathbb{R}$. Similarly for Eq. (2) by replacing the differential system by the differential inclusion. For Eq. (3), a *T-periodic solution* is a T-periodic sequence $u : \mathbb{Z} \to \mathbb{R}^N$ such that

$$\max_{n=1,\ldots,T} |\Delta u(n-1)| < 1,$$

which satisfies Eq. (3).

In 2012[7], the existence of at least $N+1$ geometrically distinct T-periodic solutions was proved for a relativistic-type Lagrangian system of the form

$$-(\phi(u'))' = \nabla_u F(t, u) + h(t),$$

where ϕ is a homeomorphism from $B(\sigma)$ onto $\mathbb{R}^N$ such that $\phi(0) = 0$, $\phi = \nabla\Phi$, with $\Phi : \overline{B}(\sigma) \to \mathbb{R}$ of class C^1 on $B(\sigma)$, continuous, strictly convex on $\overline{B}(\sigma)$ and such that $\Phi(0) = 0$, F satisfies Assumption (H_F),

$h \in L^s(0,T;\mathbb{R}^N)$ for some $s > 1$ and $\int_0^T h(t)\,dt = 0$. In order to obtain such a result, the problem was reduced to the equivalent Hamiltonian system

$$u' = \phi^{-1}(v), \quad v' = \nabla_u F(t,u) + h(t),$$

and a generalized saddle point theorem for strongly indefinite functionals due to Szulkin[13], and based upon relative category was applied. This approach was also used in 2013[8] in the corresponding problem for difference equations.

A second proof of the result was given in 2013[1], based upon an abstract multiplicity theorem for convex, lower semicontinuous perturbations of a C^1 functional, and generalizing a similar result for C^1 functionals given in[10], based upon Ljusternik-Schnirelman category.

A third proof of this result, under slightly better regularity assumptions than in[7] and[1], but in the special case of φ defined in (4), was proposed in 2015[5]. The existence of at least $N+1$ geometrically distinct T-periodic solutions for (1) was proved under the assumptions (H_F) and (H_h). Recall that two solutions u and $\widehat{u}$ of (1) are said to be *geometrically distinct* if

$$u \not\equiv \widehat{u} \ (\mathrm{mod}\ T_i e^i), \ i = 1, \ldots, N,$$

where $\{e^1, \ldots, e^N\}$ denotes the canonical basis in $\mathbb{R}^N$.

The idea of the proof in[5] was to replace φ in Eq. (1) by a homeomorphism $\psi : \mathbb{R}^N \to \mathbb{R}^N$ chosen in such a way that the T-periodic solutions of Eq. (1) and of equation

$$-(\psi(u'))' = \nabla_u F(t,u) + h(t)$$

coincide, and to apply directly the abstract multiplicity theorem for periodic functionals given in[10] to this system.

The same approach has been successfully applied to the corresponding difference problem in[9], and, combined with techniques of non-smooth analysis, has been used in[6] to treat the case where the differentiability of $F(t,x)$ with respect to x is replaced by a Lipschitz condition.

In what follows, we describe the main ideas of the papers[5], [6] and[9].

2. The mapping ψ associated to the modified problems

Given $\beta > 0$, let us fix $R = R(\beta) \in (0,1)$ in such a way that

$$\frac{R}{\sqrt{1-R^2}} \geq \beta,$$

and let us define $\psi : \mathbb{R}^N \to \mathbb{R}^N$ by

$$\psi(y) = \begin{cases} \dfrac{y}{\sqrt{1-|y|^2}} & \text{if} \quad |y| \le R; \\[2ex] \dfrac{y}{\sqrt{1-R^2}} & \text{if} \quad |y| > R. \end{cases} \tag{5}$$

Note that ψ is a homeomorphism, having as inverse

$$\psi^{-1}(y) = \begin{cases} \dfrac{y}{\sqrt{1+|y|^2}} & \text{if} \quad |y| \le \frac{R}{\sqrt{1-R^2}}; \\[2ex] y\sqrt{1-R^2} & \text{if} \quad |y| > \frac{R}{\sqrt{1-R^2}}. \end{cases}$$

The following lemma, proved in [5], shows that the operator ψ is strongly monotone; $(\cdot|\cdot)$ stands for the usual inner product in $\mathbb{R}^N$.

Lemma 2.1. *For every $z, y \in \mathbb{R}^N$, it holds*

$$(\psi(z) - \psi(y)|z - y) \ge |z - y|^2.$$

To introduce the variational formulation of the modified problems, we define $\Psi : \mathbb{R}^N \to \mathbb{R}$ by

$$\Psi(y) = \begin{cases} 1 - \sqrt{1-|y|^2} & \text{if} \quad |y| \le R; \\[2ex] 1 - \sqrt{1-R^2} + \dfrac{|y|^2 - R^2}{2\sqrt{1-R^2}} & \text{if} \quad |y| > R. \end{cases} \tag{6}$$

Observe that $\Psi' = \psi$ on $\mathbb{R}^N$. Also, a straightforward computation shows that

$$\frac{1}{2}|y|^2 \le \Psi(y) \le \frac{1}{\sqrt{1-R^2}}|y|^2 \quad \left(y \in \mathbb{R}^N\right). \tag{7}$$

3. Periodic solutions for some continuous perturbations of the relativistic operator

It follows from Assumption (H_F) that there exists a T-periodic function $\alpha : \mathbb{R} \to (0, +\infty)$ such that $\alpha \in L^1(0, T)$ and

$$|\nabla_u F(t, u)| \le \alpha(t) \tag{8}$$

for a.e. $t \in \mathbb{R}$ and all $u \in \mathbb{R}^N$. The following a priori estimate will suggest the construction of the modified system.

Lemma 3.1. *If Assumptions (H_F) and (H_h) hold, any possible T-periodic solution of Eq. (1) is such that*

$$|\varphi(u'(t))| \leq \sqrt{N}(|\alpha|_1 + |h|_1) \quad (t \in \mathbb{R}). \tag{9}$$

Proof. If u is a possible T-periodic solution of Eq. (1) and $j \in \{1, \ldots, N\}$, then, because of $u_j(0) = u_j(T)$, there exists $t_j \in (0, T)$ such that $u_j'(t_j) = 0$, and hence, using (8),

$$\begin{aligned}|\varphi_j(u'(t))| &= |\varphi_j(u'(t)) - \varphi_j(u'(t_j))| = \left| \int_{t_j}^{t} (\varphi_j(u'(s))' \, ds \right| \\ &= \left| \int_{t_j}^{t} [\partial_{u_j} F(s, u(s)) + h_j(s)] \, ds \right| \leq |\alpha|_1 + |h|_1\end{aligned}$$

for all $t \in [0, T]$, and hence for all $t \in \mathbb{R}$. Consequently,

$$|\varphi(u'(t))| \leq \sqrt{N}(|\alpha|_1 + |h|_1)$$

for all $t \in \mathbb{R}$. □

Let now $\psi : \mathbb{R}^N \to \mathbb{R}^N$ be the homeomorphism associated to

$$\beta = \sqrt{N}(|\alpha|_1 + |h|_1)$$

in definition (5), and let us consider the modified problem

$$-(\psi(u'))' = \nabla_u F(t, u) + h(t). \tag{10}$$

It follows immediately from Lemma 3.1 that any T-periodic solution of Eq. (1) is a T-periodic solution of Eq. (10). The converse was proved in[5] by showing, through a contradiction argument and a similar Rolle technique as in Lemma 3.1, that every possible solution of Eq. (10) is such that $|u'(t)| \leq R$ for the $R > 0$ used in the definition of ψ. Consequently,

Lemma 3.2. *u is a T-periodic solution of Eq. (1) if and only if u is a T-periodic solution of Eq. (10).*

If

$$H_T^1 := \{u \in H^1(\mathbb{R}, \mathbb{R}^N) : u(t + T) = u(t), \ \forall \, t \in \mathbb{R}\},$$

with the usual norm,

$$\|u\| = \left(\int_0^T |u(t)|^2 \, dt + \int_0^T |u'(t)|^2 \, dt \right)^{1/2},$$

corresponding inner product $\langle u, v \rangle$, and if

$$f(u) := \int_0^T [\Psi(u') - F(t,u) - (h(t)|u)] \, dt, \tag{11}$$

where $\Psi : \mathbb{R}^N \to \mathbb{R}$ is defined in (6), the assumptions (H_F) and (H_h), and inequalities (7) easily imply that $f \in C^1(H_T^1, \mathbb{R})$, that, for each $u, v \in H_T^1$,

$$\langle f'(u), v \rangle = \int_0^T [(\psi(u'(t))|v'(t)) - (\nabla_u F(t, u(t))|v(t)) - (h(t)|v(t))] \, dt,$$

so that the weak (and hence classical) solutions of (10) are the critical points of f.

Furthermore, f is bounded from below and satisfies the following version of *Palais-Smale condition*: every sequence (u_k) in H_T^1 such that $(f(u_k))$ is bounded, $\lim_{k\to\infty} f'(u_k) = 0$ and $(\overline{u}_k)$ is bounded, contains a convergent subsequence (see[5] for details). Here,

$$\overline{u} := T^{-1} \int_0^T u(t) \, dt.$$

Finally, we have

$$f\left(u + \sum_{j=1}^N k_j T_j e^j\right) = f(u) \tag{12}$$

for all $u \in H_T^1$ and $k = (k_1, \ldots, k_N) \in \mathbb{Z}^N$. Consequently, the functional f satisfies all the assumptions of Theorem 4.12 in[10] (an abstract version of a theorem of Rabinowitz[12]), so that f has at least $N+1$ critical orbits, which correspond to $N+1$ geometrically distinct solutions of (10). Using Lemma 3.2, we obtain, according to[5], the following multiplicity result for Eq. (1).

Theorem 3.1. *If Assumptions (H_F) and (H_h) hold, then Eq. (1) has at least $N+1$ geometrically distinct T-periodic solutions.*

Example 3.1. For any $a \in \mathbb{R}$ and any T-periodic $h \in L^1(0,T)$ such that $\int_0^T h(t) \, dt = 0$, the relativistic pendulum equation

$$-\left(\frac{u'}{\sqrt{1-u'^2}}\right)' = a \sin u + h(t),$$

has at least two geometrically distinct T-periodic solutions.

4. Periodic solutions for some discontinuous perturbations of the relativistic operator

In this section, following[6], we deal with the T-periodic solutions of systems of differential inclusions of the type (2), where φ is defined by (4), the mapping $F : \mathbb{R} \times \mathbb{R}^N \to \mathbb{R}$ satisfies Assumption (H'_F), and h satisfies Assumption (H_h).

With respect to (H_F), the differentiability condition of $F(t, \cdot)$ is replaced by the weaker Lipschitz condition, so that the differential system (1) has to be replaced by the differential inclusion (2), requiring techniques of nonsmooth critical point theory developed by K.C. Chang[2].

We introduce, like in Section 3, the modified problem

$$-(\psi(u'))' \in \partial F(t,u) + h(t), \tag{13}$$

where $\psi : \mathbb{R}^N \to \mathbb{R}^N$ is the homeomorphism associated in (5) to

$$\beta = \sqrt{N}(|\alpha|_1 + |h|_1),$$

with α given in Assumption (H'_F). Noticing that if u is a T-periodic solution of Eq. (2), and

$$v(t) := -(\varphi(u'(t)))' \in \partial F(t,u) + h(t),$$

one can show that $|v(t)| \leq \alpha(t) + |h(t)|$ for a.e. $t \in \mathbb{R}$ and then prove the equivalent of Lemma 3.2 for Eq. (2) and Eq. (13) (see[6] for details).

The functional on H^1_T corresponding to problem (13), again given by f defined in (11), is now the sum of a C^1 functional (associated to Ψ) and of a Lipschitzian one, and hence is locally Lipscitizian. For the convenience of the reader, we briefly recall the requested notions and results of non-smooth critical point theory in sense of Clarke[3].

The *generalized directional derivative* of f at $u \in H^1_T$ in the direction of $v \in H^1_T$ is defined by

$$f^0(x; v) = \limsup_{w \to u, s \downarrow 0} \frac{f(w + sv) - f(w)}{s}$$

and Clarke's *generalized gradient*[3] of f at $u \in H^1_T$ is the set

$$\partial f(u) = \left\{ u^* \in (H^1_T)^* : \ f^0(u; v) \geq \langle u^*, v \rangle, \ \ \forall\, v \in H^1_T \right\},$$

where $\langle \cdot, \cdot \rangle$ stands here for the duality pairing between $(H^1_T)^*$ and H^1_T. Since the nonempty convex set $\partial f(u)$ is weakly*- compact in $(H^1_T)^*$, we can define $\lambda : H^1_T \to \mathbb{R}$ by

$$\lambda(u) = \min_{u^* \in \partial f(u)} \|u^*\|_{(H^1_T)^*}, \quad \forall\, u \in H^1_T,$$

which is a lower semi-continuous function[2].

$u \in H_T^1$ is said to be *a critical point* of the locally Lipschitz functional f if $0 \in \partial f(u)$, i.e. if

$$f^0(u; v) \geq 0, \qquad \forall\, v \in H_T^1.$$

The proof that the critical points of f are T-periodic solutions of Eq. (13) is more involved than in the smooth case and is detailed in[6].

Like in the smooth case, one proves that f is bounded from below, satisfies the periodicity condition (12), and that the following *Palais-Smale condition* holds: any sequence (u_n) in H_T^1 such that $(f(u_n))$ is bounded, $\lim_{n\to\infty} \lambda(u_n) = 0$ and $(\overline{u}_n)$ is bounded contains a convergent subsequence.

Hence, f satisfies all conditions of Mironescu-Rădulescu's generalization[11] of Theorem 4.12 of[10] to locally Lipschitizian functionals, insuring the existence of at least $N+1$ critical orbits of f, and hence of at least $N+1$ geometrically distinct T-periodic solutions for Eq. (13). From the equivalence between the T-periodic solutions of Eq. (13) and Eq. (2), one obtains the multiplicity result of[6]:

Theorem 4.1. *If Assumptions (H_F') and (H_h) hold, then problem (2) has at least $N+1$ geometrically distinct T-periodic solutions.*

Example 4.1. For each T-periodic $h \in L^1(0,T)$ such that $\int_0^T h(t)\, dt = 0$, the differential inclusion

$$-\left(\frac{u'}{\sqrt{1-u'^2}}\right)' \in [\underline{f}(u), \overline{f}(u)] + h(t),$$

where

$$\begin{aligned} f(u) &= u - [u] - \frac{1}{2}, \\ \underline{f}(u) &= \lim_{\delta \searrow 0} \operatorname{essinf}\{f(v) : |u-v| < \delta\}, \\ \overline{f}(u) &= \lim_{\delta \searrow 0} \operatorname{esssup}\{f(v) : |u-v| < \delta\}, \end{aligned}$$

has at least two geometrically distinct Filippov-type[4] T-periodic solutions.

Notice that

$$\underline{f}(u) = \begin{cases} -\frac{1}{2} & \text{if } \; u \in \mathbb{Z} \\ f(u) & \text{if } \; u \notin \mathbb{Z} \end{cases}, \qquad \overline{f}(u) = \begin{cases} \frac{1}{2} & \text{if } \; u \in \mathbb{Z} \\ f(u) & \text{if } \; u \notin \mathbb{Z} \end{cases}.$$

5. Periodic solutions for some continuous perturbations of the relativistic difference operator

In this section, given the positive integer T, H_T denotes the space of T-*periodic $\mathbb{Z}$-sequences* in $\mathbb{R}^N$, i.e. of mappings $u:\mathbb{Z}\to\mathbb{R}^N$ such that $u(n+T)=u(n)$ for all $n\in\mathbb{Z}$. H_T is isomorphic to $(\mathbb{R}^N)^T=\mathbb{R}^{NT}$ and can be endowed with the inner product and corresponding norm

$$\langle u|v\rangle := \sum_{j=1}^{T}(u(j)|v(j)), \quad |u|_2 = \left(\sum_{j=1}^{T}|u(j)|^2\right)^{1/2},$$

which makes of it a Hilbert space. We will also use the equivalent norms

$$|u|_1 = \sum_{j=1}^{T}|u(j)|, \quad |u|_\infty = \max_{1\le j\le T}|u(j)|.$$

For any $u=(u(n))_{n\in\mathbb{Z}}\in H_T$, we set

$$\overline{u} = T^{-1}\sum_{j=1}^{T}u(j)\in\mathbb{R}^N,$$

identify $\overline{u}$ with the constant T-periodic sequence with value $\overline{u}$, and set $\widetilde{u}=u-\overline{u}$.

One has, for any $u=(u(n))_{n\in\mathbb{Z}}\in H_T$ and any $k\in\mathbb{Z}$,

$$\sum_{j=1}^{T}u(j) = \sum_{j=k+1}^{k+T}u(j)$$

the identity

$$\sum_{j=1}^{T}\Delta u(j) = 0$$

and the inequality

$$|\widetilde{u}|_\infty^2 \le T|\Delta u|_2^2 \quad (n\in\mathbb{Z}).$$

Let $F:\mathbb{Z}\times\mathbb{R}^N\to\mathbb{R}$ satisfy Assumption (H_F'') and $h:\mathbb{Z}\to\mathbb{R}^N$ satisfy Assumption (H_h''). A consequence is the existence of a positive T-periodic mapping $\alpha:\mathbb{Z}\to(0,\infty)$ such that

$$|\nabla_u F(n,u)+h(n)| \le \alpha(n)+|h(n)| \quad (n\in\mathbb{Z},\ u\in\mathbb{R}^N). \tag{14}$$

Let $\varphi: B(1)\to\mathbb{R}^N$ be given by (4), and consider the existence of T-periodic solutions of Eq. (3).

Let us take $\beta = \sqrt{N}(|\alpha|_1 + |h|_1)$ and $\psi : \mathbb{R}^N \to \mathbb{R}^N$ the corresponding homeomorphism given by (5), and let us consider the modified T-periodic problem

$$-\Delta\psi[\Delta u(n-1)] = \nabla_u F[n, u(n)] + h(n) \quad (n \in \mathbb{Z}). \tag{15}$$

As there is no Rolle's theorem in the calculus of differences, the equivalence between the T-periodic solutions of Eq. (3) and Eq. (15) requires different arguments, essentially contained in the following a priori bound result.

Lemma 5.1. *If Assumptions (H_F'') and (H_h'') hold, any possible T-periodic solution $u : \mathbb{Z} \to \mathbb{R}^N$ of Eq. (3) satisfies the inequalities*

$$\frac{|\Delta u(n-1)|}{\sqrt{1 - |\Delta u(n-1)|^2}} \leq \sqrt{N}(|\alpha|_1 + |h|_1) \;\; (n \in \mathbb{Z}).$$

Proof. Let $u = (u_1, ..., u_N)$ be a solution of Eq. (3). Because u is such that $u(0) = u(T)$, given $i \in \{1, \ldots, N\}$, either there exists $m_i \in \{1, \ldots, T\}$ with $\Delta u_i(m_i) = 0$, or $\Delta u_i(j-1)$ takes both positive and negative values when $j \in \{1, \ldots, T\}$. In the first case, summing

$$-\Delta(\phi_i(\Delta u(j-1))) = \partial_{u_i} F(j, u(j)) + h_i(j) \tag{16}$$

between m_i and $n \in \{1, \ldots, T\}$ and taking the absolute value, we obtain

$$\begin{aligned}\frac{|\Delta u_i(n-1)|}{\sqrt{1 - |\Delta u(n-1)|^2}} &= \left| \sum_{j=m_i}^{n} [\partial_{u_i} F(j, u(j)) + h_i(j)] \right| \\ &\leq \sum_{j=1}^{T} |\partial_{u_i} F(j, u(j)) + h_i(j)| .\end{aligned}$$

Inequality (14) implies that

$$\frac{|\Delta u_i(n-1)|}{\sqrt{1 - |\Delta u(n-1)|^2}} \leq |\alpha|_1 + |h|_1.$$

In the second case, given $n \in \{1, \ldots, T\}$, choose $m_i \in \{1, \ldots, T\}$ such that $\Delta u_i(n-1)$ and $\Delta u_i(m_i - 1)$ have opposite signs, sum (16) between m_i and n and take the absolute value to obtain

$$\begin{aligned}&\left| \frac{\Delta u_i(n-1)}{\sqrt{1 - |\Delta u(n-1)|^2}} - \frac{\Delta u_i(m_i - 1)}{\sqrt{1 - |\Delta u(m_i - 1)|^2}} \right| \\ &= \left| \sum_{j=m_i}^{n} [\partial_{u_i} F(j, u(j)) + h_i(j)] \right| \leq \sum_{j=1}^{T} |\partial_{u_i} F(j, u(j)) + h_i(j)| .\end{aligned} \tag{17}$$

But the left-hand member of (17) is equal to

$$\frac{|\Delta u_i(n-1)|}{\sqrt{1-|\Delta u(n-1)|^2}} + \frac{|\Delta u_i(m_i-1)|}{\sqrt{1-|\Delta u(m_i-1)|^2}},$$

so that

$$\frac{|\Delta u_i(n-1)|}{\sqrt{1-|\Delta u(n-1)|^2}} \leq \sum_{j=1}^{T} |\partial_{u_i} F(j, u(j)) + h_i(j)| \leq (|\alpha|_1 + |h|_1).$$

Consequently,

$$\frac{|\Delta u(n-1)|}{\sqrt{1-|\Delta u(n-1)|^2}} \leq \sqrt{N}(|\alpha|_1 + |h|_1).$$

□

Similar arguments (see[9] for details) imply that any possible T-periodic solution of Eq. (15) satisfies the same inequality. Consequently

Lemma 5.2. *u is a T-periodic solution of Eq. (3) if and only if u is a T-periodic solution of Eq. (15).*

Define the real function $f : H_T \to \mathbb{R}$ by

$$f(u) := \sum_{j=1}^{T} \{\Psi[\Delta u(j)] - F(j, u(j)) - (h(j)|u(j))\}.$$

From Assumptions (H''_F) and (H''_h), it is not difficult to show that f is of class C^1 on H_T with, for each u, $v \in H_T$,

$$\langle f'(u)|v\rangle = \sum_{j=1}^{T} \{(\phi[\Delta u(j)]|\Delta v(j)) - (\nabla_u F(j, u(j))|v(j)) - (h(j)|v(j))\}.$$

so that any critical point of f on H_T is a solution of Eq. (15), and hence of Eq. (3).

Let us use the same notation e^i for the ith unit element in $\mathbb{R}^N$ and for the constant sequence in H_T having all elements equal to e^i. It follows from Assumptions (H''_F) and (H'_h) that, for all $u \in H_T$ and $i = 1, 2, \ldots, N$, one has $f(u + T_i e^i) = f(u)$. Proceedings in a similar way as in the continuous case, it is not difficult to show that f is bounded from below on H_T and that any sequence $(u_k)_{k\in\mathbb{N}}$ in H_T such that $(f(u_k))_{k\in\mathbb{N}}$ is bounded, $\lim_{k\to\infty} f'(u_k) = 0$ and $(\overline{u}_k)_{k\in\mathbb{N}}$ is bounded contains a convergent subsequence. Hence, again, all assumptions of Theorem 4.12 in[10] are satisfied by f on H_T and the existence of at least $N + 1$ critical orbits for f follows.

They provide $N+1$ geometrically distinct solutions of Eq. (15), and hence of Eq. (3).

Theorem 5.1. *If Assumptions (H_F'') and (H_h'') hold, then Eq. (3) has at least $N+1$ geometrically distinct T-periodic solutions.*

All details of the proofs can be found in[9].

Example 5.1. For any $a \in \mathbb{R}$ and any $h \in H_T$ such that $\sum_{n=1}^{T} h(n) = 0$, the difference equation

$$-\Delta\left(\frac{\Delta u(n-1)}{\sqrt{1-|\Delta u(n-1)|^2}}\right) = a \sin u(n) + h(n) \quad (n \in \mathbb{Z})$$

has at least two geometrically distinct T-periodic solutions.

References

1. C. Bereanu and P. Jebelean, *Multiple critical points for a class of periodic lower semicontinuous functionals*, Discrete Contin. Dyn. Syst. Ser. A **33**, 47–66 (2013).
2. K.-C. Chang, *Variational methods for nondifferentiable functionals and their applications to partial differential equations*, J. Math. Anal. Appl. **80**, 102–129 (1981).
3. F.H. Clarke, *Optimization and Nonsmooth Analysis* (John Wiley and Sons, New York, NY, 1983).
4. A.F. Filippov, *Right hand side discontinuous differential equations*, Transl. Am. Math. Soc. Ser. 2, **42**, 199–227 (1964).
5. P. Jebelean, J. Mawhin and C. Şerban, *Multiple periodic solutions for perturbed relativistic pendulum systems*, Proc. Amer. Math. Soc. **143**, 3029–3039 (2015).
6. P. Jebelean, J. Mawhin and C. Şerban, *Periodic solutions for discontinuous perturbations of the relativistic operator*, Bull. Sci. Math. **140**, 99–117 (2016).
7. J. Mawhin, *Multiplicity of solutions of variational systems involving ϕ-Laplacians with singular ϕ and periodic nonlinearities*, Discrete Contin. Dyn. Syst. Ser. A **32**, 4015–4026 (2012).
8. J. Mawhin, *Periodic solutions of second order Lagrangian difference systems with bounded or singular ϕ-Laplacian and periodic potential*, Discrete Continuous Dynamical Systems S **6**, 1065–1076 (2013).

9. J. Mawhin, *A simple proof of multiplicity for periodic solutions of Lagrangian difference systems with relativistic operator and periodic potential*, J. Difference Equations Appl. DOI: 10.1080/10236198.2015.1089867.
10. J. Mawhin and M. Willem, *Critical Point Theory and Hamiltonian Systems* (Springer, New York, NY, 1989).
11. P. Mironescu, V. Rădulescu, *A multiplicity theorem for locally Lipschitz periodic functionals*, J. Math. Anal. Appl. **195**, 621–637 (1995).
12. P. Rabinowitz, *On a class of functionals invariant under a Z^n action*, Trans. Amer. Math. Soc. **310**, 303–311 (1988).
13. A. Szulkin, *A relative category and applications to critical point theory for strongly indefinite functionals*, Nonlinear Anal. **15**, 725–739 (1990).

On the anisotropic Caginalp phase-field system with singular nonlinear terms

Alain Miranville

Université de Poitiers,
Laboratoire de Mathématiques et Applications,
UMR CNRS 7348 - SP2MI,
Boulevard Marie et Pierre Curie - Téléport 2,
F-86962 Chasseneuil Futuroscope Cedex, France
E-mail: Alain.Miranville@math.univ-poitiers.fr

Our aim in this paper is to study the well-posedness and the dissipativity of the anisotropic Caginalp phase-field system with singular nonlinear terms.

Keywords: Caginalp phase-field system; anisotropy; singular nonlinear term; well-posedness; dissipativity.

1. Introduction

The Caginalp phase-field system,

$$\frac{\partial u}{\partial t} - \Delta u + f(u) = \theta, \tag{1}$$

$$\frac{\partial \theta}{\partial t} - \Delta \theta = -\frac{\partial u}{\partial t}, \tag{2}$$

was proposed in Ref. 7 to model phase transition phenomena, such as melting-solidification phenomena. Here, u is the order parameter, θ is the relative temperature (defined as $\theta = \tilde{\theta} - \theta_E$, where $\tilde{\theta}$ is the absolute temperature and θ_E is the equilibrium melting temperature) and f is the derivative of a double-well potential F (a typical choice of potential is $F(s) = \frac{1}{4}(s^2 - 1)^2$, hence the usual cubic nonlinear term $f(s) = s^3 - s$). Furthermore, here and below, we set all physical parameters equal to one. This system has been much studied; we refer the reader to, e.g., Ref. 4–6,10,11,13,15,16,19–22,24,27,34.

These equations can be derived as follows. One introduces the (total Ginzburg-Landau) free energy

$$\Psi_{\mathrm{GL}} = \int_\Omega (\frac{1}{2}|\nabla u|^2 + F(u) - u\theta - \frac{1}{2}\theta^2)\, dx, \tag{3}$$

where Ω is the domain occupied by the system (we assume here that it is a bounded and regular domain of $\mathbb{R}^3$, with boundary Γ), and the enthalpy

$$H = u + \theta. \tag{4}$$

As far as the evolution equation for the order parameter is concerned, one postulates the relaxation dynamics (with relaxation parameter set equal to one)

$$\frac{\partial u}{\partial t} = -\frac{D\Psi_{\mathrm{GL}}}{Du}, \tag{5}$$

where $\frac{D}{Du}$ denotes a variational derivative with respect to u, which yields (1). Then, we have the energy equation

$$\frac{\partial H}{\partial t} = -\mathrm{div} q, \tag{6}$$

where q is the heat flux. Assuming finally the usual Fourier law for heat conduction,

$$q = -\nabla\theta, \tag{7}$$

we obtain (2).

In (3), the term $|\nabla u|^2$ models short-ranged interactions. It is however interesting to note that such a term is obtained by truncation of higher-order ones (see Ref. 9); it can also be seen as a first-order approximation of a nonlocal term accounting for long-ranged interactions (see Ref. 17,18).

G. Caginalp and E. Esenturk recently proposed in Ref. 8 higher-order phase-field models in order to account for anisotropic interfaces (see also Ref. 23,31,33 for other approaches which, however, do not provide an explicit way to compute the anisotropy). More precisely, these authors proposed the following modified (total) free energy:

$$\Psi_{\mathrm{HOGL}} = \int_\Omega \Big(\frac{1}{2}\sum_{i=1}^{k}\sum_{|\alpha|=i} a_\alpha |\mathcal{D}^\alpha u|^2 + F(u) - u\theta - \frac{1}{2}\theta^2\Big)\, dx, \tag{8}$$

where, for $\alpha = (k_1, k_2, k_3) \in (\mathbb{N} \cup \{0\})^3$,

$$|\alpha| = k_1 + k_2 + k_3$$

and, for $\alpha \neq (0, 0, 0)$,

$$\mathcal{D}^\alpha = \frac{\partial^{|\alpha|}}{\partial x_1^{k_1} \partial x_2^{k_2} \partial x_3^{k_3}}$$

(we agree that $\mathcal{D}^{(0,0,0)}v = v$). This then yields the following evolution equation for the order parameter u:

$$\frac{\partial u}{\partial t} + \sum_{i=1}^{k}(-1)^i \sum_{|\alpha|=i} a_\alpha \mathcal{D}^{2\alpha} u + f(u) = \theta. \tag{9}$$

In particular, for $k = 1$ (anisotropic Caginalp phase-field system), we have an equation of the form

$$\frac{\partial u}{\partial t} - \sum_{i=1}^{3} a_i \frac{\partial^2 u}{\partial x_i^2} + f(u) = \theta \tag{10}$$

and, for $k = 2$ (fourth-order anisotropic Caginalp phase-field system), we have an equation of the form

$$\frac{\partial u}{\partial t} + \sum_{i,j=1}^{3} a_{ij} \frac{\partial^4 u}{\partial x_i^2 \partial x_j^2} - \sum_{i=1}^{3} b_i \frac{\partial^2 u}{\partial x_i^2} + f(u) = \theta. \tag{11}$$

We studied in Ref. 25 the higher-order phase-field system (2)-(9), for regular nonlinear terms f and Dirichlet boundary conditions.

It is however important to note that, in phase transition, regular nonlinear terms actually are approximations of thermodynamically relevant logarithmic ones of the form $f(s) = -\lambda_1 s + \frac{\lambda_2}{2} \ln \frac{1+s}{1-s}$, $s \in (-1, 1)$, $0 < \lambda_2 < \lambda_1$, which follow from a mean-field model (see Ref. 9,24; in particular, the logarithmic terms correspond to the entropy of mixing).

The original Caginalp phase-field system, with the aforementioned logarithmic nonlinear terms, was studied in Ref. 24; see also Ref. 19 for a more general Caginalp phase-field system, with a nonlinear coupling between u and θ.

In this paper, we consider the second-order anisotropic phase-field system (2)-(10), with general singular nonlinear terms (containing the physically relevant logarithmic ones). In particular, we prove the existence and uniqueness of classical solutions, as well as the dissipativity of the associated dynamical system.

Here, we do not address the higher-order models. Indeed, when $k \geq 2$, we are not able to prove the existence of classical solutions and have to deal with a different notion of a solution, based on a variational inequality (see Ref. 12,30). This will be addressed in Ref. 26.

2. Setting of the problem

We consider the following initial and boundary value problem:

$$\frac{\partial u}{\partial t} - \sum_{i=1}^{3} a_i \frac{\partial^2 u}{\partial x_i^2} + f(u) = \theta, \tag{12}$$

$$\frac{\partial \theta}{\partial t} - \Delta\theta = -\frac{\partial u}{\partial t}, \tag{13}$$

$$u = \theta = 0 \text{ on } \Gamma, \tag{14}$$

$$u|_{t=0} = u_0, \ \theta|_{t=0} = \theta_0. \tag{15}$$

We assume that

$$a_i > 0, \ i \in \{1, 2, 3\}, \tag{16}$$

and we introduce the elliptic operator A defined by

$$\langle Av, w\rangle_{H^{-1}(\Omega), H_0^1(\Omega)} = \sum_{i=1}^{k} a_i((\frac{\partial v}{\partial x_i}, \frac{\partial w}{\partial x_i})), \tag{17}$$

where $H^{-1}(\Omega)$ is the topological dual of $H_0^1(\Omega)$. Furthermore, $((\cdot,\cdot))$ denotes the usual L^2-scalar product, with associated norm $\|\cdot\|$; more generally, we denote by $\|\cdot\|_X$ the norm on the Banach space X. We can note that

$$(v, w) \in H_0^1(\Omega)^2 \mapsto \sum_{i=1}^{k} a_i((\frac{\partial v}{\partial x_i}, \frac{\partial w}{\partial x_i}))$$

is bilinear, symmetric, continuous and coercive, so that

$$A : H_0^1(\Omega) \to H^{-1}(\Omega)$$

is indeed well defined. It then follows from elliptic regularity results for linear elliptic operators of order 2 (see Ref. 1–3) that A is a strictly positive, selfadjoint and unbounded linear operator with compact inverse, with domain

$$D(A) = H^2(\Omega) \cap H_0^1(\Omega),$$

where, for $v \in D(A)$,

$$Av = -\sum_{i=1}^{3} a_i \frac{\partial^2 v}{\partial x_i^2}.$$

We further note that $D(A^{\frac{1}{2}}) = H_0^1(\Omega)$ and, for $(v, w) \in D(A^{\frac{1}{2}})^2$,

$$((A^{\frac{1}{2}}v, A^{\frac{1}{2}}w)) = \sum_{i=1}^{3} a_i \| \frac{\partial v}{\partial x_i} \|^2.$$

We finally note that (see, e.g., Ref. 32) $\|A \cdot \|$ (resp., $\|A^{\frac{1}{2}} \cdot \|$) is equivalent to the usual H^2-norm (resp., H^1-norm) on $D(A)$ (resp., $D(A^{\frac{1}{2}})$).

Remark 2.1. Note that similar properties hold for the operator $-\Delta$, with obvious changes.

Having this, we rewrite (1) as

$$\frac{\partial u}{\partial t} + Au + f(u) = \theta. \tag{18}$$

As far as the nonlinear term f is concerned, we assume that

$$f \in \mathcal{C}^1(-1, 1), \ f(0) = 0, \tag{19}$$

$$\lim_{s \to \pm 1} f(s) = \pm\infty, \ \lim_{s \to \pm 1} f'(s) = +\infty, \tag{20}$$

$$f' \geq -c_0, \ c_0 \geq 0. \tag{21}$$

In particular, it follows from (20)-(21) that

$$-c_1 \leq F(s) \leq f(s)s + c_2, \ c_1, \ c_2 \geq 0, \ s \in (-1, 1), \tag{22}$$

where $F(s) = \int_0^s f(\xi)\,d\xi$. Here, the only difficulty is to prove that $F(s) \leq f(s)s + c$, $c \geq 0$, $s \in (-1, 1)$. To do so, it suffices to study the variations of the function $s \mapsto f(s)s - F(s) + \frac{c_0}{2}s^2$, whose derivative has, owing to (21), the sign of s.

Remark 2.2. In particular, the above properties hold for the thermodynamically relevant logarithmic nonlinear terms.

Next, we define, for $N \in \mathbb{N}$, the approximated function $f_N \in \mathcal{C}^1(\mathbb{R})$ by

$$f_N(s) = \begin{cases} f(-1 + \frac{1}{N}) + f'(-1 + \frac{1}{N})(s + 1 - \frac{1}{N}), \ s < -1 + \frac{1}{N}, \\ f(s), \ |s| \leq 1 - \frac{1}{N}, \\ f(1 - \frac{1}{N}) + f'(1 - \frac{1}{N})(s - 1 + \frac{1}{N}), \ s > 1 - \frac{1}{N}. \end{cases}$$

We thus have $f_N(0) = 0$,

$$f'_N \geq -c_0 \tag{23}$$

and, setting $F_N(s) = \int_0^s f_N(\xi)\,d\xi$,

$$-c_1 \leq F_N(s), \tag{24}$$

$$f_N(s)s \geq c_3 F_N(s) - c_4,\ c_3 > 0,\ c_4 \geq 0. \tag{25}$$

Furthermore, the constants c_3 and c_4 can be chosen independently of N (note that the other constants are independent of N). The above properties follow from the fact that similar properties hold for f and from the explicit expression of f_N (see Ref. 30 for more details).

We then consider the following approximated problem, for $N \in \mathbb{N}$:

$$\frac{\partial u^N}{\partial t} + Au^N + f_N(u^N) = \theta^N, \tag{26}$$

$$\frac{\partial \theta^N}{\partial t} - \Delta\theta^N = -\frac{\partial u^N}{\partial t}, \tag{27}$$

$$u^N = \theta^N = 0 \text{ on } \Gamma, \tag{28}$$

$$u^N|_{t=0} = u_0,\ \theta^N|_{t=0} = \theta_0. \tag{29}$$

The existence, uniqueness and regularity of the solution to (12)-(14) was proved in Ref. 25. In particular, we have the

Theorem 2.1. *We assume that* $(u_0, \theta_0) \in (H^2(\Omega) \cap H_0^1(\Omega))^2$. *Then, (12)-(14) possesses a unique solution* (u^N, θ^N) *such that,* $\forall T > 0$,

$$(u^N, \theta^N) \in L^\infty(\mathbb{R}^+; (H^2(\Omega) \cap H_0^1(\Omega))^2)$$

and

$$(\frac{\partial u^N}{\partial t}, \frac{\partial \theta^N}{\partial t}) \in L^2(0, T; L^2(\Omega)^2).$$

Throughout the paper, the same letters c, c' and c'' denote (generally positive) constants which may vary from line to line and are independent of N. Similarly, the same letter Q denotes (positive) monotone increasing (with respect to each argument) and continuous functions which may vary from line to line and are independent of N.

3. Uniform a priori estimates

Our aim in this section is to derive estimates on u^N and θ^N which are uniform with respect to N. These estimates are fully justified, owing to Theorem 2.1.

We assume in what follows that

$$\|u_0\|_{L^\infty(\Omega)} < 1. \tag{30}$$

We multiply (12) by $\frac{\partial u^N}{\partial t}$ and (27) by θ^N, sum the two resulting equalities and integrate over Ω and by parts. This gives

$$\frac{d}{dt}(\|A^{\frac{1}{2}}u^N\|^2+2\int_\Omega F_N(u^N)\,dx+\|\theta^N\|^2)+2\|\frac{\partial u^N}{\partial t}\|^2+2\|\nabla\theta^N\|^2=0. \tag{31}$$

We then multiply (12) by u^N and have, owing to (11),

$$\frac{d}{dt}\|u^N\|^2+c(\|u^N\|^2_{H^1(\Omega)}+\int_\Omega F_N(u^N)\,dx)\le c'\|\theta^N\|^2+c'',\ c>0. \tag{32}$$

Summing (17) and δ_1 times (32), where $\delta_1>0$ is small enough, we obtain a differential inequality of the form

$$\frac{dE_{1,N}}{dt}+c(E_{1,N}+\|\frac{\partial u^N}{\partial t}\|^2+\|\theta^N\|^2_{H^1(\Omega)})\le c',\ c>0, \tag{33}$$

where

$$E_{1,N}=\|A^{\frac{1}{2}}u^N\|^2+2\int_\Omega F_N(u^N)\,dx+\|\theta^N\|^2+\delta_1\|u^N\|^2$$

satisfies

$$E_{1,N}\ge c(\|u^N\|^2_{H^1(\Omega)}+\int_\Omega F_N(u^N)\,dx+\|\theta^N\|^2)-c',\ c>0. \tag{34}$$

Next, we multiply (12) by Au^N and find, employing (10),

$$\frac{d}{dt}\|A^{\frac{1}{2}}u^N\|^2+c\|u^N\|^2_{H^2(\Omega)}\le c'(\|u^N\|^2_{H^1(\Omega)}+\|\theta^N\|^2),\ c>0. \tag{35}$$

Summing (33) and δ_2 times (35), where $\delta_2>0$ is small enough, we have a differential inequality of the form

$$\frac{dE_{2,N}}{dt}+c(E_{2,N}+\|u^N\|^2_{H^2(\Omega)}+\|\frac{\partial u^N}{\partial t}\|^2+\|\theta^N\|^2_{H^1(\Omega)})\le c',\ c>0, \tag{36}$$

where

$$E_{2,N}=E_{1,N}+\delta_2\|A^{\frac{1}{2}}u^N\|^2$$

satisfies

$$E_{2,N} \geq c(\|u^N\|^2_{H^1(\Omega)} + \int_\Omega F_N(u^N)\,dx + \|\theta^N\|^2_{H^1(\Omega)}) - c', \; c > 0. \tag{37}$$

We then multiply (27) by $-\Delta\theta^N$ and $\frac{\partial\theta^N}{\partial t}$ and sum the two resulting equalities to obtain

$$\frac{d}{dt}\|\nabla\theta^N\|^2 + c(\|\theta^N\|^2_{H^2(\Omega)} + \|\frac{\partial\theta^N}{\partial t}\|^2) \leq c'\|\frac{\partial u^N}{\partial t}\|^2, \; c > 0. \tag{38}$$

Summing finally (2.4) and δ_3 times (38), where $\delta_3 > 0$ is small enough, we find a differential inequality of the form

$$\frac{dE_{3,N}}{dt} + c(E_{3,N} + \|u^N\|^2_{H^2(\Omega)} + \|\frac{\partial u^N}{\partial t}\|^2 + \|\theta^N\|^2_{H^2(\Omega)} + \|\frac{\partial\theta^N}{\partial t}\|^2) \leq c', \; c > 0, \tag{39}$$

where

$$E_{3,N} = E_{2,N} + \delta_3\|\nabla\theta^N\|^2$$

satisfies

$$E_{3,N} \geq c(\|u^N\|^2_{H^1(\Omega)} + \int_\Omega F_N(u^N)\,dx + \|\theta^N\|^2_{H^1(\Omega)}) - c', \; c > 0. \tag{40}$$

We now differentiate (12) with respect to time and have

$$\frac{\partial}{\partial t}\frac{\partial u^N}{\partial t} + A\frac{\partial u^N}{\partial t} + f'_N(u^N)\frac{\partial u^N}{\partial t} = \frac{\partial\theta^N}{\partial t}, \tag{41}$$

$$\frac{\partial u^N}{\partial t} = 0 \text{ on } \Gamma, \tag{42}$$

$$\frac{\partial u^N}{\partial t}|_{t=0} = -Au_0 - f_N(u_0) + \theta_0. \tag{43}$$

In particular, if $u_0 \in H^2(\Omega) \cap H^1_0(\Omega) \, (= D(A))$ and $\theta_0 \in L^2(\Omega)$, then $\frac{\partial u^N}{\partial t}(0) \in L^2(\Omega)$ and

$$\|\frac{\partial u^N}{\partial t}(0)\| \leq Q(\|u_0\|_{H^2(\Omega)}, \|u_0\|_{L^\infty(\Omega)}, \|\theta_0\|). \tag{44}$$

Indeed, it follows from the continuity of f, the continuous embedding $H^2(\Omega) \subset \mathcal{C}(\overline{\Omega})$ and (15) that, for N large enough, $f_N(u_0) = f(u_0)$ and

$$\|f(u_0)\| \leq Q(\|u_0\|_{H^2(\Omega)}, \|u_0\|_{L^\infty(\Omega)}). \tag{45}$$

Multiplying (41) by $\frac{\partial u^N}{\partial t}$, we obtain, owing to (10),

$$\frac{d}{dt}\|\frac{\partial u^N}{\partial t}\|^2 + c\|\frac{\partial u^N}{\partial t}\|^2_{H^1(\Omega)} \leq c'(\|\frac{\partial u^N}{\partial t}\|^2 + \|\frac{\partial\theta^N}{\partial t}\|^2), \; c > 0. \tag{46}$$

We then multiply (27) by $-\Delta\frac{\partial\theta^N}{\partial t}$ and find

$$\frac{d}{dt}\|\Delta\theta^N\|^2+\|\nabla\frac{\partial\theta^N}{\partial t}\|^2\le\|\nabla\frac{\partial u^N}{\partial t}\|^2. \tag{47}$$

Summing (46) and δ_4 times (47), where $\delta_4>0$ is small enough, we have the differential inequality

$$\frac{d}{dt}(\|\frac{\partial u^N}{\partial t}\|^2+\delta_4\|\Delta\theta^N\|^2)\le c(\|\frac{\partial u^N}{\partial t}\|^2+\|\frac{\partial\theta^N}{\partial t}\|^2). \tag{48}$$

Summing finally (39) and δ_5 times (48), where $\delta_5>0$ is small enough, we obtain an inequality of the form

$$\frac{dE_{4,N}}{dt}+c(E_{4,N}+\|u^N\|^2_{H^2(\Omega)}+\|\frac{\partial\theta^N}{\partial t}\|^2)\le c',\ c>0, \tag{49}$$

where

$$E_{4,N}=E_{3,N}+\delta_5(\|\frac{\partial u^N}{\partial t}\|^2+\delta_4\|\Delta\theta^N\|^2)$$

satisfies

$$E_{4,N}\ge c(\|u^N\|^2_{H^1(\Omega)}+\int_\Omega F_N(u^N)\,dx+\|\frac{\partial u^N}{\partial t}\|^2+\|\theta^N\|^2_{H^2(\Omega)})-c',\ c>0. \tag{50}$$

In particular, it follows from (44), (49)-(50) and Gronwall's lemma that

$$\|u^N(t)\|_{H^1(\Omega)}+\|\frac{\partial u^N}{\partial t}(t)\|+\|\theta^N(t)\|_{H^2(\Omega)} \tag{51}$$

$$\le e^{-ct}Q(\|u_0\|_{H^2(\Omega)},\|u_0\|_{L^\infty(\Omega)},\|\theta_0\|_{H^2(\Omega)})+c',\ c>0,\ t\ge 0.$$

Rewriting then (12) as an elliptic equation, for $t>0$ fixed,

$$Au^N+f_N(u^N)=-\frac{\partial u^N}{\partial t}+\theta^N,\ u^N=0 \text{ on } \Gamma, \tag{52}$$

we find, multiplying (52) by Au^N and employing (10),

$$\|u^N\|^2_{H^2(\Omega)}\le c(\|u^N\|^2_{H^1(\Omega)}+\|\frac{\partial u^N}{\partial t}\|^2+\|\theta^N\|^2),$$

whence, owing to (51),

$$\|u^N(t)\|_{H^2(\Omega)}\le e^{-ct}Q(\|u_0\|_{H^2(\Omega)},\|u_0\|_{L^\infty(\Omega)},\|\theta_0\|_{H^2(\Omega)})+c',\ c>0,\ t\ge 0. \tag{53}$$

Let now $\eta \in (0,1)$ be such that $\|u_0\|_{L^\infty(\Omega)} \leq \eta$ and $\|\theta^N(t)\|_{L^\infty(\Omega)} - f(\eta) \leq 0$, $\forall t \geq 0$. Such an η exists, owing to (20) and (15), noting that it follow from (51) that

$$\|\theta^N(t)\|_{L^\infty(\Omega)} \leq c\|\theta^N(t)\|_{H^2(\Omega)} \leq Q(\|u_0\|_{H^2(\Omega)}, \|u_0\|_{L^\infty(\Omega)}, \|\theta_0\|_{H^2(\Omega)}). \tag{54}$$

We then take $N \geq N_0$, where N_0, depending only on $Q(\|u_0\|_{H^2(\Omega)}, \|u_0\|_{L^\infty(\Omega)}, \|\theta_0\|_{H^2(\Omega)})$, is such that $1 - \frac{1}{N_0} \geq \eta$. In particular, for $N \geq N_0$, $f_N = f$.

Setting $U_N = u^N - \eta$, we have, for $N \geq N_0$,

$$\frac{\partial U_N}{\partial t} + AU_n + f(u^N) - f(\eta) = \theta^N - f(\eta)\,(\leq 0), \tag{55}$$

$$U_N \leq 0 \text{ on } \Gamma, \tag{56}$$

$$U_N|_{t=0} = u_0 - \eta\,(\leq 0). \tag{57}$$

Multiplying (55) by $U_N^+ = \max(U_N, 0)$, we have, noting that $\theta^N - f(\eta) \leq 0$ and $U_N^+ = 0$ on Γ and employing (21),

$$\frac{d}{dt}\|U_N^+\|^2 \leq c\|U_N^+\|^2, \tag{58}$$

which yields, since $U_N^+(0) = 0$,

$$\|U_N^+(t)\| \leq 0,\ t \geq 0, \tag{59}$$

whence

$$u^N(t) \leq \eta,\ t \geq 0. \tag{60}$$

Proceeding in a similar way for a lower bound, we deduce that there exists

$$\eta = \eta(\|u_0\|_{H^2(\Omega)}, \|u_0\|_{L^\infty(\Omega)}, \|\theta_0\|_{H^2(\Omega)}) \in (0,1)$$

such that

$$\|u^N(t)\|_{L^\infty(\Omega)} \leq \eta,\ t \geq 0, \tag{61}$$

hence a strict separation from the pure phases ± 1.

4. The dissipative semigroup

We have the

Theorem 4.1. *We assume that* $(u_0, \theta_0) \in (H^2(\Omega) \cap H_0^1(\Omega))^2$*. Then, (1)-(5) possesses a unique solution* (u, θ) *such that,* $\forall T > 0$,

$$(u, \theta) \in L^\infty(\mathbb{R}^+; (H^2(\Omega) \cap H_0^1(\Omega))^2),$$

$$(\frac{\partial u}{\partial t}, \frac{\partial \theta}{\partial t}) \in L^2(0, T; L^2(\Omega)^2)$$

and

$$\|u(t)\|_{L^\infty(\Omega)} \leq \eta, \ t \geq 0, \tag{62}$$

where

$$\eta = \eta(\|u_0\|_{H^2(\Omega)}, \|u_0\|_{L^\infty(\Omega)}, \|\theta_0\|_{H^2(\Omega)}) \in (0, 1).$$

Proof. The proof of existence follows from the a priori estimates derived in the previous section. In particular, it follows from (61) that (u^N, θ^N), $N \geq N_0$, actually is a solution to (1)-(5), since, for $N \geq N_0$, f_N coincides with f.

Let now (u_1, θ_1) and (u_2, θ_2) be two solutions to (1)-(3) with initial data $(u_{0,1}, \theta_{0,1})$ and $(u_{0,2}, \theta_{0,2})$, respectively. We set $(u, \theta) = (u_1 - u_2, \theta_1 - \theta_2)$ and $(u_0, \theta_0) = (u_{0,1} - u_{0,2}, \theta_{0,1} - \theta_{0,2})$ and have

$$\frac{\partial u}{\partial t} + Au + f(u_1) - f(u_2) = \theta, \tag{63}$$

$$\frac{\partial \theta}{\partial t} - \Delta\theta = -\frac{\partial u}{\partial t}, \tag{64}$$

$$u = \theta = 0 \text{ on } \Gamma, \tag{65}$$

$$u|_{t=0} = u_0, \ \theta|_{t=0} = \theta_0. \tag{66}$$

Multiplying (63) by u, we obtain, owing to (21),

$$\frac{d}{dt}\|u\|^2 + c\|u\|^2_{H^1(\Omega)} \leq c'(\|u\|^2 + \|\theta\|^2), \ c > 0. \tag{67}$$

Next, multiplying (64) by $u + \theta$, we find

$$\frac{d}{dt}\|u + \theta\|^2 + c\|\theta\|^2_{H^1(\Omega)} \leq c'\|u\|^2_{H^1(\Omega)}, \ c > 0. \tag{68}$$

Summing then (67) and δ_6 times (68), where $\delta_6 > 0$ is small enough, we have a differential inequality of the form

$$\frac{dE_5}{dt} \leq cE_5, \tag{69}$$

where

$$E_5 = \|u\|^2 + \delta_6\|u+\theta\|^2$$

satisfies

$$E_5 \geq c(\|u\|^2 + \|\theta\|^2),\ c > 0. \tag{70}$$

It follows from (69)-(70) and Gronwall's lemma that

$$\|u(t)\|^2 + \|\theta(t)\|^2 \leq ce^{c't}(\|u_0\|^2 + \|\theta_0\|^2),\ t \geq 0, \tag{71}$$

hence the uniqueness, as well as the continuous dependence with respect to the initial data in the $L^2 \times L^2$-norm. □

Remark 4.1. We can also prove the continuous dependence with respect to the initial data in the $H^2 \times H^2$-norm (for solutions which satisfy (62)).

We set

$$\Phi = \{(v,\Theta) \in (H^2(\Omega) \cap H_0^1(\Omega))^2,\ \|v\|_{L^\infty(\Omega)} < 1\}.$$

It follows from Theorem 4.1 that we can define the continuous semigroup $S(t) : \Phi \to \Phi$, $(u_0, \theta_0) \mapsto (u(t), \theta(t))$, $t \geq 0$ (i.e., $S(0) = I$ (identity operator) and $S(t+\tau) = S(t) \circ S(\tau)$, t, $\tau \geq 0$).

We now consider the ODE's

$$y_\pm' + f(y_\pm) = \pm\|\theta\|_{L^\infty(\mathbb{R}^+\times\Omega)},\ y_\pm(0) = \pm\|u_0\|_{L^\infty(\Omega)}. \tag{72}$$

It then follows from the properties of f that (see Ref. 28, Corollary A.1)

$$|y_\pm(t)| \leq \eta,\ t \geq 1, \tag{73}$$

where

$$\eta = \eta(\|\theta\|_{L^\infty(\mathbb{R}^+\times\Omega)}) \in (0,1)$$

is continuous. In particular, η is independent of $y_\pm(0)$. It then follows from the comparison principle for second-order parabolic equations that

$$y_-(t) \leq u(t,x) \leq y_+(t),\ \text{a.e. } (t,x) \in \mathbb{R}^+ \times \Omega. \tag{74}$$

It finally follows from (51) and the continuity of η that

$$\frac{1}{1-\eta} \leq e^{-ct}Q(\|u_0\|_{H^2(\Omega)}, \|u_0\|_{L^\infty(\Omega)}, \|\theta_0\|_{H^2(\Omega)}) + c',\ c > 0,\ t \geq 0, \tag{75}$$

whence

$$\frac{1}{1-\|u(t)\|_{L^\infty(\Omega)}} \leq e^{-ct}Q(\|u_0\|_{H^2(\Omega)}, \|u_0\|_{L^\infty(\Omega)}, \|\theta_0\|_{H^2(\Omega)})+c',\ c>0,\ t\geq 1. \tag{76}$$

We deduce from (51), (53) and (76) the

Theorem 4.2. *The semigroup $S(t)$ is dissipative in Φ, in the sense that it possesses a bounded absorbing set $\mathcal{B}_0 \subset \Phi$ (i.e., $\forall B \subset \Phi$ bounded, $\exists t_0 = t_0(B) \geq 0$ such that $t \geq t_0 \implies S(t)B \subset \mathcal{B}_0$).*

Remark 4.2. It follows from (71) that we can extend, by continuity and in a unique way, $S(t)$ to the closure of Φ in the $L^2 \times L^2$-topology, i.e., to the space

$$\Phi_1 = \{(v,\Theta) \in L^\infty(\Omega) \times L^2(\Omega),\ \|v\|_{L^\infty(\Omega)} \leq 1\},$$

meaning that the initial data can now reach the pure phases ± 1. We can also prove that the solutions instantaneously regularize, in the sense that $S(t): \Phi_1 \to \Phi$, $t > 0$.

Remark 4.3. The dissipativity is a first step in view of the study of the (temporal) asymptotic behavior of the associated dynamical system. In particular, an important issue is to prove the existence of finite-dimensional attractors: such objects describe all possible dynamics of the system; furthermore, the finite-dimensionality means, very roughly speaking, that, even though the initial phase space Φ has infinite dimension, the reduced dynamics can be described by a finite number of parameters (we refer the interested reader to, e.g., Ref. 29,32 for discussions on this subject). This will be studied elsewhere.

References

1. S. Agmon, *Lectures on elliptic boundary value problems*, Mathematical Studies, Van Nostrand, New York 1965.
2. S. Agmon, A. Douglis and L. Nirenberg, *Estimates near the boundary for solutions of elliptic partial differential equations, I*, Commun. Pure Appl. Math. **12**, 623 (1959).
3. S. Agmon, A. Douglis and L. Nirenberg, Estimates near the boundary for solutions of elliptic partial differential equations, II, *Commun. Pure Appl. Math.* **17**, 35 (1964).
4. S. Aizicovici and E. Feireisl, Long-time stabilization of solutions to a phase-field model with memory, *J. Evol. Eqns.* **1**, 69 (2001).

5. S. Aizicovici, E. Feireisl and F. Issard-Roch, Long-time convergence of solutions to a phase-field system, *Math. Methods Appl. Sci.* **24**, 277 (2001).
6. D. Brochet, X. Chen and D. Hilhorst, Finite dimensional exponential attractors for the phase-field model, *Appl. Anal.* **49**, 197 (1993).
7. G. Caginalp, An analysis of a phase field model of a free boundary, *Arch. Rational Mech. Anal.* **92**, 205 (1986).
8. G. Caginalp and E. Esenturk, *Anisotropic phase field equations of arbitrary order, Discrete Contin. Dyn. Systems S* **4**, 311 (2011).
9. J.W. Cahn and J.E. Hilliard, Free energy of a nonuniform system I. Interfacial free energy, *J. Chem. Phys.* **2**, 258 (1958).
10. L. Cherfils and A. Miranville, Some results on the asymptotic behavior of the Caginalp system with singular potentials, *Adv. Math. Sci. Appl.* **17**, 107 (2007).
11. L. Cherfils and A. Miranville, On the Caginalp system with dynamic boundary conditions and singular potentials, *Appl. Math.* **54**, 89 (2009).
12. L. Cherfils, A. Miranville and S. Peng, Higher-order Allen-Cahn models with logarithmic nonlinear terms, submitted.
13. R. Chill, E. Fašangovà and J. Prüss, Convergence to steady states of solutions of the Cahn-Hilliard equation with dynamic boundary conditions, *Math. Nachr.* **279**, 1448 (2006).
14. M. Conti, S. Gatti and A. Miranville, A generalization of the Caginalp phase-field system with Neumann boundary conditions, *Nonlinear Anal.* **87**, 11 (2013).
15. C.G. Gal and M. Grasselli, The nonisothermal Allen-Cahn equation with dynamic boundary conditions, *Discrete Contin. Dyn. Systems A* **22**, 1009 (2008).
16. S. Gatti and A. Miranville, *Asymptotic behavior of a phase-field system with dynamic boundary conditions*, in *Differential equations: inverse and direct problems* (Proceedings of the workshop "Evolution Equations: Inverse and Direct Problems", Cortona, June 21-25, 2004, A series of Lecture notes in pure and applied mathematics, Vol. 251, A. Favini and A. Lorenzi eds., Chapman & Hall, pp. 149–170, 2006).
17. G. Giacomin and J.L. Lebowitz, Phase segregation dynamics in particle systems with long range interaction I. Macroscopic limits, *J. Statist. Phys.* **87**, 37 (1997).
18. G. Giacomin and J.L. Lebowitz, Phase segregation dynamics in particle systems with long range interaction II. Interface motion, *SIAM J. Appl.*

Math. **58**, 1707 (1998).
19. M. Grasselli, A. Miranville, V. Pata and S. Zelik, Well-posedness and long time behavior of a parabolic-hyperbolic phase-field system with singular potentials, *Math. Nachr.* **280**, 1475 (2007).
20. M. Grasselli, A. Miranville and G. Schimperna, The Caginalp phase-field system with coupled dynamic boundary conditions and singular potentials, *Discrete Contin. Dyn. Systems* **28**, 67 (2010).
21. M. Grasselli, H. Petzeltová and G. Schimperna, Long time behavior of solutions to the Caginalp system with singular potential, *Z. Anal. Anwend.* **25**, 51 (2006).
22. M. Grasselli and V. Pata, Existence of a universal attractor for a fully hyperbolic phase-field system, *J. Evol. Eqns.* **4**, 27 (2004).
23. R. Kobayashi, Modelling and numerical simulations of dendritic crystal growth, *Phys. D* **63**, 410 (1993).
24. A. Miranville, Some mathematical models in phase transition, *Discrete Contin. Dyn. Systems Series S* **7**, 271 (2014).
25. A. Miranville, Higher-order anisotropic Caginalp phase-field systems, submitted.
26. A. Miranville, in preparation.
27. A. Miranville and S. Zelik, Robust exponential attractors for singularly perturbed phase-field type equations, *Electronic J. Diff. Eqns.* **2002**, 1 (2002).
28. A. Miranville and S. Zelik, Robust exponential attractors for Cahn-Hilliard type equations with singular potentials, *Math. Methods Appl. Sci.* **27**, 545 (2004).
29. A. Miranville and S. Zelik, Attractors for dissipative partial differential equations in bounded and unbounded domains, in *Handbook of Differential Equations, Evolutionary Partial Differential Equations* (Vol. 4, C.M. Dafermos and M. Pokorny eds., Elsevier, Amsterdam, pp. 103–200, 2008).
30. A. Miranville and S. Zelik, The Cahn-Hilliard equation with singular potentials and dynamic boundary conditions, *Discrete Cont. Dyn. Systems* **28**, 275 (2010).
31. J.E. Taylor, Mean curvature and weighted mean curvature, *Acta Metall. Mater.* **40**, 1475 (1992).
32. R. Temam, *Infinite-dimensional dynamical systems in mechanics and physics*, Second edition (Applied Mathematical Sciences, Vol. 68, Springer-Verlag, New York, 1997).
33. A.A. Wheeler and G.B. McFadden, On the notion of ξ-vector and stress

tensor for a general class of anisotropic diffuse interface models, *Proc. R. Soc. London Ser. A* **453**, 1611 (1997).

34. Z. Zhang, Asymptotic behavior of solutions to the phase-field equations with Neumann boundary conditions, *Comm. Pure Appl. Anal.* **4**, 683 (2005).

Space, time, similarity

Umberto Mosco
Worcester Polytechnic Institute,
100 Worcester Road, Worcester, MA 01609, USA

The article contains an overview of our recent results about initial value problems with fractal filling space attractors and about fully discrete models for nonlinear anomalous diffusions on synchronized space-time grids.

Keywords: Elliptic operators, singular homogenization, fractals, composite media.

1. Introduction

The background for this paper goes back to the constructive theory of fractals. Fractals are meant here to be mathematical objects representing physical bodies of complicated geometry and unusual dynamic properties. The constructions involve the iteration by similarity of discrete finite grids that become finer and finer at each iteration. The refinement of the spacial grids as the iteration proceeds can be viewed as an evolution process in discrete time. This brings fractal constructions in connections with a variety of scaled discrete dynamical systems on structures that also becomes finer and finer as the system evolves. In this paper the lines that connect different problems are driven by similarity and symmetry and are displayed over finite synchronized space-time grids of increasing cardinality. As we shall see, the coupling of short range interactions with fast running local times is the basic tool that allows to capture the essential features of very diverse mathematical objects.

A discrete approach of this kind is a powerful method for the study of properties that would otherwise escape a continuous analysis. The spectral properties of a fractal can only be revealed by discovering the scaling rule of the energy and cannot be derived from a differential setup. The combination of power-laws and experiments serve to this purpose in physics. Constructions on infinite grids as in this paper offer a mathematical alternative that can be applied to non-uniform structures non accessible to differential equations. Increasing progress in fast computing can be expected to make the large numerical calculations involved in these models affordable and comparable with the experimental data.

In this paper we focus on two examples.

Our first example is the construction, outlined in Section 2, of certain second-order initial value problems for ODEs with attractors of fractal kind. We introduce a family of synchronized countable space-time grids, that incorporate fractal similarity and local times. As space discretization becomes finer and finer, time discretization becomes faster and faster. On such space-time grids we define suitable kinetic vector fields. We introduce initial value Cauchy problems for the kinetic vector fields and study the attractors generated by the orbits asymptotically at infinite time. We find that the attractors have unusual topological and dimensional features of a fractal nature.

Our second example, outlined in Section 3, is the construction of a fully discrete nonlinear evolution process of self-organized-criticality (SOC) type, also based on synchronized space-time grids. The equations of the process on each finite grid are coupled in time with an impulsive change of initial data at each refinement of the grid. Grid size and time frequency are synchronized and short range space interactions are observed at fast ticking local times. This leads to a quantitative estimate of the extinction time of the process, new in discrete SOC theory.

The problems stated in Section 2 and Section 3 are the object of the papers[18], [19]. For the fractal background and its connections with the power laws of physics we refer in particular to[13], [15], [16].

2. Filling attractors

In this section we show that fractal constructions of the kind mentioned in the Introduction can be transformed into a family of ODEs with asymptotic fractal attractors. More precisely, we construct a family of second order initial value Cauchy problems in the plane with orbits that stabilize at infinite time into filling and quasi-filling attractors. The basic tool is the construction of suitable *kinetic vector fields* on infinite countable grids in the cylindrical space-time region $\mathbb{R}^2 \times [0, +\infty)$.

Kinetic vector fields are maps

$$\gamma: \ \{(P, \tau) : P \in G^{\beta,\infty}, \tau \in \mathcal{T}_P^{\infty}\} \mapsto \mathbb{R}^2$$

which assign a vector of $\mathbb{R}^2$ to each pair (P, τ), where P is a point of an infinite countable grid $G^{\beta,\infty} \subset \mathbb{R}^2$ and τ belongs to an infinite countable set $\mathcal{T}_P^\infty \subset \mathcal{T}^\infty$ of local times at P. In our construction, the field $\gamma = \gamma^\beta$ depends on a parameter $1 \leq \beta \leq 2$ through the angle

$$\theta = \theta^\beta := \arccos\left(\frac{2-\beta}{\beta}\right), \qquad 0 \leq \theta \leq \pi/2\,,$$

and is obtained as follows.

There is a space-time synchronization driven by the dictionary set

$$\mathcal{W} = \{n\kappa i_1 \cdots i_n : n \in \{0, \mathbb{N}\}, \kappa \in \{0,1,2,3\}, (i_1,\ldots,i_n) \in \{0,1,2,3\}^n\} \tag{1}$$

whose elements are the multi-indices (words) $n\kappa i/n$, with $n \in \{0, \mathbb{N}\}$, $\kappa \in \{0,1,2,3\}$, $i/n = i_1 \ldots i_n := (i_1,\ldots,i_n) \in \{0,1,2,3\}^n$, ordered lexicographically. For $n \geq 1$ we define $0^n := (0,\ldots,0) \in \mathbb{R}^n$ and for $n = 0$ we define $i/0 = 0^0 := \emptyset$ and $0\kappa i/0 = 0\kappa\emptyset$. Synchronization consists in that, that with every multi-index $n\kappa i/n \in \mathcal{W}$ we assign simultaneously the point $P = P^\beta_{n\kappa i/n} \in G^{\beta,\infty}$ and the local time $\tau = \tau_{n\kappa i/n} \in \mathcal{T}_P^\infty \in \mathcal{T}^\infty$, as we now specify.

The location of P in the plane is regulated by the iterations $\psi^\beta_{i/n} := \psi^\beta_{i_1} \circ \psi^\beta_{i_2} \circ \cdots \circ \psi^\beta_{i_n}$ of a suitable family $\{\psi^\beta_0,\ldots,\psi^\beta_3\}$ of $\beta/4$-contractive similarity maps of the plane – as in section 2 of[20] – and by the maps z_κ, $\kappa = 0,\ldots,3$ that, in succession as κ increases from 0 to 3, bring the unit right-open segment $I = \{w = \xi + i\eta : \eta = 0, 0 \leq \xi < 1\}$ onto one of the right-open sides that decompose the boundary of a square

$$\mathcal{Q} = (-L,L) \times (-L,L), \quad L > 0$$

as

$$\partial\mathcal{Q} = [Z_0,Z_1) \cup [Z_1,Z_2) \cup [Z_2,Z_3) \cup [Z_3,Z_0)$$

where $Z_0 = (-L,-L)$, $Z_1 = (L,-L)$, $Z_2 = (L,L)$, $Z_3 = (-L,L)$. The closure of $\mathcal{Q}$ in $\mathbb{R}^2$ is $\overline{\mathcal{Q}} = [-L,L]\times[-L,L]$. For $n = 0$ we define $\psi^\beta_{i/0} = Id_{\mathbb{R}^2}$, where $Id_{\mathbb{R}^2}$ denotes the identity map of $\mathbb{R}^2$ in itself. With every $n\kappa i/n \in \mathcal{W}$, we associate the point

$$P^\beta_{n\kappa i/n} := z_\kappa \circ \psi^\beta_{i/n}(0)\,,$$

$P^\beta_{0\kappa\emptyset} = Z_\kappa$ for $k = 0,1,2,3$. As $n\kappa i/n$ varies in $\mathcal{W}$, the points $P^\beta_{n\kappa i/n}$ form a countable set, $G^{\beta,\infty}$. The map $\mathcal{W} \mapsto G^{\beta,\infty}$ is not injective. With the same multi-index $n\kappa i/n \in \mathcal{W}$ we associate the time

$$\tau_{n\kappa\, i/n} := 4n + \kappa + \sum_{m=1}^{n} i_m\, 4^{-m}\,,$$

$\tau_{0\kappa\, i/0} = 4n + \kappa$. As $n\kappa i/n$ varies in $\mathcal{W}$, the times $\tau_{n\kappa i/n}$ form a countable set, $\mathcal{T}^{\infty}$. For a given point $P \in G^{\beta,\infty}$, all $\tau = \tau_{n\kappa\, i/n} \in \mathcal{T}^{\infty}$ such that $P^{\beta}_{n\kappa i/n} = P$ are the *local times* at P and the set of all local times at P is denoted by $\mathcal{T}^{\infty}_P$.

The field γ, for each fixed β, is obtained as follows. For all indices $n\,\kappa\, i/n \in \mathcal{W}$ different from $0\,0\,\emptyset$, we define the vector $\gamma(P^{\beta}_{n\,\kappa\, i/n}, \tau_{n\,\kappa\, i/n}) \in \mathbb{C}$ by setting

$$\gamma(P^{\beta}_{n\,\kappa\, i/n}, \tau_{n\,\kappa\, i/n}) := \gamma^{\beta}_{n\kappa i/n}, \quad n\kappa i/n \in \mathcal{W},\, n\kappa i/n \neq 0\,0\,\emptyset\,, \tag{2}$$

where $\gamma^{\beta}_{n\kappa i/n} \in \mathbb{C}$ are vectors suitably chosen. In our notation complex vectors act multiplicatively in $\mathbb{C}$ and are identified with maps of $\mathcal{L}(\mathbb{R}^2, \mathbb{R}^2)$, in particular $e^{i0} \in \mathbb{C}$ is identified with the identity map $Id_{\mathbb{R}^2}$. The vector $\gamma^{\beta}_{n\kappa i/n} \in \mathbb{C}$ is defined for the various indices $n\kappa i/n \in \mathcal{W}, n\kappa i/n \neq 0\,0\,\emptyset$, by:

$$\begin{aligned}
& n \geq 1,\ \kappa = 0,\ i/n = 0^n : \\
& \quad \gamma^{\beta}_{n\,\kappa\,0^n} := \ 2L\beta^n \ e^{i\,arctan(\frac{1}{\beta})} \sqrt{1 + (1/\beta)^2}\,; \\
& n \geq 0,\ \kappa = 1,2,3,\ i/n = 0^n \ : \\
& \quad \gamma^{\beta}_{n\,\kappa\,0^n} := \ 2L\beta^n \ e^{i\,(\kappa\frac{\pi}{2} + \frac{\pi}{4})} \ \sqrt{2}\,; \\
& n = 1,\ \kappa = 0,1,2,3,\ i_1 = 1,2,3 : \\
& \quad \gamma^{\beta}_{1\,\kappa\, i_1} := \ 2L\beta\ e^{i\,\kappa\,\frac{\pi}{2}} \cdot e^{i\,(\delta_{1\,i_1}\,\frac{\pi+\theta}{2} - \delta_{2\,i_1}\,\frac{\pi}{2} + \delta_{3\,i_1}\,\frac{\pi-\theta}{2})}\,\cdot \\
& \quad \cdot\, (\delta_{1 i_1}\, 2\sin(\theta/2) \,+ \delta_{2\,i_1}\ 2\sin\theta\, +\, \delta_{3\,i_1}\ 2\sin(\theta/2))\,; \\
& n \geq 2,\ \kappa = 0,1,2,3,\ i/n \neq 0^n : \\
& \quad \gamma^{\beta}_{n\,\kappa\, i/n} := 2L\beta^n\, e^{i\,\kappa\,\frac{\pi}{2}} \cdot e^{i\,\theta\,\sum_{h=1}^{\overline{n}-1}(\delta_{1\,\overline{i}_h} - \delta_{2\,\overline{i}_h})} \cdot e^{i\,(\delta_{1\,\overline{i}_{\overline{n}}}\,\frac{\pi+\theta}{2} - \delta_{2\,\overline{i}_{\overline{n}}}\,\frac{\pi}{2} + \delta_{3\,\overline{i}_{\overline{n}}}\,\frac{\pi-\theta}{2})} \\
& \quad \cdot\, (\ \delta_{1\,\overline{i}_{\overline{n}}}\, 2\sin(\tfrac{\theta}{2}) \,+\, \delta_{2\,\overline{i}_{\overline{n}}}\ 2\sin(\theta) \,+ \delta_{3\,\overline{i}_{\overline{n}}}\ 2\sin(\tfrac{\theta}{2}))
\end{aligned}$$

where $\overline{n}\,\kappa\,\overline{i}/\overline{n} \in \mathcal{W}$ is the reduced index of $n\,\kappa\, i/n \in \mathcal{W}$ at the point $P = P_{n\,\kappa\, i/n}$ of $G^{\beta,\infty}$.

For $n \geq 2$, the *reduced index* of $n\,\kappa\, i/n \in \mathcal{W}$ at the point $P = P_{n\,\kappa\, i/n} \in G^{\beta,\infty}$ is defined to be the *smallest index* $\overline{n}\,\kappa\,\overline{i}/\overline{n} = \overline{n}\,\kappa\,\overline{i}_1 \ldots \overline{i}_{\overline{n}} \in \mathcal{W}$ in the lexicographic order of $\mathcal{W}$ such that $P = P_{\overline{n}\,\kappa\,\overline{i}/\overline{n}}$. The index $\overline{n}\,\kappa\,\overline{i}/\overline{n}$ is uniquely determined in $\mathcal{W}$ and, in turn, it determines the *smallest local time* $\overline{\tau} \in \mathcal{T}^{\infty}_P$ at P, which is given by

$$\overline{\tau} = \tau_{\overline{n}\,\kappa\,\overline{i}/\overline{n}} = 4\overline{n} + \kappa + \sum_{m=1}^{\overline{n}} \overline{i}_m\, 4^{-m}.$$

The construction of the kinetic vector field $\gamma = \gamma^{\beta}$ for every $1 \leq \beta \leq 2$ is fully accomplished.

We now extend the kinetic field $\gamma = \gamma^{\beta}$ to the whole space-time cylinder $\mathbb{R}^2 \times [0, +\infty)$ by defining $g(P,t) \in \mathbb{R}^2$ at every $(P,t) \in \mathbb{R}^2 \times [0, +\infty)$

according to

$$\begin{cases} g(P,t) := \gamma(P,t) \text{ if } P \in G^{\beta,\infty} \text{ and } t \in [\tau, \tau_+) \text{ with } \tau \in \mathcal{T}_P^\infty \\ g(P,t) := (0,0) \quad \text{otherwise} \end{cases}$$

where τ_+ denotes the local time that follows τ in $\mathcal{T}_P^\infty$. With this highly irregular vector field $g : \mathbb{R}^2 \times [0, +\infty) \mapsto \mathbb{R}^2$ at hand, we consider the Cauchy initial value problem formally stated as

$$\begin{cases} \ddot{y}^\beta(t) = g(y(t), t) \quad (y(t), t) \in \mathbb{R}^2 \times [0, \infty) \\ y^\beta(0) = x \\ \dot{y}_+^\beta(0) = \overrightarrow{v}_0 \end{cases}$$

where $x := (-L, -L)$ and $\overrightarrow{v}_0 := (L, -L) - x$.

A rigorous solution of this problem is defined as follows. To the time-grid $\mathcal{T}^\infty$ we attach the space $Y_{\mathcal{T}^\infty}$ of the vector-valued functions $y : [0, +\infty) \mapsto \mathbb{R}^2$ that have the following properties:

(i) $y(t)$ is continuous on $(0, +\infty)$ and right-continuous at $t = 0$ with value denoted by $y_+(0)$, $y_+(0) = y(0)$;

(ii) $y(t)$ has a continuous derivative y in each open interval $(\tau_{n\kappa i/n}, \tau_{n\kappa i/n+})$ with $\tau_{n\kappa i/n} \in \mathcal{T}^\infty$, $\dot{y}$ is right-continuous at each $\tau_{n\kappa i/n} \in \mathcal{T}^\infty$, that is, $\dot{y}(\tau_{n\kappa i/n}) = \dot{y}_+(\tau_{n\kappa i/n})$ where

$$\dot{y}_+(\tau_{n\kappa i/n}) := \lim_{0<\epsilon\to 0} \frac{y(\tau_{n\kappa i/n} + \epsilon) - y(\tau_{n\kappa i/n})}{\epsilon}$$

and it possesses the left-limit

$$\dot{y}_-(\tau_{n\kappa i/n}) := \lim_{0<\epsilon\to 0} \frac{y(\tau_{n\kappa i/n}) - y(\tau_{n\kappa i/n} - \epsilon)}{\epsilon}$$

at each $\tau_{n\kappa i/n} \in \mathcal{T}^\infty$, $\tau_{n\kappa i/n} > 0$, the limits being taken in $\mathbb{R}^2$;

(iii) $\dot{y}$ has a bounded continuous derivative $\ddot{y}$ in each open interval $(\tau_{n\kappa i/n}, \tau_{n\kappa i/n+})$ with $\tau_{n\kappa i/n} \in \mathcal{T}^\infty$ and a measure-valued derivative $\ddot{y} := d\dot{y}^\beta/dt \in \mathcal{M}_{\mathcal{T}^\infty}$ over $(0, +\infty)$, where $\mathcal{M}_{\mathcal{T}^\infty}$ denotes the space of regular Borel $\mathbb{R}^2$-vector-valued measures on $(0, +\infty)$ supported on $\mathcal{T}^\infty$.

A *solution* of the Cauchy problem stated before is a function $y = y^\beta \in Y_{\mathcal{T}^\infty}$ that satisfies the initial conditions $y^\beta(0) = x$, $\dot{y}_+^\beta(0) = \overrightarrow{v}_0$ and the equation $\ddot{y}^\beta(t) = g(y^\beta(t), t)$ on $(0, +\infty)$ in the measure sense of $\mathcal{M}_{\mathcal{T}^\infty}$.

We summarize the main results obtained in[18].

Theorem 2.1. *For every fixed $1 \le \beta \le 2$, the Cauchy initial value problem stated before has a unique solution $y^\beta \in Y_{\mathcal{T}^\infty}$. The solution y^β is given for*

$t \in [0, +\infty)$ by

$$y^\beta(t) = \sum_{\tau_{n\kappa i/n} \in \mathcal{T}^\infty} \chi_{[\tau_{n\kappa i/n}, \tau_{n\kappa i/n+})}(t)(P_{n\kappa i/n} + \frac{t - \tau_{n\kappa i/n}}{\tau_{n\kappa i/n+} - \tau_{n\kappa i/n}}(P_{n\kappa i/n+} - P_{n\kappa i/n})$$

where $n\kappa i/n+ \succ n\kappa i/n$ denotes the index that follows $n\kappa i/n$ in the lexicographic order of $\mathcal{W}$. Moreover, $y^\beta(t) \in \overline{\mathcal{Q}}$ for all $t \in [0, +\infty)$.

The orbits described by the trajectory $y^\beta(t)$ as t runs in $[0, +\infty)$ have a surprising peculiar structure. This structure depends critically on the parameter β. In the case $1 \leq \beta < 2$, orbits and attractors as $t \to \infty$ are analyzed in the following

Theorem 2.2. *Let $1 \leq \beta < 2$. For every $n \in \{0, \mathbb{N}\}$, the orbit*

$$\Gamma^{\beta,n} := \{y \in \mathbb{R}^2 : y = y^\beta(t), 4n \leq t < 4(n+1), y^\beta(4n) = x, y^\beta(4(n+1)-) = x\}$$

described by the trajectory $y^\beta(\cdot)$ in the interval of time $4n \leq t < 4(n+1)$ is a closed Jordan curve homeomorphic to the boundary $\partial\mathcal{Q}$ and it is itself the boundary $\Gamma^{\beta,n} = \partial D^{\beta,n}$ of an open connected domain $D^{\beta,n} \subset \mathbb{R}^2$, moreover, $\Gamma^{1,n} = \partial\mathcal{Q}$ for every n. In the Hausdorff metric of compact sets, the limit attractor

$$\Gamma^\beta := \lim_{n\to\infty} \Gamma^{\beta,n}$$

is also a closed Jordan curve homeomorphic to the boundary $\partial\mathcal{Q}$ and the boundary $\Gamma^\beta = \partial D^\beta$ of an open connected domain $D^\beta \subset \mathcal{Q}$. The attractor Γ^β has Hausdorff dimension

$$1 \leq d_H(\Gamma^\beta) = \frac{\log 4}{\log(4/\beta)} < 2$$

and, in the Hausdorff metric,

$$\lim_{\beta\to 2} \Gamma^\beta = \overline{\mathcal{Q}}.$$

This result shows that for $\beta = 1$ the orbits belong the the boundary of the square $\mathcal{Q}$ and the attractor is $\partial\mathcal{Q}$. For $1 < \beta < 2$ the orbits intrude into the interior of the square and become closed Jordan interfaces that split $\mathcal{Q}$ into an interior domain and an exterior domain. As t grows to infinity the orbit stabilizes into a fractal Jordan interface, with no multiple points, which still splits $\mathcal{Q}$ into an interior domain and an exterior domain. The dimension of this interface increases with β, reaching 2 as $\beta \to 2$. The interface topologically quasi-fills the whole square, that is, the (disconnected) open region $\mathcal{Q}\setminus(D^\beta\cup\Gamma^\beta)$ of $\mathcal{Q}$ and the (connected) domain D^β have both a small area for β close to 2 that vanishes as $\beta\to 2$.

The limit case $\beta = 2$ is topologically quite different from the case $1\leq\beta<2$. For $\beta=2$, the attractor at infinite time actually fills the whole closed square, losing in the limit the property of being homeomorphic to the boundary. This of course should be expected, as we cannot set a homeomorphism of an interval onto a square. Nevertheless, the trajectory $y^\beta(\cdot)=y^2(\cdot)$ still provides a surjective, though non-injective, continuous map of the boundary of $\mathcal{Q}$ onto $\overline{\mathcal{Q}}$. Since this map is not injective, the trajectory has multiple points. In fact, it can be shown that in each time interval $[4n,4(n+1))$ the trajectory revisits any given point of the open square at 4 different local times; it revisits any point of the boundary, except the vertices, at 2 different local times; and it visits every vertex only once. Together with the results for $1\leq\beta<2$, the result for $\beta=2$ shows that the multiple singularities of the 2D-attractor $\overline{\mathcal{Q}}$ can be removed by just moving back the dimensional parameter β from the value 2 to any smaller value $\beta-\varepsilon$. Then, the multiple points disappear, the ones with multiplicity 4 split into quadruplets of nearby points, the points with multiplicity 2 split into pairs of nearby points, and the attractor becomes a simple Jordan curve with the properties stated before. This process, from filling to quasi-filling, is of great potential interest in the study of boundary value problems with dynamic interfaces.

The model described so far can be generalized to more general geometries. This can be done by replacing the square with a regular $\mathcal{K}\geq 2$ sided polygon inscribed in a circle of radius $R=L\sqrt{2}$ and by introducing suitable families of $\mathcal{N}\geq 2$ similarities, with the polygon and the similarities both depending on the dimensional parameter $1\leq\beta\leq 2$. Space-time synchronization is driven by the dictionary $\mathcal{W}=\{n\kappa k : n\in\{0,\mathbb{N}\},\kappa\in\{0,1,\ldots,\mathcal{K}-1\},k\in\{0,1,\ldots,\mathcal{N}^n-1\}\}$. Orbits and attractors can be confined in a narrow neighborhood of the unit circle with a transversal diameter that decreases as $\mathcal{K}$ increases. When $1<\beta<2$, the fractal attractors are interfaces separating an interior domain from an exterior one. In this more general setting, the internal domains can be very small, while their boundary can be very rich and rippled. Further generalizations could take into account alternating similarity families in the construction of the kinetic fields, giving more flexibility to the models.

The results described before open new interesting lines of research in various analytic and geometric applications. These include for example boundary value problems with dynamic fractal interfaces, as those recently studied in[12], [9], [20], [17]; numerical approximation of Laplace operators; geometric motion of curves; optimal design of boundaries and interfaces.

3. Self organized criticality

In this section we present a new, fully discrete model for the so called SOC (self organized criticality) process. Problems of this kind are widely studied in the physical literature and recently also in the mathematical literature. The general idea underlying these works is that certain complicated dynamic systems like sand-piles and avalanches have the striking intrinsic property that, if left unperturbed, they reach a critical equilibrium in a finite time. This peculiar feature emerges qualitatively from the early automata-like models of the physics literature, as in Bak, Tang, Wiesenfeld[1], [2], Bantay, Ianosi[3], Carlson, Swindle[8], where the evidence of scale-invariance in the sandpile models is put in relation with fractal geometries. SOC continuous models have necessarily only an indirect connection with the original particle models of physics. The mathematical theory of certain nonlinear deterministic and stochastic PDE models has been developed recently by Barbu[4], [5] Barbu, Da Prato, Röckner[6], Barbu, Röckner[7], Gess[10]. The model presented here is inspired by the constructive theory of dynamic fractals mentioned in the Introduction. It develops initial ideas shared by Viorel Barbu with the author on the occasion of his visit to WPI in 2013. The aim of the model is to provide a rigorous proof of the existence of a finite extinction time of the process in a purely discrete setting close to the automata particle models of physics.

The model is set as an impulsive evolution process that evolves over a space-time synchronized infinite grid. A local time is defined at each site and the increasingly short-range spacial interactions are observed at increasingly fast-running intervals of time. Delicate actualization factors for the initial conditions at each change of the grid from one time cycle to the next are put into effect. The domain where the process takes place is the square $\mathcal{O} = (-L, L) \times (-L, L)$ of the plane, $L > 0$, with time running in $[0, \infty)$. The continuous structures of $\mathcal{O}$, $\overline{O}$, $\partial\mathcal{O} = (-L \times [-L, L]) \cup (L \times [-L, L])$ and of $[0, \infty)$ are replaced by suitable discrete space-time grids of increasing cardinality. Both space and time discretizations are regulated by the set W of all multi-integer-indices $ms\ell k \in \{0, \mathbb{N}\}^4$, ordered lexicographically. The discrete time-grid over the real half line $[0, +\infty)$ is given by

$$
\begin{aligned}
&t^{\ell,k}_{m,s} = 4m + 4^{-m}(\ell + k\,4^{-s}) \\
&0 \le m < +\infty, \quad 0 \le s < +\infty, \\
&0 \le \ell \le 4^{m+1} - 1, \quad 0 \le k \le 4^s - 1 \\
&m,\, s,\, \ell,\, k \in \{0, \mathbb{N}\}\,.
\end{aligned}
$$

For each given s, the set of the indexes $m\ell k \in \{0,\mathbb{N}\}^3$ is ordered lexicographically as $m\ell k$ increase in their ranges. The index which in this order is successive to an index $ms\ell k$ is denoted by $m's\ell'k'$. As the multi-index $ms\ell k$ increases according to the the lexicographic order of $m\ell k$, the corresponding times $t^{\ell,k}_{m,s}$ increase over the non-negative real line $\mathbb{R}_+ = [0, +\infty)$. For given m and s, the set

$$
\mathcal{T}_{m,s} := \{t^{\ell,k}_{m,s} : 0 \le \ell \le 4^{m+1} - 1,\ 0 \le k \le 4^s - 1\}
$$

is a partition of the time interval $[4m, 4(m+1))$ into the subintervals $[t^{\ell,k}_{\overline{m},\overline{s}}, t^{\ell',k'}_{\overline{m},\overline{s}})$ of equal length $4^{-(m+s)}$. We set $t^{LAST}_{m,s} := t^{4^{m+1}-1,4^s-1}_{m,s}$. As $s \to +\infty$, $t^{LAST}_{m,s} = 4(m+1) - 4^{-(m+s)}$ converges to $4(m+1)$ from the left. We now discretize space. For each $m \in \{0,\mathbb{N}\}$, the time period $[4m, 4(m+1))$ is synchronized with the discrete grids $G_m \subset \mathcal{O}$, $\partial G_m \subset \partial\mathcal{O}$, $\overline{G}_m \subset \overline{\mathcal{O}}$ of a mesh size $h_m = 2L/2^m$ constructed as follows. For $m = 0$, $G_0 = \emptyset$ and $\partial G_0 = \overline{G}_0$ is the set of the 4 vertexes of $\mathcal{O}$. For $m \in \mathbb{N}$, we set

$$
\begin{aligned}
G_m \; := \; & \left\{ (q_1 \frac{2L}{2^m}, q_2 \frac{2L}{2^m}) : (q_1, q_2) \in \mathbb{Z}, |q_1| \le 2^{m-1} - 1, |q_2| \le 2^{m-1} - 1 \right\} \\
\partial G_m := & \left\{ (q_1 \frac{2L}{2^m}, q_2 \frac{2L}{2^m}) : (q_1, q_2) \in \mathbb{Z}, |q_1| = 2^{m-1}, |q_2| \le 2^{m-1} - 1 \right\} \\
\cup & \left\{ (q_1 \frac{2L}{2^m}, q_2 \frac{2L}{2^m}) : (q_1, q_2) \in \mathbb{Z}, |q_1| = 2^{m-1} - 1, |q_2| = 2^{m-1} \right\} \\
\cup & \left\{ (q_1 \frac{2L}{2^m}, q_2 \frac{2L}{2^m}) : (q_1, q_2) \in \mathbb{Z}, |q_1| \le 2^{m-1}, |q_2| = 2^{m-1} \right\} \\
\overline{G}_m := & \; G_m \cup \partial G_m.
\end{aligned}
$$

For every fixed $m \in \{0,\mathbb{N}\}$, we write equivalently $ij = (ih_m, jh_m) = (q_1h_m, q_2h_m) \in \overline{G}_m$ and for every function u on $\overline{G}_m$, $u_{ij} = u(ih_m, jh_m) = u(q_1h_m, q_2h_m)$ as well as $u = (u_{ij})_{ij\in\overline{G}_m}$ or $u = (u_{ij})$. For every function u on $\overline{G}_m$, the centered difference Laplacian $\Delta_m u$ is defined at every site $ij \in G_m$

$$
(\Delta_m u)_{ij} = \frac{1}{h_m^2}[4u_{ij} - u_{i-1j} - u_{i+1j} - u_{ij-1} - u_{ij+1}] \quad \forall\; ij \in G_m.
$$

$\mathbf{H}$ is the multivalued Heaviside operator given by the relation $\eta \in \mathbf{H}(z)$ for arbitrary functions (vectors) η and z given on $\overline{G}_m$, where $\eta \in \mathbf{H}(z)$ is equivalent to the following set of conditions for the components η_{ij} of η:

$$\eta_{ij} = 0 \text{ if } z_{ij} < 0\,; \quad 0 \leq \eta_{ij} \leq 1 \text{ if } z_{ij} = 0\,; \quad \eta_{ij} = 1 \text{ if } z_{ij} > 0\,, \quad \forall\, ij \in \overline{G}_m\,.$$

We introduce the countable space-grids

$$G_\infty := \cup_{m=0}^{\infty} G_m, \quad \partial G_\infty := \cup_{m=0}^{\infty} \partial G_m, \quad \overline{G}_\infty := G_\infty \cup \partial G_\infty\,.$$

We have $G_\infty \subset \mathcal{O}$, $\partial G_\infty \subset \partial\mathcal{O}$, $\overline{G}_\infty \subset \overline{\mathcal{O}}$. For a function u defined on $\overline{G}_\infty$, u_m is the restriction of u to $\overline{G}_m$.

Now we set the model. Our data are given functions $u^c = (u^c_{ij})_{ij \in \overline{G}_\infty}$ – the critical state – and $u^0 = (u^0_{ij})_{ij \in \overline{G}_\infty}$ – the initial configuration – both defined on $\overline{G}_\infty$. The functions u^c and u^0 are assumed to satisfy the conditions

$$\begin{aligned} &u^c = (u^c_{ij})_{ij \in \overline{G}_\infty}, \quad u^0 = (u^0_{ij})_{ij \in \overline{G}_\infty} \text{ satisfying} \\ &u^0_{ij} \geq u^c_{ij} \ \forall\ ij \in \overline{G}_\infty, \quad u^0_{ij} = u^c_{ij} \ \forall\ ij \in \partial G_\infty\,. \end{aligned} \tag{3}$$

The first set of equations in the model refers to solutions $u^{(s)}(t^{\ell,k}_{m,s})$ defined for $(t^{\ell,k}_{m,s}, ij) \in \mathcal{T}_{m,s} \times \overline{G}_m$. For fixed $ms \in \mathbb{R}^+ \times \{0, \mathbb{N}\}$, they are obtained iteratively in ℓk by the following finite-difference scheme with assigned initial conditions u^*_m on the grid $\overline{G}_m$:

$$\begin{aligned} u^{(s)}(t^{0,0}_{m,s}) &= u^*_m \quad \text{on } \overline{G}_m \\ u^{(s)}(t^{\ell',k'}_{m,s}) &= u^{(s)}(t^{\ell,k}_{m,s}) - 4^{-(m+s)}\Delta_m(\eta^{\ell',k'}_{m,s}) \text{ with} \\ \eta^{\ell',k'}_{m,s} &\in \mathbf{H}(u^{(s)}(t^{\ell',k'}_{m,s}) - u^c_m) \text{ on } \overline{G}_m \\ u^{(s)}(t^{\ell',k'}_{m,s}) &= u^{(s)}(t^{\ell,k}_{m,s}) = u^c_m\,, \eta^{\ell',k'}_{m,s} = 0 \quad \text{on } \partial G_m \end{aligned} \tag{4}$$

for $0 \leq \ell < 4^{m+1} - 1,\ 0 \leq k \leq 4^s - 1$.

The sequence of initial states u^*_m on the grids $\overline{G}_m$ appearing in the preceding equations is iteratively determined by a second set of equations, with an initial state $u^*_m = u^*_0$ assigned for $m = 0$ by setting

$$u^*_0 := u^0_0 \text{ on } \overline{G}_0\,. \tag{5}$$

We note that for $m = 0$, $\overline{G}_0 = \partial G_0 = \{ij : i, j = 2^{m-1}\}$ consists only of the 4 vertexes of the square $[-L, L] \times [-L.L]$. Moreover, $u^0_0 = u^c_0$ on $\overline{G}_0$.

For given indices

$$1 \leq m \leq n\,, \quad n \in \{0, \mathbb{N}\}$$

the u_m^* are determined iteratively by the equations (5), (6), (7), (8) that follow. The state u_m^* is defined on the grid G_m in three steps. For every $s \in \{0, \mathbb{N}\}$,

$$u_m^*(ij) = u_m^c(ij) + 4(\, u_{m-1}^{(s)}(t_{m-1,s}^{4^m-1,4^s-1}, ij) - u_{m-1}^c(ij)\,) \qquad (6)$$
$$\forall n \geq 1,\, \forall 1 \leq m \leq n,\, \forall ij \in G_m \cap G_{m-1} = G_{m-1}\,.$$

The value $u_{m-1}^{(s)}(t_{m-1,s}^{4^m-1,4^s-1}, ij)$ is the last value reached by the solution of the diffusion-type process on the grid G_{m-1} in the time-period $[t_{m-1}^{0,0}, t_m)$. We also set

$$u_m^*(ij) := u_m^c(ij) + \frac{1}{4} 4^{-(n-m)}(u_m^0(ij) - u_m^c(ij)) \qquad (7)$$
$$\forall n \geq 1,\, \forall 1 \leq m \leq n,\ \forall ij \in G_m - G_{m-1}$$

Finally, we define

$$u_m^*(ij) = u_m^c(ij),\ \ \forall\, ij \in \partial G_m\,. \qquad (8)$$

This completes the definition of u_m^* on the grids $\overline{G}_m$, $\overline{G}_m = G_m \cup \partial G_m$, for all m under consideration.

The set of equations (5), (6), (7), (8) is coupled with the equations (4) that define the process on each grid G_m and together the two sets of equations describe the model.

A remark about space-time synchronization in the model and about the dependence on n of the initial state u_m^* for $1 \leq m \leq n$ is here in order. For all $m \geq 0$, the time intervals $[4m, 4(m+1))$ are synchronized with the grids $\overline{G}_m$, in the sense that there is a 1-1 correspondence of each time interval $[4m, 4(m+1))$ with the space grid $\overline{G}_m$ indexed by the same m. The grids $\overline{G}_m$ are set-increasing as m increases, that is $\overline{G}_{m-1} \subset \overline{G}_m$ for all $m \geq 0$. For every fixed $n \geq 1$, the grid G_n is then layered in n disjoint subsets,

$$G_n = G_0 \cup (\bigcup_{m=1}^{n} (G_m - G_{m-1}))\,.$$

Because of synchronization, the space stratification of G_n into the layers $G_m - G_{m-1}$, $1 \leq m \leq n$, is also a time stratification, because each layer $G_m - G_{m-1}$ enters the grid G_n at a different time, due to the fact that G_m enters G_n at the initial time $4m$ of the corresponding time interval $[4m, 4(m+1))$: the smaller is m, the older is the layer $G_m - G_{m-1}$ of the sites in G_n. The numerical factors $4^{-(n-m)}$ occurring in the impulsive equations (5), (6), (7), (8) are thus actualization factors for the values of the solutions u_{ij} of the process (4) at the sites ij in each layer $G_m - G_{m-1}$.

With the actualization factors taken into account, the values taken by the solutions $u^{(s)}$ of Theorem 3.1 below are all actualized at the initial time $t = 0$ of the process. Such an actualization is necessary in order to establish the finite-time extinction property as an *a-priori* feature of the model, as in Theorem 3.2 and Theorem 3.3 that follow.

The first result in[19] is about the existence and uniqueness of the solutions of the model set before and the maximum-principle properties satisfied by them.

Theorem 3.1. *We are given an initial state u^0 and a critical state u^c on $\overline{G}_\infty$ satisfying* (3)*. We fix an arbitrary $s \in \{0, \mathbb{N}\}$. Then, for all $m \in \{0, \mathbb{N}\}$, $0 \le \ell \le 4^{m+1} - 1$, $0 \le k \le 4^s - 1$ and for all discrete times $t^{\ell,k}_{m,s} \ge 0$ there exists a unique solution $u^{(s)}$ with values $u^{(s)}(t^{\ell,k}_{m,s})$ of the impulsive system* (4), (5), (6), (7), (8)*. Moreover*

$$u^c \le u^{(s)}(t^{\ell,k}_{m,s}) \le u^c + M_m, \quad \text{on } \overline{G}_m \quad \forall\, t^{\ell,k}_{m,s} \ge 0\,, \tag{9}$$

where

$$M_m := \max_{ij \in \overline{G}_m} \{u^0_{ij} - u^c_{ij}\}\,. \tag{10}$$

The proof of this theorem is based on maximal monotone operators in finite-dimensional Hilbert space and on an adaptation to our discrete setting of the classical De Giorgi - Stampacchia truncation method for weak solutions of elliptic equations.

Our next result takes into account the long time behavior of the process and establishes a weak form of the finite time extinction property. The data of the model are the two real-valued functions u^0 and u^c, point-wise defined on the infinite (discrete) spacial grid $\overline{G}_\infty$, satisfying condition (3). We now assume in addition that u^0 and u^c have the property

$$0 < M_\infty < +\infty, \;\; M_1 < +\infty\,, \tag{11}$$

where

$$M_\infty := M_\infty(u^0 - u^c) = \sup_{ij \in G_\infty} (u^0(ij) - u^c(ij)) \le +\infty$$

$$M_1 := M_1(u^0 - u^c) = \sup_{n \in \mathbb{N}} \sum_{ij \in \overline{G}_n} (\, u^0_n(ij) - u^c_n(ij))\, h_n^2 \le +\infty.$$

For $\tau = 4m + \ell 4^{-m}$, $m \ge 0$, $\ell \in \{0, \dots, 4^{m+1} - 1\}$ we define

$$\Phi_\infty(\tau) := \liminf_{s \to +\infty} \sum_{ij \in \overline{G}_m} (\, u^{(s)}(t^{\ell,k}_{m,s}, ij) - (u^c)_{ij}\,)\, h_m^2\,, \tag{12}$$

possibly equal to $-\infty$, where $u^{(s)}(t_{m,s}^{\ell,k})$ are the solutions of the impulsive system (4), (5)(6), (7), (8). We then define $T^* \in [0, +\infty]$ by

$$T^* := \sup\{4n : n \in \{0, \mathbb{N}\},\ \text{such that}\ \Phi_\infty(\tau) > 0\ \forall\, 0 \le m \le n \quad (13)$$
$$\forall\, \tau \in [4m, 4(m+1)),\ \tau = 4m + \ell 4^{-m},\ \ell \in \{0, \dots, 4^{m+1} - 1\}\}\,.$$

Theorem 3.2. *Let u^0 and u^c be two functions defined on the infinite grid $\overline{G}_\infty$, satisfying the assumptions* (3), (11). *Then, there exists a grid $\overline{G}_{m^*}$ and a finite time $\tau^* = 4m^* + \ell^* 4^{-m^*}$, $m^* \ge 0$, $0 \le \ell^* \le 4^{m*+1} - 1$, such that the solutions $u^{(s)}$ of the system* (4), (5), (6), (7), (8) *satisfy the property*

$$\Phi_\infty(\tau^*) = 0\,,$$

implying

$$T^* < +\infty\,. \tag{14}$$

The proof of this result is rather intricate. It involves in-grid estimates for the decay of the nonlinear diffusion on each finite grid G_n. These estimates are based on suitable discrete Poincaré and Sobolev inequalities and on a delicate discrete weak chain-rule; on across-grid estimates for the jumps occurring recursively in time at each synchronized change of the grid, where the actualization factors play a crucial role; on a conclusive argument by contradiction, that combines the in-grid and across-grid estimates to show that T^* is in fact finite.

By strengthening the assumptions on the data u^0 and u^c, we obtain a more precise, quantitative estimate of the extinction time. This is achieved by estimating not only the limsup of the integrated sum in (12), as in the preceding theorem, but also the liminf of that sum, that involves the fine-grid solutions $u^{(s)}(t_{m,s}^{\ell,k}, ij)$ of the process. We additionally assume that u^0 and u^c are the restrictions to $\overline{G}_\infty$ of two continuous functions $u^0 \ge u^c$ on $\overline{\mathcal{O}}$, such that $u^0 - u^c$ is Lipschitz continuous on $\overline{\mathcal{O}}$. We define

$$B = \iint_{\overline{\mathcal{O}}} (u^0 - u^c)\,dxdy < +\infty\,,$$

we denote the Lipschitz semi-norm of $u^0 - u^c$ on $\overline{\mathcal{O}}$ by $c_{Lip} = c_{Lip}(u^0 - u^c)$, $0 \le c_{Lip} < +\infty$, and introduce the constant

$$C := c_{Lip}\, 2L\sqrt{2}\,. \tag{15}$$

We then focus on a coarser time grid. For every $m \in \{0, \mathbb{N}\}$ and $\ell = 0, \dots, 4^m - 1$ we consider the times

$$t_m^\ell = 4m + \ell 4^{-m}\,. \tag{16}$$

The set

$$\mathcal{T}_m := \{t_m^\ell : 0 \le \ell \le 4^{m+1} - 1\} \tag{17}$$

is ordered lexicographically by the index $m\ell$. The set $\mathcal{T}_m$, for each fixed s, is an ordered subset of the set $\mathcal{T}_{m,s}$ introduced before. We set

$$\mathcal{T}^\infty := \cup_{m=1}^n \mathcal{T}_m \,. \tag{18}$$

For $\tau \in \mathcal{T}^\infty$ there is a unique $m \in \{0, \mathbb{N}\}$ and a unique $\ell \in \{0, \ldots, 4^m - 1\}$ such that $\tau = t_m^\ell \in \mathcal{T}_m$. We then define

$$\Phi^\infty(\tau) := \limsup_{s \to +\infty} \sum_{ij \in \overline{G}_m} (\, u^{(s)}(t_{m,s}^{\ell,k}, ij) - (u^c)_{ij}\,)\, h_m^2. \tag{19}$$

The conclusive result in [19] is the following theorem that provides a quantitative estimate of the extinction time:

Theorem 3.3. *Under the assumptions of Theorem* 3.2 *and the additional assumptions stated before, we have*

$$\Phi^\infty(\tau) = 0$$

for all $\tau \in \mathcal{T}^\infty$, $\tau \ge \tau^*$, *where*

$$\tau^* := cM_\infty\, [\frac{|B|}{4} + (\frac{C}{4})^{\frac{1}{2}}]^2 \,, c < +\infty.$$

We add some comments. Theorem 3.2 and Theorem 3.3 capture the essential feature of SOC, that is the finite-time extinction property, in a purely discrete framework for both space and time. The proof of this property in a fully discrete setting is a distinctive feature of the model presented in this paper. It is so because the discrete equations in our model are not obtained by discretizing a PDE model of the kind referred to before, they are instead set up intrinsically, as automata-like equations on infinite countable grids describing the a-priori long-time behavior of a particle dynamics.

The a-priori, discrete character of the model suggests another interesting investigation, namely the identification and the study of the continuous differentiable equations that could emerge spontaneously by taking suitable limits of the synchronized space-time grids and of the equations of the present model. The limit continuous models can be expected to inherit the a-priori finite-time extinction property, and this would bring them in connection with the differential models of the literature mentioned before.

As time discretization is concerned, limits from discrete to continuous settings have been studied recently for nonlinear dynamics by Giga, Mitake and Tran [11]. Similar methods are applicable in the synchronized setting, that might reveal new interesting features of SOC-like nonlinear dynamics.

The fractal model of Section 2 suggests an interesting generalization of the SOC model presented before. The 2D grid of the SOC model is the same as the grid $G^{2,\infty}$ constructed in Section 2 for $\beta = 2$. SOC models similar to the one described in this section could also be constructed on the fractional grids $G^{\beta,\infty}$ of Section 2 with $1 \leq \beta < 2$. The finite extinction property of SOC would then be established in a fractional continuous dimensional setting. Conceptually, this would add universality to the SOC paradigm.

Acknowledgement

The author is grateful to the referee for the useful remarks and suggestions and for bringing reference[11] to his attention. This work is supported by the NSF grant No. 1109356.

References

1. P. Bak, C. Tang, K. Wiesenfeld, *Self-organized criticality: an explanation of the* $1/f$ *noise*, Physical Review Letters, 59 (1987), 381-394.
2. P. Bak, C. Tang, K. Wiesenfeld, Physical Review, A 38 (1988), 364-375.
3. P. Bantay, M. Ianosi, *Self-organization and anomalous diffusion*, Physica, A 185 (1992), 11-189.
4. V. Barbu, *Self-organized criticality and convergence to equilibrium of solutions to nonlinear diffusion problems*, Annual Review in Control, JARAP, vol. 340 (2010), 52-61.
5. V. Barbu, *Self-organized criticality of cellular automata model; absorbtion in finite-time of supercritical region into the critical one*, Math. Methods in Appl. Sci., vol. 13, 15 (2013), 1726-1733.
6. V. Barbu, G. Da Prato, M. Röckner, *Stochastic porous media equation and self-organized criticality*, Communications in Mathematical Physics, 285 (2009), 901-923.
7. V. Barbu, M. Röckner, *Stochastic porous media equation and self-organized criticality: convergence to the critical state in all dimensions*, Communications in Mathematical Physics, 311 (2012), 539-555.
8. J.M. Carlson, G.H. Swindle, *Self-organized criticality: sandpiles, singularities and scaling*, Proceedings of the National Academy of Sciences of the United states of America, 92 (1995), 6710-6719.
9. M. Cefalo, M.R. Lancia, H. Liang, *Heat flow problems across fractal*

mixtures: regularity results and numerical approximation, Differential and Integral Equations **26** (2013) no. 9-10, 1027-1054.
10. B. Gess, *Finite time extinction for stochastic sign and fast diffusion and self-organized criticality* (to appear).
11. Y. Giga, H. Mitake, H. V. Tran, *On asymptotic speed of solutions to level-set mean curvature flow equations with driving and source term,* arXiv:1512.03742, 2016.
12. M.R. Lancia, P. Vernole, *Irregular heat flow problems,* SIAM J. Math. Anal. **42** (4) (2010), 1539-1567.
13. U. Mosco, *Invariant field metrics and dynamical scaling on fractals,* Phys. Rev. Letters 79, n. 21, Nov. 1997, p. 4067.
14. U. Mosco, *Variational fractals,* Annali Scuola Normale Sup. Pisa, Cl. Sci. (4), XXV (1997, in Memory of Ennio De Giorgi), 683 - 712.
15. U. Mosco, *Dirichlet forms and self-similarity,* New directions in Dirichlet forms, 117155, AMS/IP Stud. Adv. Math. **8** Amer. Math. Soc., Providence, RI, 1998.
16. U. Mosco, *Energy functionals on certain fractal structures,* J. Convex Analysis, 9 (2002), n.2, 581-600.
17. U. Mosco, *Analysis and numerics of some fractal boundary value problems,* in "Analysis and Numerics of Partial Differential Equations - In Memory of Enrico Magenes", F. Brezzi, P. Colli Franzone, U. Gianazza, G. Gilardi eds., Springer INDAM Series, Vol. 4, 2013, pag. 237. ISBN: 978-88-470-2591-2.
18. U. Mosco, *Filling attractors,* to appear.
19. U. Mosco, *Finite-time self-organized-criticality on synchronized infinite grids,* to appear.
20. U. Mosco, M.A. Vivaldi, *Layered fractal fibers and potentials,* Journal des Mathematiques Pures et Appliquées, 103 (2015) 1198-1227.

Singularly perturbed problems for abstract differential equations of second order in Hilbert spaces

Andrei Perjan* and Galina Rusu**

Department of Fundamental Mathematics, Moldova State University, Chisinau, Republic of Moldova

E-mail: *aperjan1248@gmail.com, **rusugalinamoldova@gmail.com

In a real Hilbert space H we study the behavior of solutions $u_{\varepsilon\delta}$ to the following Cauchy problem

$$\begin{cases} \varepsilon u''_{\varepsilon\delta}(t) + \delta u'_{\varepsilon\delta}(t) + A\,u_{\varepsilon\delta}(t) + B(u_{\varepsilon\delta}(t)) = f_\varepsilon(t),\ t\in(0,T), \\ u_{\varepsilon\delta}(0) = u_{0\varepsilon},\ u'_{\varepsilon\delta}(0) = u_{1\varepsilon}, \end{cases} \qquad (P_{\varepsilon\delta})$$

where $A : D(A) \subset H \to H$ is a linear, self-adjoint and positive definite operator; $B : D(B) \subset H \to H$ is a nonlinear operator. The case when $\varepsilon \to 0$, $\delta = 1$, B is Lipschitz or monotone and the case when $\varepsilon \to 0, \delta \to 0,\ B = 0$ are analyzed.

Keywords: Singular perturbation; abstract second order Cauchy problem; boundary layer function; a priori estimate.

1. Introduction

Let H and V be two real Hilbert spaces endowed with norms $|\cdot|$ and $||\cdot||$, respectively, such that $V \subset H$ continuously and densely. Denote by $(\cdot,\cdot)$ the scalar product in H. Let $A : V = D(A) \subset H \to H$ be a self-adjoint positive definite operator and let $B : D(B) \subset H \to H$ be a nonlinear operator.

The main goal is to study the behavior of solutions $u_{\varepsilon\delta}$ to the following Cauchy problem

$$\begin{cases} \varepsilon u''_{\varepsilon\delta}(t) + \delta u'_{\varepsilon\delta}(t) + A\,u_{\varepsilon\delta}(t) + B\big(u_{\varepsilon\delta}(t)\big) = f_\varepsilon(t),\ t\in(0,T), \\ u_{\varepsilon\delta}(0) = u_{0\varepsilon}, \quad u'_{\varepsilon\delta}(0) = u_{1\varepsilon}. \end{cases} \qquad (P_{\varepsilon\delta})$$

We analyze two cases: (i) $\varepsilon \to 0$, $\delta = 1$, B is Lipschitz or monotone; (ii) $\varepsilon \to 0, \delta \to 0,\ B = 0$.

Namely, in (i) case we study the behavior of solutions to the perturbed problem

$$\begin{cases} \varepsilon u''_{\varepsilon}(t) + u'_{\varepsilon}(t) + Au_\varepsilon(t) + B\big(u_\varepsilon(t)\big) = f_\varepsilon(t), \\ u_\varepsilon(0) = u_{0\varepsilon}, \quad u'_\varepsilon(0) = u_{1\varepsilon}. \end{cases} \qquad (P_\varepsilon)$$

relative to the corresponding unperturbed problem:

$$\begin{cases} v'(t) + Av(t) + B\big(v(t)\big) = f(t), & t \in (0,T), \\ v(0) = u_0; \end{cases} \qquad (P_0)$$

in (ii) case we study the behavior of solutions to the perturbed problem

$$\begin{cases} \varepsilon\, u''_{\varepsilon\delta}(t) + \delta\, u'_{\varepsilon\delta}(t) + Au_{\varepsilon\delta}(t) = f(t), & t \in (0,T), \\ u_{\varepsilon\delta}(0) = u_0, \quad u'_{\varepsilon\delta}(0) = u_1, \end{cases} \qquad (P_{\varepsilon\delta})$$

when $\varepsilon \to 0$ and $\delta \to 0$, relative to solutions to the following unperturbed problem:

$$v(t) = A^{-1}f(t), \quad t \in [0,T). \qquad (P_{00})$$

The problem $(P_{\varepsilon\delta})$ is the abstract model of singularly perturbed problems of hyperbolic-parabolic type in the case (i) and of hyperbolic-parabolic-elliptic type in the case (ii). Such kind of problems arises in the mathematical modeling of elasto-plasticity phenomena. These abstract results can be applied to singularly perturbed problems of hyperbolic-parabolic-elliptic type with stationary part defined by strongly elliptic operators.

For example in[2], the equation

$$\rho v_{tt} + \gamma v_t = \sigma \Delta v$$

is considered (where ρ, γ, σ are the mass density per unit area of the membrane, the coefficient of viscosity of the medium, and the tension of the membrane, respectively), which characterizes the vibration of a membrane in a viscous medium, which can be rewritten as

$$\varepsilon^2 u_{tt} + u_t = \Delta u,$$

with $\varepsilon = (\rho\sigma)^{1/2}/\gamma$.

In the case when the medium is highly viscous ($\gamma \gg 1$), or the density ρ is very small, we have $\varepsilon \to 0$ and the formal "limit" of this equation will be the following first order equation

$$u_t = \Delta u.$$

Without pretending to make a complete analysis, let us mention some works dedicated to the study of singularly perturbed Cauchy problems for linear or nonlinear differential equations of second order of type $(P_{\varepsilon\delta})$. The case when $\delta = 1$ was widely studied by various mathematicians (see, e.g.[3,4,6,8] and the bibliography therein). In[5] the asymptotic behavior of solutions to singular perturbation problems for second order equations, as $\varepsilon \to 0$ and $\delta \to 0$, is studied. In[1,10,11], some numerical results about singular behaviour of solutions to the problem $(P_{\varepsilon\delta})$ for some ordinary differential equations and their applicability in modeling of different physical and engineering processes are presented.

In what follows we need some notations. For $k \in \mathbb{N}^\star$, $1 \le p \le +\infty$, $(a,b) \subset (-\infty,+\infty)$ and Banach space X we denote by $W^{k,p}(a,b;X)$ the Banach space of all vectorial distributions $u \in D'(a,b;X)$, $u^{(j)} \in L^p(a,b;X)$, $j = 0,1,\ldots,k$, endowed with the norm

$$\|u\|_{W^{k,p}(a,b;X)} = \begin{cases} \left(\sum_{j=0}^{k} \|u^{(j)}\|^p_{L^p(a,b;X)} \right)^{1/p}, & p \in [1,\infty), \\ \max_{0\le j\le k} \|u^{(j)}\|_{L^\infty(a,b;X)}, & p = \infty. \end{cases}$$

If $p = 2$, and X is a Hilbert space, then $W^{k,2}(a,b;X)$ is also a Hilbert space with the inner product

$$(u,v)_{H^k(a,b;X)} = \sum_{j=0}^{k} \int_a^b \left(u^{(j)}(t), v^{(j)}(t) \right)_X dt.$$

For $s \in \mathbb{R}$, $k \in \mathbb{N}$ and $p \in [1,\infty]$, we define the Banach space

$$W^{k,p}_s(a,b;X) = \{ f : (a,b) \to H; f^{(l)}(\cdot)e^{-st} \in L^p(a,b;X),\, l = 0,\ldots,k \},$$

with the norm $\|f\|_{W^{k,p}_s(a,b;X)} = \|fe^{-st}\|_{W^{k,p}(a,b;X)}$.

2. The monotone case

The framework of this section will be determined by the following conditions:

(H1) *The operator $A : D(A) \subseteq H \to H$ is linear, self-adjoint and positive definite, i. e. there exists $\omega > 0$ such that*

$$(Au,u) \ge \omega\, |u|^2, \quad \forall u \in D(A);$$

(HM1) *Operator $B : D(B) \subseteq H \to H$ is $A^{1/2}$ locally Lipschitz, i. e. $D(A^{1/2}) \subset D(B)$ and for every $R > 0$ there exists $L(R) \ge 0$ such that*

$$\begin{aligned} |B(u_1) - B(u_2)| \le L(R)|A^{1/2}(u_1 - u_2)|,\ \forall u_i \in D(A^{1/2}), \\ |A^{1/2}u_i| \le R,\ i = 1,2; \end{aligned} \tag{1}$$

(HM2) *Operator B is the Fréchet derivative of some convex and positive functional $\mathcal{B}$ with $D(A^{1/2}) \subset D(\mathcal{B})$.*

(HM3) *Operator B possesses the Fréchet derivative B' in $D(A^{1/2})$ and there exists constant $L_1(R) \geq 0$ such that*

$$\big|\big(B'(u_1) - B'(u_2)\big)v\big| \leq L_1(R)\,\big|A^{1/2}(u_1 - u_2)\big|\,|A^{1/2}v|,$$

$$\forall u_1,\, u_2,\, v \in D(A^{1/2}),\ |A^{1/2}u_i| \leq R,\ i = 1, 2.$$

The hypothesis **(HM2)**, in particular, implies that operator B is monotone and verifies condition

$$\frac{d}{dt}\mathcal{B}(u(t)) = \big(B(u(t)), u'(t)\big), \quad \forall t \in [a, b] \subset \mathbb{R},$$

in the case when $u \in C([a, b], D(A_0^{1/2})) \cap C^1([a, b], H)$.

Definition 2.1. Let $T > 0$ and $f \in L^2(0, T; H)$, $A : D(A) \subseteq H \to H$, $B : D(B) \subseteq H \to H$. The function $u \in L^2(0, T; D(A) \cap D(B))$ with $u' \in L^2(0, T; H)$ and $u'' \in L^2(0, T; H)$ is called strong solution to the Cauchy problem

$$u''(t) + u'(t) + Au(t) + B\big(u(t)\big) = f(t), \quad \forall t \in (0, T), \tag{2}$$

$$u(0) = u_0, \quad u'(0) = u_1, \tag{3}$$

if u satisfies the equality (2) in the sense of distributions a.e. $t \in (0, T)$ and the initial conditions (3).

Definition 2.2. Let $T > 0$ and $f \in L^2(0, T; H)$, $A : D(A) \subseteq H \to H$, $B : D(B) \subseteq H \to H$. The function $v \in L^2(0, T; D(A) \cap D(B))$ with $v' \in L^2(0, T; H)$ is called strong solution to the Cauchy problem

$$v'(t) + Av(t) + B\big(v(t)\big) = f(t), \quad \forall t \in (0, T), \tag{4}$$

$$v(0) = u_0. \tag{5}$$

if v verifies the equality (4) in the sense of distributions a.e. $t \in (0, T)$ and the initial condition (5).

Theorem 2.1 (Perjan&Rusu, 2012). *Let $T > 0$. Assume that the operator $A : D(A) \subset H \to H$ is linear, self-adjoint and positive definite and the operator $B : D(B) \subset H \to H$ satisfies conditions* (**HM1**) *and* (**HM2**). *If $u_0 \in D(A)$, $u_1 \in D(A^{1/2})$ and $f \in W^{1,1}(0, T; H)$, then there exists the* *unique strong solution to problem* (2), (3), *such that* $u \in C^2([0,T];H)$, $A^{1/2}u \in C^1([0,T];H)$, $Au \in C([0,T];H)$.

If, in addition, $u_1 \in D(A)$, $f(0) - B(u_0) - Au_0 - u_1 \in D(A^{1/2})$, $f \in W^{2,1}(0,T;H)$ *and condition* (**HM3**) *is fulfilled, then* $A^{1/2}u \in W^{2,\infty}(0,T;H)$ *and* $u \in W^{3,\infty}(0,T;H)$.

Theorem 2.2 (Perjan&Rusu, 2012). *Let* $T > 0$. *Assume that* $A : D(A) \subset H \to H$ *is linear, self-adjoint, positive definite operator and* B *verifies* (**HM1**) *and* (**HM2**). *If* $u_0 \in D(A)$ *and* $f \in W^{1,1}(0,T;H)$, *then there exists a unique strong solution to the problem* (4), (5), *such that* $v \in C^1([0,T];H)$, $Av \in C([0,T];H)$. *The following estimates*

$$||v||_{C([0,t];H)} + ||A^{1/2}v||_{L^2(0,t;H)} \leq C\,\mathbf{M}_0(t), \quad \forall t \in [0,T],$$

$$||A^{1/2}v||_{C([0,t];H)} + ||v'||_{C([0,t];H)}$$

$$+||A^{1/2}v'||_{L^2(0,t;H)} \leq C(\omega)\,\mathbf{M}_1(t), \quad \forall t \in [0,T],$$

$$\mathbf{M}_0(t) = |u_0| + \int_0^t \big(|f(s)| + |B(0)|\big)\,ds,$$

$$\mathbf{M}_1(t) = |Au_0| + ||f||_{W^{1,1}(0,t;H)} + |B(0)| + |f(0)|.$$

are valid.

Theorem 2.3 (Perjan&Rusu, 2012). *Let* $T > 0$ *and* $p \in (1,\infty]$. *Assume that the operator* A *is linear, self-adjoint and positive definite and the operator* B *verifies conditions* (**HM1**) *and* (**HM2**). *If*

$$\begin{cases} u_0, u_{0\varepsilon} \in D(A), \quad u_{1\varepsilon} \in D(A^{1/2}), \\ f_\varepsilon \in W^{1,p}(0,T;H), \ ||f_\varepsilon||_{W^{1,p}(0,T;H)} \leq \mathbf{M}, \\ ||Au_{0\varepsilon}||_H + ||A^{1/2}u_{1\varepsilon}||_H + ||B(u_{0\varepsilon})||_H \leq \mathbf{M}, \end{cases}$$

then there exist constants $C = C(T,p,\omega,\mathbf{M}) > 0$ *and* $\varepsilon_0 \in (0,1)$, *such that*

$$||u_\varepsilon - v||_{C([0,T];H)} + ||A^{1/2}u_\varepsilon - A^{1/2}v||_{L^2(0,T;H)}$$

$$\leq C\left(\varepsilon^\beta + ||u_{0\varepsilon} - u_0||_H + ||f_\varepsilon - f||_{L^p(0,T;H)}\right),$$

where u_ε *and* v *are strong solutions to the problems* (P_ε) *and* (P_0), *respectively,* $\beta = \min\{1/4,\ (p-1)/2p\}$.

Theorem 2.4 (Perjan&Rusu, 2012). *Let $T > 0$ and $p \in (1, \infty]$. Assume that A is linear, self-adjoint and positive definite and B verify conditions* (**HM1**), (**HM2**) *and* (**HM3**). *If*

$$\begin{cases} Au_0, Au_{0\varepsilon}, Bu_{0\varepsilon}, u_{1\varepsilon}, f(0), f_\varepsilon(0) \in D(A), \\ f, f_\varepsilon \in W^{2,p}(0,T;H), \ ||f_\varepsilon||_{W^{2,\,p}(0,\,T;\,H)} \leq \mathbf{M}, \\ \big|\big|Au_{0\varepsilon}\big|\big|_H + \big|\big|Au_{1\varepsilon}\big|\big|_H + \big|\big|B(u_{0\varepsilon})\big|\big|_H \leq \mathbf{M}, \end{cases}$$

then there exist constants $C = C(T, p, \omega, \mathbf{M}) > 0$, $\varepsilon_0 \in (0,1)$, such that

$$||u_\varepsilon' - v' + h_\varepsilon(t)||_{C([0,\,T]\,;H)} + \big|A^{1/2}\big(u_\varepsilon' - v' + h_\varepsilon(t)\big)\big|\big|_{L^2(0,T;H)}$$

$$\leq C\left(\varepsilon^\beta + \mathbf{D}_\varepsilon\right), \quad \forall \varepsilon \in (0, \varepsilon_0],$$

where u_ε and v are strong solutions to the problems (P_ε) and (P_0), respectively, $\beta = \min\{1/4, (p-1)/2p\}$,

$$\mathbf{D}_\varepsilon = ||f_\varepsilon - f||_{W^{1,\,p}(0,\,T;\,H)} + ||A(u_{0\varepsilon} - u_0)||_H + ||B(u_{0\varepsilon}) - B(u_0)||_H,$$

and the boundary function h_ε is defined by

$$h_\varepsilon(t) = \left(f_\varepsilon(0) - Au_{0\varepsilon} - B(u_{0\varepsilon}) - u_{1\varepsilon}\right)e^{-\frac{t}{\varepsilon}}.$$

From this theorem we can state that the derivative of u_ε has the boundary layer function h_ε which describes the singular behavior of derivatives of solutions u_ε as $\varepsilon \to 0$ in the boundary layer, and the boundary layer is the neighborhood of $t = 0$.

Our approach is based on two key points:

- *a priori* estimates of solutions to problem (P_ε), which are uniform with respect to the small parameter ε;
- the relationship between solutions to the problems (P_ε) and (P_0) in the linear case.

These key points are presented in what follows.

Lemma 2.1 (Perjan&Rusu, 2012). *Let $T > 0$. Assume that the operator $A : D(A) \subset H \to H$ is linear, self-adjoint, positive definite and the operator B verifies conditions* (**HM1**) *and* (**HM2**). *If $u_0 \in D(A)$, $u_1 \in D(A^{1/2})$ and $f \in W^{1,1}(0, \infty; H)$, then there exists constant $C = C(\omega) > 0$ such that for every strong solution u_ε to the problem (P_ε) the following estimates hold:*

$$||A^{1/2}u_\varepsilon\,||_{C([0,\,t];H)} + ||u_\varepsilon'||_{L^2(0,\,t;H)} + \Big(\mathcal{B}\big(u_\varepsilon(t)\big)\Big)^{1/2} \leq \mathbf{m},$$

$$\varepsilon||u''_\varepsilon||_{C([0,t];H)} + ||u'_\varepsilon||_{C([0,t];H)} + \big|\big|A^{1/2}u'_\varepsilon\big|\big|_{L^2(0,t;H)} \leq Ce^{12L^2(\mathbf{m})t}\mathbf{m_1},$$

$\forall\varepsilon \in (0, 1/2], \forall t \geq 0,$

$$||Au_\varepsilon(t)||_{C([0,t];H)} \leq C\mathbf{m_2}e^{\left(6L^2(\mathbf{m})+1\right)t},$$

where

$$\mathbf{m} = |A^{1/2}u_0| + |u_1| + |\mathcal{B}(u_0)|^{1/2} + ||f||_{L^2(0,\infty;H)},$$

$$\mathbf{m_1} = |Au_0| + |A^{1/2}u_1| + |B(u_0)| + |\mathcal{B}(u_0)|^{1/2} + ||f||_{W^{1,1}(0,\,\infty;H)},$$

$$\mathbf{m_2} = \big(L(\mathbf{m}) + 1\big)\mathbf{m_1}.$$

For $\varepsilon > 0$ denote

$$K(t,\tau,\varepsilon) = \frac{1}{2\sqrt{\pi\varepsilon}}\Big(K_1(t,\tau,\varepsilon) + 3K_2(t,\tau,\varepsilon) - 2K_3(t,\tau,\varepsilon)\Big),$$

where

$$K_1(t,\tau,\varepsilon) = \exp\Big\{\frac{3t-2\tau}{4\varepsilon}\Big\}\lambda\Big(\frac{2t-\tau}{2\sqrt{\varepsilon t}}\Big),$$

$$K_2(t,\tau,\varepsilon) = \exp\Big\{\frac{3t+6\tau}{4\varepsilon}\Big\}\lambda\Big(\frac{2t+\tau}{2\sqrt{\varepsilon t}}\Big),$$

$$K_3(t,\tau,\varepsilon) = \exp\Big\{\frac{\tau}{\varepsilon}\Big\}\lambda\Big(\frac{t+\tau}{2\sqrt{\varepsilon t}}\Big), \quad \lambda(s) = \int_s^\infty e^{-\eta^2}\,d\eta.$$

Theorem 2.5 (Perjan, 2003). *Let $B = 0$. Assume that $A : D(A) \to H$, $D(A) \subset H$, is a positive definite operator and $f \in L^\infty_\gamma(0,\infty;H)$ for some $\gamma \geq 0$. If u_ε is the strong solution to the problem (2), (3) with $u_\varepsilon \in W^{2,\infty}_\gamma(0,\infty;H) \cap L^\infty_\gamma(0,\infty;H)$, $Au_\varepsilon \in L^\infty_\gamma(0,\infty;H)$ then for every $0 < \varepsilon < (4\gamma)^{-1}$ the function w_ε, definite by*

$$w_\varepsilon(t) = \int_0^\infty K(t,\tau,\varepsilon)\,u_\varepsilon(\tau)\,d\tau,$$

is the strong solution in H to the problem

$$\begin{cases} w'_\varepsilon(t) + Aw_\varepsilon(t) = F_0(t,\varepsilon), & a.e. \quad t > 0, \quad in \quad H, \\ w_\varepsilon(0) = \varphi_\varepsilon, \end{cases}$$

$$F_0(t,\varepsilon) = \frac{1}{\sqrt{\pi}}\Big[2\exp\Big\{\frac{3t}{4\varepsilon}\Big\}\lambda\Big(\sqrt{\frac{t}{\varepsilon}}\Big) - \lambda\Big(\frac{1}{2}\sqrt{\frac{t}{\varepsilon}}\Big)\Big]v_1 + \int_0^\infty K(t,\tau,\varepsilon)\,f(\tau)\,d\tau,$$

$$\varphi_\varepsilon = \int_0^\infty e^{-\tau}\,u_\varepsilon(2\varepsilon\tau)\,d\tau.$$

3. The Lipschitz case

The framework of this section will be determined by the following conditions:

(HL1) *The operator $B : D(B) \subseteq H \to H$ is $A^{1/2}$ Lipschitz, i.e. there exists $L > 0$ such that*

$$|B(u) - B(v)| \leq L\,|A^{1/2}(u-v)|, \quad \forall u, v \in D(A^{1/2}) \cap D(B);$$

(HL2) *The operator B possesses the Fréchet derivative in H and there exist $L_0 \geq 0$, $L_1 \geq 0$ such that*

$$\big|\big(B'(u_1) - B'(u_2)\big)v\big| \leq L_1\,\big|A^{1/2}(u_1-u_2)\big|\,|A^{1/2}v|,$$

$$\forall u_1,\, u_2,\, v \in D(A^{1/2}) \cap D(B),$$

$$\big|B'(u)\,v\big| \leq L_0\,|v|, \quad \forall u, v \in D(B),$$

in the case when the operator A is linear, self-adjoint and positive definite.

Similar to the main results of the previous section the following two theorems hold.

Theorem 3.1 (Perjan&Rusu, 2011). *Let $T > 0$ and $p \in (1,\infty]$. Suppose that the operator A is linear, self-adjoint and positive definite and the operator B verifies condition* (**HL1**). *If*

$$\begin{cases} u_0, u_{0\varepsilon} \in D(A), u_{1\varepsilon} \in D(A^{1/2}), f, f_\varepsilon \in W^{1,p}(0,T;H) \\ \big|\big|Au_{0\varepsilon}\big|\big|_H + \big|\big|A^{1/2}u_{1\varepsilon}\big|\big|_H + \big|\big|B(u_{0\varepsilon})\big|\big|_H \leq \mathbf{M}, \end{cases}$$

then there exist constants $C = C(T,p,\omega,\mathbf{M}) > 0$, $\varepsilon_0 \in (0,1)$, such that

$$||u_\varepsilon - v||_{C([0,T];H)} + ||A^{1/2}\big(u_\varepsilon - v\big)||_{L^2(0,T;H)}$$

$$\leq C\,\Big(\varepsilon^\beta + |u_{0\varepsilon} - u_0| + ||f_\varepsilon - f||_{L^p(0,T;H)}\Big),\ \varepsilon \in [0,\varepsilon_0],$$

where u_ε and v are strong solutions to the problems (P_ε) and (P_0), respectively, $\beta = min\{1/4,\,(p-1)/2p\}$.

Theorem 3.2 (Perjan&Rusu, 2011). *Let $T > 0$ and $p \in (1,\infty]$. Suppose that the operator A is linear, self-adjoint and positive definite and the operator B satisfy* (**HL1**) *and* (**HL2**). *If u_0, $u_{0\varepsilon}$, Au_0, $Au_{0\varepsilon}$, $Bu_{0\varepsilon}$, $u_{1\varepsilon}$, $f(0)$, $f_\varepsilon(0) \in D(A)$ and f_ε, $f \in W^{2,p}(0,T;H)$ then*

$$||u_\varepsilon' - v' + h_\varepsilon e^{-t/\varepsilon}||_{C([0,T];H)} + ||A^{1/2}\big(u_\varepsilon' - v' + h_\varepsilon e^{-t/\varepsilon}\big)||_{L^2(0,T;H)}$$

$$\leq C\Big(\mathcal{M}\varepsilon^\beta + \mathbf{D}_\varepsilon\Big), \varepsilon \in (0,\varepsilon_0],$$

where u_ε and v are strong solutions to the problems (P_ε) and (P_0), respectively,

$$h_\varepsilon = f_\varepsilon(0) - Au_{0\varepsilon} - B(u_{0\varepsilon}) - u_{1\varepsilon}, \ \beta = min\{1/4, (p-1)/2p\},$$

$$\mathcal{M} = \mathcal{M}\Big(|Au_{0\varepsilon}|, |A_1 u_{0\varepsilon}|, |Au_{1\varepsilon}|, |Ah_\varepsilon|, |B(u_{0\varepsilon})|, ||f_\varepsilon||_{W^{2,\,p}(0,T;H)}\Big),$$

$$\mathbf{D}_\varepsilon = ||f_\varepsilon - f||_{W^{1,\,p}(0,T;H)} + |A(u_{0\varepsilon} - u_0)|.$$

4. The two parameters case

In this section we study the behavior of solutions to the perturbed problem

$$\begin{cases} \varepsilon\, u''_{\varepsilon\delta}(t) + \delta\, u'_{\varepsilon\delta}(t) + Au_{\varepsilon\delta}(t) = f(t), \quad t \in (0,T), \\ u_{\varepsilon\delta}(0) = u_0, \quad u'_{\varepsilon\delta}(0) = u_1, \end{cases} \tag{$P_{\varepsilon\delta}$}$$

when $\varepsilon \to 0$ and $\delta \to 0$, relative to solutions to the following unperturbed problem:

$$v(t) = A^{-1}f(t), \quad t \in [0,T). \tag{P_{00}}$$

Theorem 4.1 (Perjan&Rusu, 2014). *Let $T > 0$ and $p \in (1,\infty)$. Let us assume that operator A is linear, self-adjoint and positive definite. If $u_0 \in V$, $u_1 \in H$, $f \in W^{1,p}(0,T;H)$, then there exist constants $C = C(p,T,\omega) > 0$ and $\varepsilon_0 \in (0,1)$, such that*

$$||u_{\varepsilon\delta} - v - h_\delta||_{C([0,T];H)} \leq C\,\mathbf{M}\,\Theta(\varepsilon,\delta), \ \forall\varepsilon \in (0,\varepsilon_0], \ \forall\delta \in (0,1],$$

where $u_{\varepsilon\delta}$ and v are strong solutions to the problems $(P_{\varepsilon\delta})$ and (P_{00}), respectively,

$$\Theta(\varepsilon,\delta) = \frac{\varepsilon^\beta}{\delta^{1+1/p}} + \sqrt{\delta},$$
$$\beta = \min\{1/4, (p-1)/2p\}, \mathbf{M} = |A^{1/2}u_0| + |u_1| + ||f||_{W^{1,p}(0,T;H)}.$$

The function h_δ is the solution to the problem

$$\begin{cases} \delta h'_\delta(t) + Ah_\delta(t) = 0, \quad t \in (0,T), \\ h_\delta(0) = u_0 - A^{-1}f(0) \end{cases}$$

and

$$|h_\delta(t)| \leq |u_0 - A^{-1}f(0)|e^{-\delta t/\omega}, \quad t \in [0,T].$$

If, in addition, $u_1 \in D(A^{1/2})$, then

$$||u_{\varepsilon\delta} - v - h_\delta||_{C([0,T];H)}$$

$$\leq C\left(|A^{1/2}u_0| + |A^{1/2}u_1| + ||f||_{W^{1,p}(0,T;H)}\right)\Theta_1(\varepsilon,\delta),$$

$\forall\varepsilon \in (0,\varepsilon_0], \forall\delta \in (0,1]$,

$$\Theta_1(\varepsilon,\delta) = \frac{\varepsilon^{\frac{p-1}{2p}}}{\delta^{2+\frac{1}{p}}} + \sqrt{\delta}.$$

5. Examples

Example 1. Let $\Omega \subset \mathbb{R}^n$ be an open bounded set with C^m boundary $\partial\Omega$. In the real Hilbert space $L^2(\Omega)$, with the scalar product $(u,v) = \displaystyle\int_\Omega u(x)\,v(x)\,dx$, we consider the following Cauchy problem

$$\begin{cases} \varepsilon\partial_t^2 u_\varepsilon(x,t) + \partial_t u_\varepsilon(x,t) + Au_\varepsilon(x,t) + B\,u_\varepsilon(x,t) = f(x,t), x \in \Omega, t > 0, \\ u_\varepsilon(x,0) = u_{0\varepsilon}(x), \quad \partial_t\, u_\varepsilon(x,0) = u_{1\varepsilon}(x), \quad x \in \overline{\Omega}, \\ \dfrac{\partial^j u_\varepsilon}{\partial\nu^j}\Big|_{\partial\Omega} = 0, \quad j = 0, 1, \ldots, m-1, \quad t \geq 0, \end{cases} \tag{6}$$

The unperturbed Cauchy problem associated to (6) is

$$\begin{cases} \partial_t v(x,t) + A(x,\partial_x)v(x,t) + B\,v = f(x,t), x \in \Omega, t > 0, \\ v(x,0) = u_0(x), \quad x \in \overline{\Omega}, \\ \dfrac{\partial^j v}{\partial\nu^j}\Big|_{\partial\Omega} = 0, \quad j = 0, 1, \ldots, m-1, \quad t \geq 0, \end{cases} \tag{7}$$

where $D(A) = H^{2m}(\Omega) \cap H_0^m(\Omega)$,

$$A(x,\partial_x)u(x) = \sum_{|\alpha|\leq m} (-1)^{|\alpha|}\partial^\alpha\Big(a_\alpha(x)\partial^\alpha u(x)\Big), u \in D(A),$$

$$a_\alpha \in C^m(\overline{\Omega}), \alpha = (\alpha_1,\ldots,\alpha_n), \alpha_i \in \mathbb{N}, \quad \partial^\alpha = \Big(\frac{\partial}{\partial x_1}\Big)^{\alpha_1}\cdots\Big(\frac{\partial}{\partial x_n}\Big)^{\alpha_n}.$$

We suppose that operator A is self-adjoint, i.e.

$$\int_\Omega \Big(A(x,\partial_x)u(x)\Big)v(x)\,dx$$

$$= \int_\Omega u(x)\Big(A(x,\partial_x)v(x)\Big)dx, \ \forall u,v \in D(A). \tag{8}$$

Moreover, we suppose that

$$\sum_{|\alpha|\leq m}\Big(a_\alpha(x)\,\xi^\alpha,\,\xi^\alpha\Big)_{\mathbb{R}^n} \geq a_0\,||\xi||^{2m}, \ \forall x \in \bar{\Omega}, \ a_0 > 0. \tag{9}$$

Conditions (8) and (9) assure the strong ellipticity of the operator A and that A is self-adjoint and positive definite operator.

Define the operator B by the relationships:

$$D(B) = L^2(\Omega) \cap L^{2(q+1)}(\Omega), \quad Bu = b\,|u|^q u.$$

If $b > 0$, then B is a Fréchet derivative of convex and positive functional $\mathcal{B}$, which is defined as follows:

$$D(\mathcal{B}) = L^{q+2}(\Omega) \cap L^2(\Omega), \quad \mathcal{B}u = \frac{b}{q+2} \int_\Omega |u(x)|^{q+2}\, dx,$$

and the Fréchet's derivative of operator B is defined by the relationships:

$$D\big(B'(u)\big) = \{v \in L^2(\Omega) : u^q\, v \in L^2(\Omega)\},$$

$$B'(u)v = b\,(q+1)|u|^q\, v.$$

If Ω is bounded with the C^m boundary $\partial\Omega$, the condition (9) is fulfilled and q verifies the following condition

$$\begin{cases} q \in [0, 2m/(n-2m)], & if \quad n > 2m, \\ q \in [(m-1)/2m), \infty), & if \quad n = 2m, \quad m > 1, \\ q \in [0, \infty), & if \quad n = 2, \ m = 1, \\ q \in [0, \infty), & if \quad n < 2m, \end{cases} \tag{10}$$

then the operator B verifies conditions (**HM1**). If Ω is bounded with C^m boundary $\partial\Omega$, condition (9) is fulfilled and q verifies condition

$$\begin{cases} q \in [1, 2m/(n-2m)], & if \quad n > 2m, \\ q \in [1, \infty), & if \quad n \leq 2m, \end{cases} \tag{11}$$

then the operator B verifies condition (**HM3**). Using the above-mentioned facts, Theorems 2.1 and 2.2 we get:

Theorem 5.1 (Perjan&Rusu, 2012). *Let $\Omega \subset \mathbb{R}^n$ be an open bounded set with C^m boundary $\partial\Omega$. Let $T > 0$, $p \in (1, \infty]$, $b > 0$, q verifies* (10) *and* (8) – (9) *are fulfilled. If $u_0, u_{0\varepsilon} \in H^{2m}(\Omega) \cap H_0^m(\Omega)$, $u_{1\varepsilon} \in H_0^m(\Omega)$ and $f, f_\varepsilon \in W^{1,p}(0,T; L^2(\Omega))$ then there exist $C = C(T, p, a_0, b, n, m, q, \Omega)$ and $\varepsilon_0 = \varepsilon_0(a_0, n, m, \Omega)$, $\varepsilon_0 \in (0,1)$, such that:*

$$||u_\varepsilon - v||_{C([0,T];L^2(\Omega))} + ||u_\varepsilon - v||_{L^2(0,T;\, H_0^m(\Omega))}$$

$$\leq C\left(\mathbf{M}\,\varepsilon^\beta + ||f_\varepsilon - f||_{L^p(0,T;L^2(\Omega))} + ||u_{0\varepsilon} - u_0||_{L^2(\Omega)}\right), \ \forall \varepsilon \in [0, \varepsilon_0],$$

where u_ε and v are strong solutions to the problems (6) *and* (7), *respectively,*

$$\mathbf{M} = ||u_{1\varepsilon}||_{H_0^m(\Omega)} + ||Au_{0\varepsilon}||_{L^2(\Omega)} + ||f_\varepsilon||_{W^{1,p}(0,T;L^2(\Omega))},$$

$$\beta = \min\{1/4,\, (p-1)/2p\}.$$

Theorem 5.2 (Perjan&Rusu, 2012). *Let $\Omega \subset \mathbb{R}^n$ be an open bounded set with C^m boundary $\partial\Omega$. Let $T > 0$, $p \in (1,\infty]$, $b > 0$, q verifies* (11) *and* (8), (9) *are fulfilled. If u_ε, $u_{0\varepsilon}$, A_0u_0, $h_\varepsilon \in H^{2m}(\Omega)\cap H_0^m(\Omega)$, $u_{1\varepsilon} \in H_0^m(\Omega)$ and $f, f_\varepsilon \in W^{2,p}(0,T;L^2(\Omega))$ then there exist $C = C(T,p,\ a_0,\ b,\ n,\ m,q,\Omega) > 0$ and $\varepsilon_0 = \varepsilon_0(a_0,n,m,\Omega)$, $\varepsilon_0 \in (0,1)$, such that*

$$||u'_\varepsilon - v' + h_\varepsilon e^{-t/\varepsilon}||_{C([0,T];L^2(\Omega))} + ||u'_\varepsilon - v' + h_\varepsilon e^{-t/\varepsilon}||_{L^2(0,T;\,H_0^m(\Omega))}$$

$$\leq C\Big(\mathbf{M}(T,u_{0\varepsilon},u_{1\varepsilon},f_\varepsilon)\,\varepsilon^\beta + \mathbf{D}_\varepsilon\Big), \quad \forall\varepsilon \in [0,\varepsilon_0],$$

where u_ε and v are strong solutions to the problems (6) *and* (7), *respectively,*

$$h_\varepsilon = f_\varepsilon(0) - Au_{0\varepsilon} - B(u_{0\varepsilon}) - u_{1\varepsilon},$$

$$\mathbf{M} = ||u_{1\varepsilon}||_{H_0^m(\Omega)} + ||Au_{0\varepsilon}||_{L^2(\Omega)} + ||f_\varepsilon||_{W^{1,p}(0,T;L^2(\Omega))},$$

$$\beta = \min\{1/4,\, (p-1)/2p\},$$

$$\mathbf{D}_\varepsilon = ||f_\varepsilon - f||_{W^{1,p}(0,T;L^2(\Omega))} + ||u_{0\varepsilon} - u_0||_{L^2(\Omega)} + ||B(u_{0\varepsilon}) - B(u_0)||_{L^2(\Omega)}.$$

In particular, one can consider the Cauchy problem

$$\begin{cases} \varepsilon u_{tt}(x,t) + u_t(x,t) - \Delta u(x,t) + |u(x,t)|^q\, u(x,t) = f(x,t), x \in \Omega, t > 0, \\ u(x,0) = u_0(x), \quad u_t(x,0) = u_1(x), \quad x \in \overline{\Omega}, \\ u\Big|_{\partial\Omega} = 0, \quad t \geq 0, \end{cases}$$

and the associated unperturbed Cauchy problem:

$$\begin{cases} v_t(x,t) - \Delta v(x,t) + |v(x,t)|^q\, v(x,t) = f(x,t), x \in \Omega, t > 0, \\ v(x,0) = u_0(x), \quad x \in \overline{\Omega}, \\ v\Big|_{\partial\Omega} = 0, \quad t \geq 0, \end{cases}$$

with

$$\begin{cases} q \in [0, 2/(n-2)], & \text{if} \quad n > 2, \\ q \in [0,\infty), & \text{if} \quad n = 1, 2, \end{cases}$$

and get:

$$||u-v||_{C([0,T];L^2(\Omega))}+||u-v||_{L^2(0,T;\,H_0^1(\Omega))}$$

$$\leq C\,\mathbf{M}(T,u_{0\varepsilon},u_{1\varepsilon},f_\varepsilon)\,\varepsilon^{\beta},\quad \forall\varepsilon\in[0,\varepsilon_0].$$

For

$$\begin{cases} q\in[1,2/(n-2)], & \text{if} \quad n>2,\\ q\in[1,\infty), & \text{if} \quad n=1,2\end{cases}$$

we get:

$$||u'-v'+he^{-t/\varepsilon}||_{C([0,T];L^2(\Omega))}+||u'-v'+he^{-t/\varepsilon}||_{L^2(0,T;\,H_0^1(\Omega))}$$

$$\leq C\,\mathbf{M}(T,u_{0\varepsilon},u_{1\varepsilon},f_\varepsilon)\,\varepsilon^{\beta},\quad \forall\varepsilon\in[0,\varepsilon_0].$$

$$h=f(0)+\Delta u_0-|u_0|^q\,u_0-u_1,\quad \beta=\min\{1/4,\,(p-1)/2p\}.$$

Example 2. Let us define in (6) and (7) the operator B as follows:

$$D(B)=H^r(\Omega),\quad Bu(x)=b(x,u(x),\ldots,D^r u(x)),$$

where $D^k u=\big\{\partial^\alpha u:|\alpha|=k\big\}$ and b is a nonlinear function such that there exist $r_0\in\mathbb{N},\ r_0\leq m$ and

$$p_0\in\begin{cases}\Big(\dfrac{1}{2},\dfrac{n}{n-2(m-r_0)}\Big], \text{ if } r_0\geq m-\dfrac{n}{2},\\ (0,\infty),\ \text{ if } r_0<m-\dfrac{n}{2},\end{cases} \tag{12}$$

which verifies the inequality

$$\Big|b(x,u_1(x),\ldots,D^r u_1(x))-b(x,u_2(x),\ldots,D^r u_2(x))\Big|$$

$$\leq b_0\max_{|\alpha|\leq r_0}\big|\partial^\alpha\left(u_1(x)-u_2(x)\right)\big|^{p_0}\quad x\in\overline{\Omega}. \tag{13}$$

Then, for $r_0\in\mathbb{N},\ m-\dfrac{n}{2}\leq r_0\leq m$ and for $\dfrac{1}{2}<p_0\leq\dfrac{n}{n-2(m-r_0)}$, due to (13) and the Galiardo-Nirenberg inequality, we can state

$$||b(x,u_1(x),\ldots,D^r u_1(x))-b(x,u_2(x),\ldots,D^r u_2(x))||_{L_2(\Omega)}$$

$$\leq b_0\Big(\int_\Omega\max_{|\alpha|\leq r_0}\big|\partial^\alpha\left(u_1(x)-u_2(x)\right)\big|^{2p_0}\Big)^{1/2}\leq C\,||u_1-u_2||^{p_0}_{H_0^m(\Omega)}$$

$$=C\,||A_0(u_1-u_2)||^{p_0}_{L_2(\Omega)},\forall u_1,u_2\in H_0^m(\Omega). \tag{14}$$

For $r_0 \in \mathbb{N}$, $0 \leq r_0 < m - \dfrac{n}{2}$ and $p_0 > 0$, due to (13) and the Galiardo-Nirenberg inequality, we can state

$$||b(x, u_1(x), \ldots, D^r u_1(x)) - b(x, u_2(x), \ldots, D^r u_2(x))||_{L_2(\Omega)}$$

$$\leq b_0 \left(\int_\Omega \max_{|\alpha| \leq r_0} \left| \partial^\alpha \left(u_1(x) - u_2(x)\right)\right|^{2p_0} \right)^{1/2} dx$$

$$\leq (mes\, \Omega)^{1/2}\, C\, ||u_1 - u_2||^{p_0}_{C^{r_0}(\bar{\Omega})} \leq (mes\, \Omega)^{1/2}\, C\, ||u_1 - u_2||^{p_0}_{H_0^m(\Omega)}$$

$$= (mes\, \Omega)^{1/2}\, C\, ||A_0(u_1 - u_2)||^{p_0}_{L_2(\Omega)}, \quad \forall u_1, u_2 \in H_0^m(\Omega). \tag{15}$$

Due to (12) - (15) we conclude that, if Ω is bounded with C^m boundary $\partial\Omega$, then the operator B verifies (**HL1**).

Using Theorem 3.1, we get:

Theorem 5.3 (Perjan&Rusu, 2011). *Let $\Omega \subset \mathbb{R}^n$ be an open bounded set with C^m boundary $\partial\Omega$. Let $T > 0$, $p \in (1, \infty]$, $q \leq m$ and (12), (13) are fulfilled. If u_0, $u_{0\varepsilon}$, $u_{1\varepsilon} \in H^{2m}(\Omega) \cap H_0^m(\Omega)$, $u_{1\varepsilon} \in H_0^m(\Omega)$ and $f, f_\varepsilon \in W^{1,p}(0, T; L^2(\Omega))$ then there exist $C = C(T, p, a_0, b_0, r_0, p_0, n, m, \Omega) > 0$ and $\varepsilon_0 = \varepsilon_0(a_0, b_0, r_0, p_0, n, m, \Omega)$, $\varepsilon_0 \in (0, 1)$, such that*

$$||u_\varepsilon - v||_{C([0,T]; L^2(\Omega))} + ||u_\varepsilon - v||_{L^2(0,T;\, H_0^m(\Omega))}$$

$$\leq C\left(\mathbf{M}\, \varepsilon^\beta + D_\varepsilon\right), \quad \varepsilon \in [0, \varepsilon_0],$$

where u_ε and v are strong solutions to the problems (6) and (7), respectively,

$$\mathbf{M}(T, u_{0\varepsilon}, u_{1\varepsilon}, f_\varepsilon) = ||A^{1/2} u_{1\varepsilon}||_{L_2(\Omega)} + ||A u_{0\varepsilon}||_{L_2(\Omega)}$$

$$+ ||B(u_{0\varepsilon})||_{L_2(\Omega)} + ||f_\varepsilon||_{W^{1,p}(0,\, T\,; L_2(\Omega))},$$

$$\beta = min\{1/4,\, (p-1)/2p\},$$

$$D_\varepsilon = ||f_\varepsilon - f||_{L^p(0,T; L_2(\Omega))} + ||u_{0\varepsilon} - u_0||_{L_2(\Omega)}.$$

References

1. M. Cakir, Uniform Second-Order Difference Method for a Singularly Perturbed Three-Point Boundary Value Problem, *Advances in Difference Equations*, (2010).
2. H. O. Fattorini, The hyperbolic singular perturbation problem: an operator approach, *J. Differential Equations* **70**, 1 (1987).
3. M. Ghisi and M. Gobbino, Global-in-time uniform convergence for linear M. hyperbolic-parabolic singular perturbations, *Acta Math. Sinica* (English Series), **22**, 4 (2006).
4. M. Gobbino, Singular perturbation hyperbolic - parabolic for degenerate nonlinear equations of Kirchhoff type, *Nonlinear Anal.*, **44**, 3 (2001).
5. R. E. O'Malley, Two parameter singular perturbation problems for second order equations, *J. Math. Mech.*, **16** (1967).
6. B. Najman, Convergence estimate for second order Cauchy problems with a small parameter, *Czechoslovak Mathematical Journal,* **48**, 123 (1998).
7. A. Perjan, Linear singular perturbations of hyperbolic-parabolic type, *Bul. Acad. Stiinte Repub. Mold. Mat.*, **2**, 42 (2003).
8. A. Perjan and G. Rusu, Convergence estimates for abstract second-order singularly perturbed Cauchy problems with Lipschitz nonlinearities, *Asymptotic Analysis,* **74**, 3-4 (2011).
9. A. Perjan and G. Rusu Convergence estimates for abstract second order singularly perturbed Cauchy problems with monotone nonlinearities. *Annals of the Academy of Romanian Scientists. Series on Mathematics and its Applications*, **4**, 2 2012.
10. E. O'Rodin, L. M. Pickett and G. I. Shishkin, Singularly perturbed problems modeling reaction-convection-diffusion processes, *Computational methods in applied mathematics,* **3**, 3 (2003).
11. W. K. Zahra and A. M. El Mhlawy, Numerical solution of two-parameter singularly perturbed boundary value problems via exponential spline, *Journal of King Saud University Science,* **25** (2013).

Global controllability and mixing for the Burgers equation with localised finite-dimensional external force

Armen Shirikyan

Department of Mathematics, University of Cergy–Pontoise
CNRS UMR 8088, 2 avenue Adolphe Chauvin
95302 Cergy–Pontoise, France
and National Research University MPEI, Moscow, Russia
E-mail: Armen.Shirikyan@u-cergy.fr

The aim of this short note to give a summary of some recent results on global controllability of the viscous Burgers equation and mixing properties of the associated stochastic problem. We also outline the main ideas of the proofs. The details are given in [1,2].

Keywords: Burgers equation, exact and approximate controllability, mixing, localised external force

1. Introduction

We consider the problem

$$\partial_t u - \nu\partial_x^2 u + u\partial_x u = f(t,x), \tag{1}$$

$$u(t,0) = u(t,1) = 0, \tag{2}$$

$$u(0,x) = u_0(x), \tag{3}$$

where $x \in I = (0,1)$ and $t \geq 0$, $\nu > 0$ is a fixed parameter, $u = u(t,x)$ is an unknown function with range in $\mathbb{R}$, u_0 is a given initial condition, and f is an external force. Let us fix a closed interval $[a,b] \subset I$ and denote by E the two-dimensional vector space spanned by the functions

$$e_1(x) = \sin\Bigl(\pi\frac{x-a}{b-a}\Bigr), \quad e_2(x) = \sin\Bigl(2\pi\frac{x-a}{b-a}\Bigr), \quad x \in [a,b], \tag{4}$$

extended to $I \setminus [a,b]$ by zero. We deal with following two situations:

Control problem. *The external force has the form*

$$f(t,x) = h(x) + \zeta(t,x), \tag{5}$$

where[a] $h \in H^1(I)$ *is a given function and* $\zeta(t,x)$ *is a control with range in* E.

Stochastic problem. *The external force is stochastic and is given by the relation*

$$f(t,x) = h(x) + \eta(t,x), \tag{6}$$

in which $h \in H^1(I)$ *is a deterministic function and* η *is an* E*-valued white noise of the form*

$$\eta(t,x) = \frac{\partial}{\partial t}\sum_{j=1}^{2} b_j\beta(t)e_j(x), \tag{7}$$

where $b_j > 0$ *are some numbers and* $\{\beta_j\}$ *are independent standard Brownian motions.*

In the first case, we are interested in global approximate controllability to trajectories, while the second problem concerns the uniqueness of a stationary distribution for the Markov process associated with (1), (6), (7) and the large-time behaviour of its trajectories. These two questions were intensively studied in the literature, and we now describe some results concerning the Burgers equation.

The problem of asymptotic behaviour of trajectories for stochastic Burgers equation was first studied by Sinai[3] who established the convergence of the laws of solutions to a measure independent of the initial condition. This result was refined and developed in a number of works, and it is now well known that the Markov process corresponding to problem (1), (6) with $h \equiv 0$ has a unique stationary distribution, and all other solutions converge to it weakly as $t \to +\infty$; see the paper[4] and the references therein for the exact result and discussion. The control problem was investigated by Fursikov and Imanuvilov[5,6], who proved the local exact controllability to trajectories and the absence of global approximate controllability. Agrachev and Sarychev[7,8] (see also[9]) developed a general approach, applicable to the Burgers equation, that enables one to study the approximate controllability of nonlinear PDE's by a low-dimensional control. Chapouly[10] applied Coron's return method to study the exact controllability of the Burgers equation by two boundary controls combined with one distributed control. The reader is referred to the books[11–15] for more details on these two subjects and further results for other equations.

[a]We refer the reader to the Notation below for the definition of functional spaces.

The aim of this short note is to give a summary of the results established in[1,2]. We first discuss the problem of global exponential stabilisation of non-stationary solutions for (1), (5) by a localised control. This property combined with the local exact controllability implies that the Burgers equation is exactly controllable to trajectories by a control localised in the physical space. Moreover, invoking a result of Phan and Rodrigues[16], one can easily deduce that the Burgers equation is approximately controllable to trajectories by a two-dimensional localised control. We next turn to the stochastic counterpart of these results. Namely, the contraction property for the resolving operator of the Burgers equation combined with the global approximate controllability to trajectories implies that problem (1), (6), (7) has a unique stationary distribution, which is asymptotically stable (in the weak topology of measures) as $t \to +\infty$. We confine ourselves to giving some ideas of the proofs of these results, referring the reader to[1,2] for details.

Notation

We write $I = (0,1)$ and $\mathbb{R}_+ = [0, +\infty)$ and denote by $L^p(I)$ and $H^s(I)$ the usual Lebesgue and Sobolev spaces on I, endowed with the standard norms $\|\cdot\|_{L^p}$ and $\|\cdot\|_s$, respectively, and by $H_0^s(I)$ the closure in $H^s(I)$ of the space of infinitely smooth functions on I with compact support. Very often, we shall omit the interval I from the notation and write L^p, H^s, and H_0^s. Given a closed interval $J \subset \mathbb{R}$ and a separable Banach space X, let $C(J, X)$ be the space of continuous functions from J to X and $L^p(J, X)$ be the space of Borel-measurable functions $f : J \to X$ such that

$$\|f\|_{L^p(J,X)}^p := \int_J \|f(t)\|_X^p \mathrm{d}t < \infty,$$

with obvious modification in the case $p = \infty$.

2. Controllability to trajectories

In this section, we consider problem (1)–(3), in which $u_0 \in L^2$, and the external force f has the form (5). We shall always assume that $h \in H^1$ is a fixed function, and ζ is a control taking values in the closed subspace

$$F = \{v \in H_0^1(I) : \operatorname{supp} v \subset [a, b]\},$$

where $[a, b] \subset I$ is an interval. Let us recall that the initial-boundary value problem (1)–(3) is well posed in the space $H := L^2$. Namely, for any $u_0 \in H$

and $f \in L^1_{\rm loc}(\mathbb{R}_+, H)$, there is a unique function

$$u \in L^2_{\rm loc}(\mathbb{R}_+, H^1_0) \cap W^2_{\rm loc}(\mathbb{R}_+, H^{-1})$$

satisfying (1) and (3). We denote by $\mathcal{R}_t(u_0, f)$ the mapping that takes the pair (u_0, f) to $u(t)$. The following theorem is the main result of[2].

Theorem 2.1. *Let $\nu > 0$, $[a, b] \subset I$, and $h \in H^1$ be given. Then there are positive numbers C and γ such that the following property holds: for arbitrary $u_0, \hat{u}_0 \in H$ there is a piecewise continuous function $\zeta : \mathbb{R}_+ \to F$ such that*

$$\|\mathcal{R}_t(u_0, h+\zeta) - \mathcal{R}_t(\hat{u}_0, h)\|_{H^1} + \|\zeta(t)\|_{H^1} \le Ce^{-\gamma t} \min\big(\|u_0 - \hat{u}_0\|_{L^1}^{2/5}, 1\big), \quad (8)$$

where $t \ge 1$.

In other words, any solution of (1), (2) corresponding to the zero control $\zeta \equiv 0$ can be exponentially stabilized by an H^1-smooth control localized in a given interval $[a, b]$. Before giving some ideas of the proof of this result, we state two corollaries. The first one concerns the exact controllability to trajectories for problem (1), (2).

Corollary 2.1. *Under the hypotheses of Theorem 2.1, there is $T > 0$ such that, given initial conditions $u_0, \hat{u}_0 \in H$, one can find a control $\zeta \in L^2([0, T], F)$ for which*

$$\mathcal{R}_T(u_0, h + \zeta) = \mathcal{R}_T(\hat{u}_0, h).$$

To prove this property, it suffices to use the control constructed in Theorem 2.1 to steer a solution of (1), (2) sufficiently close to the reference trajectory $\hat{u}(t) = \mathcal{R}_t(\hat{u}_0, h)$ and then to apply the Fursikov–Imanuvilov result about local exact controllability to trajectories; see Section I.6 in[6].

The second corollary concerns the global approximate controllability to trajectories for the Burgers equation by a localised control of low dimension. Recall that we denote by E the vector span of the functions $e_1, e_2 \in H^1_0$ defined by (4) on $[a, b]$ and extended by zero outside $[a, b]$.

Corollary 2.2. *Under the hypotheses of Theorem 2.1, there is $T > 0$ such that, given $u_0, \hat{u}_0 \in H$ and $\varepsilon > 0$, one can find a control $\zeta \in C^\infty([0, T], E)$ for which*

$$\|\mathcal{R}_T(u_0, h + \zeta) - \mathcal{R}_T(\hat{u}_0, h)\|_{L^2} \le \varepsilon.$$

Thus, the global approximate controllability to trajectories holds, provided that the time of control is sufficiently large. To establish this property, it suffices to combine Corollary 2.1 with the results of Section 5 in[16].

Scheme of the proof of Theorem 2.1. We begin with two remarks:

Regularisation. The Burgers equation is a globally well-posed nonlinear parabolic PDE and, hence, possesses a regularising property. Namely, if $f(t,x)$ is a bounded function of time with range in H^1, then for any initial condition $u_0 \in H$ the corresponding solution $u(t,x)$ of (1)–(3) is bounded in H^2 on the half-line $[1,+\infty)$ by universal constant R not depending on u_0. Thus, there is no loss of generality in assuming that both initial functions u_0 and $\hat{u}_0$ belong to the ball of radius R centred at zero in the space $H_0^1 \cap H^2$.

Interpolation. The following inequality is well known:

$$\|v\|_{H^1} \leq C\|v\|_{L^1}^{2/5}\|v\|_{H^2}^{3/5} \quad \text{for any } v \in H^2.$$

Therefore, if we construct a bounded control $\zeta : \mathbb{R}_+ \to F$ such that

$$\|\mathcal{R}_t(u_0, h+\zeta) - \mathcal{R}_t(\hat{u}_0, h)\|_{L^1} \leq C_1 e^{-\gamma_1 t}\|u_0 - \hat{u}_0\|_{L^1}, \quad t \geq 0, \tag{9}$$

then the required estimate for the difference between two solutions will follow from the above interpolation inequality and the boundedness of the H^2-norms.

A key for the proof of (9) is the following result about the linear PDE

$$\partial_t w - \nu\partial_x^2 w + \partial_x(a(t,x)w) = 0, \tag{10}$$

where $a(t,x)$ is a given sufficiently regular function.

Proposition 2.1. *For any closed interval $I' \subset I$, one can find positive numbers ε and $q < 1$ such that any solution $w(t,x)$ of (10) satisfies one of the inequalites*

$$\|w(1)\|_{L^1} \leq q\|w(0)\|_{L^1} \quad \text{or} \quad \|w(1)\|_{L^1(I')} \geq \varepsilon\|w(0)\|_{L^1}. \tag{11}$$

In other words, either the L^1-norm w contracts by a factor strictly less than 1, or a nontrivial mass is concentrated on I'. Now note that if v and $\hat{u}$ are two solutions of (1), (2) with $\zeta \equiv 0$, then their difference $w = v - \hat{u}$ satisfies Eq. (10) and, hence, one of the inequalities in (11). In both cases, we can modify $v(t,x)$ on the domain $[0,1] \times [a,b]$ so that the resulting function $u(t,x)$ satisfies the inequality

$$\|u(1) - \hat{u}(1)\|_{L^1} \leq \theta\,\|u(0) - \hat{u}(0)\|_{L^1}, \tag{12}$$

where $\theta < 1$ depends only on the size of initial conditions. Since v and u coincide on $[0,1] \times (I \setminus [a,b])$, we see that $u(t,x)$ is a solution of the controlled equation (1) in which $\mathrm{supp}\zeta(t,\cdot) \subset [a,b]$. Iteration of (12) gives the required inequality (9). □

3. Mixing of the stochastic flow

In this section, we consider problem (1)–(3), in which the right-hand side is a stochastic process of the form (6), (7). The existence and uniqueness of a solution is well known, and our goal is to study the large-time behaviour of trajectories. The family of all solutions for (1), (2) form a Markov process in the phase space $H = L^2(I)$, and we denote by $P_t(u, \Gamma)$ its transition function and by

$$\mathfrak{P}_t : C_b(H) \to C_b(H), \quad \mathfrak{P}_t^* : \mathcal{P}(H) \to \mathcal{P}(H)$$

the corresponding Markov operators. Here $C_b(H)$ stands for the space of bounded continuous functions $g : H \to \mathbb{R}$, endowed with the L^∞-norm, and $\mathcal{P}(H)$ denotes the set of all probability Borel measures on H. The latter is endowed with the topology of weak convergence, which is denotes by $\rightharpoonup$; see Chapter 1 in[15] for the definition of all these concepts.

Theorem 3.1. *Let $\nu > 0$, $[a,b] \subset I$, and $h \in H^1$ be given and let the coefficients b_j defining the random force (7) be positive. Then there is a unique measure $\mu \in \mathcal{P}(H)$ such that $\mathfrak{P}_t^*\mu = \mu$ for all $t \geq 0$. Moreover, μ is asymptotically stable in the sense that, for any $g \in C_b(H)$ and $\lambda \in \mathcal{P}(H)$, we have*

$$\mathfrak{P}_t g \to (g, \mu) \quad \text{in } C_b(H) \text{ as } t \to +\infty, \tag{13}$$

$$\mathfrak{P}_t^*\lambda \rightharpoonup \mu \quad \text{in } \mathcal{P}(H) \text{ as } t \to +\infty, \tag{14}$$

where (g, μ) stands for the integral of g over H against the measure μ.

Scheme of the proof. The existence of a stationary measure is a simple consequence of the Bogolyubov–Krylov argument (e.g., see Section 14 in[11]), and therefore we confine ourselves to outlining the proof of uniqueness and asymptotic stability. To this end, let us recall a sufficient condition for the validity of these properties. Namely, as is proved in Section 3.1 of[15], the uniqueness of a stationary measure μ and the convergence relations (13) and (14) are valid if the following two properties hold:

Stability. *There is a function $\gamma(r) \geq 0$ going to zero as $r \to 0$ such that, for any 1-Lipschitz function $g \in C_b(H)$ satisfying $\|g\|_{L^\infty} \leq 1$, we have*

$$\sup_{t\geq 1} \big|\big(g, P_t(u,\cdot)\big) - \big(g, P_t(v,\cdot)\big)\big| \leq \gamma(r) \quad \text{for } \|u - v\|_H \leq r. \tag{15}$$

Given an initial condition $u_0 \in H$ and a closed ball $B \subset H$, let $\tau_{u_0}(B)$ be the first instant of time when the trajectory $\mathcal{R}_t(u_0, h + \eta)$ hits B.

Recurrence. *There is a sequence of balls $B_m \subset H$, whose diameters go zero as $m \to \infty$, and functions $p_m(t)$, going to zero as $t \to +\infty$, such that*

$$\mathbb{P}\{\tau_v(B_m) > t\} \leq p_m(t) \quad \textit{for any } v \in H,\, m \geq 1,\, t > 0. \tag{16}$$

The stability is a straightforward consequence of the regularisation for the viscous Burgers equation and the following *contraction property*:

$$\|\mathcal{R}_t(u_0, h+\eta) - \mathcal{R}_t(v_0, h+\eta)\|_{L^1} \leq \|u_0 - v_0\|_{L^1} \quad \text{for } t \geq 0,\, u_0, v_0 \in H. \tag{17}$$

Let us consider the recurrence. A regularisation property for the stochastic Burgers equation enables one to find positive numbers $s \in (1,2)$, T_1, R, and p such that

$$\mathbb{P}\Big\{\|\mathcal{R}_{T_1}(u_0, h+\eta)\|_{H^s} \leq R\Big\} \geq p \quad \text{for any } u_0 \in H.$$

Furthermore, the controllability property established in Corollary 2.2 implies that, if $\hat{u} \in H_0^1 \cap H^2$ is a time-independent solution of (1) with $f = h$, then there is $T_2 > 0$ such that, for any $u_0 \in H_0^1 \cap H^s$ satisfying $\|u_0\|_s \leq R$, we have

$$\mathbb{P}\Big\{\|\mathcal{R}_{T_2}(u_0, h+\eta) - \hat{u}\|_{L^2} \leq 1/m\Big\} \geq \varepsilon_m \quad \text{for } m \geq 1,$$

where $\varepsilon_m > 0$ are some numbers not depending on u_0. The above two inequalities imply the required estimate (16), in which $B_m \subset H$ is the ball of radius $1/m$ centred at $\hat{u}$; e.g., see the proof of Proposition 3.3.6 in[15]. □

Acknowledgments

This research was supported by the RSF grant 14-49-00079.

References

1. A. Shirikyan, Mixing for the Burgers equation driven by a localised two-dimensional stochastic forcing, *preprint* (2016).
2. A. Shirikyan, Global exponential stabilisation for the Burgers equation with localised control, *preprint* (2016).
3. Y. G. Sinaĭ, Two results concerning asymptotic behavior of solutions of the Burgers equation with force, *J. Statist. Phys.* **64**, 1 (1991).
4. A. Boritchev, Sharp estimates for turbulence in white-forced generalised Burgers equation, *Geom. Funct. Anal.* **23**, 1730 (2013).

5. A. V. Fursikov and O. Y. Imanuvilov, On controllability of certain systems simulating a fluid flow, in *Flow control (Minneapolis, MN, 1992)*, ed. M. D. Gunzburger, IMA Vol. Math. Appl., Vol. 68 (Springer, New York, 1995) pp. 149–184.
6. A. V. Fursikov and O. Y. Imanuvilov, *Controllability of Evolution Equations* (Seoul National University, Research Institute of Mathematics, Global Analysis Research Center, Seoul, 1996).
7. A. A. Agrachev and A. V. Sarychev, Navier–Stokes equations: controllability by means of low modes forcing, *J. Math. Fluid Mech.* **7**, 108 (2005).
8. A. A. Agrachev and A. V. Sarychev, Controllability of 2D Euler and Navier–Stokes equations by degenerate forcing, *Comm. Math. Phys.* **265**, 673 (2006).
9. A. Shirikyan, Approximate controllability of the viscous Burgers equation on the real line, in *Geometric control theory and sub-Riemannian geometry*, (Springer, Cham, 2014) pp. 351–370.
10. M. Chapouly, Global controllability of nonviscous and viscous Burgers-type equations, *SIAM J. Control Optim.* **48**, 1567 (2009).
11. G. Da Prato and J. Zabczyk, *Ergodicity for Infinite Dimensional Systems* (Cambridge University Press, Cambridge, 1996).
12. A. V. Fursikov, *Optimal control of Distributed Systems. Theory and Applications* (American Mathematical Society, Providence, RI, 2000).
13. J.-M. Coron, *Control and Nonlinearity* (American Mathematical Society, Providence, RI, 2007).
14. F. Flandoli, *Random perturbation of PDEs and fluid dynamic models*, Lecture Notes in Mathematics, Vol. 2015 (Springer, Heidelberg, 2011).
15. S. Kuksin and A. Shirikyan, *Mathematics of Two-Dimensional Turbulence* (Cambridge University Press, Cambridge, 2012).
16. D. Phan and S. S. Rodrigues, Approximate controllability for equations of fluid mechanics with a few body controls, in *Control Conference (ECC), 2015 European*, (IEEE, 2015) pp. 2682–2687.

Boundary observation in shape optimization

Dan Tiba

Institute of Mathematics (Romanian Academy)
and Academy of Romanian Scientists, Bucharest
E-mail: dan.tiba@imar.ro

We discuss the case of shape optimization problems with boundary cost functional. We investigate theoretical properties and an important tool is the implicit parametrization theorem.

Keywords: optimal design, boundary cost, implicit representation

1. Introduction

A typical example of shape optimization problems has the form:

$$\operatorname*{Min}_{\Omega \in \mathcal{O}} \int_{\Lambda} j(x, y_\Omega(x), \nabla y_\Omega(x))dx, \tag{1}$$

$$-\Delta y_\Omega = f \quad in\ \Omega, \tag{2}$$

$$y_\Omega = 0 \quad on\ \partial\Omega \tag{3}$$

with other supplementary constraints, if necessary. See Sokolowsky and Zolesio[17], Pironneau[15], Haslinger and Neittaanmaki[4], Neittaanmaki, Sprekels and Tiba[9]. Here, $\mathcal{O}$ is a family of admissible domains in R^d, $d \in N$, satisfying certain regularity hypotheses and conditions like

$$E \subset \Omega \subset D,\ \forall\, \Omega \in \mathcal{O} \tag{4}$$

where E, D are given (bounded) domains, E may be even void, etc.

Function $f \in L^2(D)$ and Λ may be either E (if nonvoid) or Ω, or some part of $\partial\Omega$. The integrand $j(\cdot,\cdot,\cdot) : D \times R \times R^d \to R$ is of Carathéodory type and convexity, growth and/or regularity hypotheses have to be imposed, according to the applications. More general differential operators, other boundary conditions, more general integrand functionals, other restrictions (for instance constraints on the state y_Ω), etc., may be considered as well.

Spectral cost functionals are discussed in Sokolowsky and Zolesio[17], Henrot and Pierre[5], Bucur and Butazzo[1].

In optimal control theory, boundary observation is an important and realistic case. This case is less discussed in shape optimization problems due to the natural difficulties related to the regularity or to the handling of unknown boundaries.

This article is devoted to the examination of boundary cost functionals in optimal design. In the literature, special cases are already discussed, for instance in Pironneau[15], Neittaanmaki and Haslinger[4], Sokolowsky and Zolesio[17], etc. By using the recent implicit parametrization theorem[18], [11] we may consider here general families of domains and combined boundary and topological variations. Our approach is of fixed domain type and avoids well known drawbacks like remeshing and recomputing of the mass matrix in each iteration, strong regularity assumptions, etc.

There are numerous ways to reduce certain shape optimization problems to optimal control problems via geometric controllability properties or the use of characteristic functions, etc., Lions[6], Pironneau[15], Neittaanmäki, Sprekels and Tiba[9]. This paper uses the new fixed domain approach based on *functional variations* of the geometry. They have been introduced in Neittaanmäki, Pennanen and Tiba[8]. See as well the recent papers Neittaanmäki and Tiba[10], Philip and Tiba[13]. Functional variations apply implicit representations of domains, but are different from the level set method of Osher and Sethian[12] (for instance, no time variable and no "evolution" of domains is used, no Hamilton-Jacobi equation is needed).

In the next section, we briefly recall a significant example, the solution of (1)-(4) for distributed observation, via the functional variations approach and we formulate the problem to be studied. In section 3 we give a short review in dimension two of the recent implicit parametrization theorem[18], which plays a key role in the sequel. The last section is devoted to the examination of shape optimization problems with boundary cost functional via functional variations/optimal control methods.

2. Example and Problem Formulation

A first property, specific to Dirichlet boundary conditions, ensures the approximate extension of the boundary value problem (2), (3) from the unknown and variable domain Ω to the given domain D. We associate to (2), (3) the approximating problem ($\varepsilon > 0$):

$$-\Delta \widehat{y} + \frac{1}{\varepsilon}(1 - H_\Omega)\widehat{y} = f \quad \text{in } D, \tag{5}$$

$$\widehat{y} = 0 \quad \text{on } \partial D. \tag{6}$$

Here, $H_\Omega : D \to R$ is the characteristic function of Ω in D.

In order to use functional variations, we define the class $\mathcal{O}$ of admissible open sets by:

$$\Omega = \Omega_g = \text{int}\{x \in D; g(x) \geq 0\}, \tag{7}$$

where $g \in X(D) \subset C(\overline{D})$ and $X(D)$ is some functional space on D. Examples of such spaces $X(D)$ may be finite element spaces, etc. In practice, it is enough to take g piecewise continuous and bounded, Philip and Tiba[14,13].

Under representation formula (7), if we impose

$$g \geq 0 \quad \text{in } E, \tag{8}$$

then constraint (4) is fulfilled. If $H : R \to R$ is the Heaviside function, then $H(g) : D \to R$ is the characteristic function of $\overline{\Omega}_g$. Notice that in this setting, the set $\{x \in D; g(x) = 0\}$ may have even positive measure, but Ω_g is always a Carathéodory set, i.e. $\Omega_g = \text{int } \overline{\Omega}_g$.

Denote by $H^\varepsilon : R \to R$ a smoothing of the Yosida approximation of the maximal monotone extension of H to $R \times R$. Then $H^\varepsilon(g) : D \to R$ is an approximation of the characteristic function H_Ω. Such procedures have been first introduced in Makinen, Neittaanmäki and Tiba[7]. More recently, they are frequently used in image reconstruction problems and are sometimes called the Chan-Vese regularization,[2].

We further approximate (5) by

$$-\Delta y_\varepsilon + \frac{1}{\varepsilon}(1 - H^\varepsilon(g))y_\varepsilon = f \quad \text{in } D, \tag{9}$$

$$y_\varepsilon = 0 \quad \text{on } \partial D. \tag{10}$$

Proposition 2.1. *If $\Omega = \Omega_g$ is of class C, then $y_\varepsilon|_{\Omega_g} \to y$ (the solution of (2), (3)) weakly in $H^1(\Omega_g)$ and strongly in $L^2(\Omega_g)$.*

This is a consequence of the Hedberg - Keldys stability property for domains of class C (i.e. with continuous boundary),[9,8]. Notice the very weak regularity assumptions on $\Omega \in \mathcal{O}$.

Based on Prop. 2.1, we approximate the shape optimization problem (1) - (4) by (1), (9), (10), (4). We underline that this is a control by the coefficients problem with the control g entering the lower order term of the differential operator and the state equation (9) being a simple penalization of (2). This is a very direct approach compared, for instance, with the mapping method[15]. Its numerical implementation is very efficient and some examples are reported in[8, 14].

The term "functional variations"[8, 10] comes from the natural variations of type $g + \lambda h$, $g, h \in X(D)$, $\lambda \in R$ that may generate very complex geometric variations of $\Omega = \Omega_g$, combining boundary, interior and topological variations in (1) - (4).

Remark 2.1. An important question that we address in the sequel is related to boundary cost functionals, which are seldom discussed in shape optimization. For instance, to minimize the flow through $\partial\Omega$

$$\text{Min} \int_{\partial\Omega} \left(\frac{\partial \mathrm{y}}{\partial \mathrm{n}}\right)^2 d\sigma \tag{11}$$

is desirable in many applications. Clearly, more regularity hypotheses have to be imposed on the admissible $\Omega \in \mathcal{O}$ in order that (11) makes sense.

3. The Implicit Parametrization Theorem

We limit our discussion to dimension two, but in Tiba[18], Tiba[19], Nicolai-Tiba[11], the higher dimensional situation is discussed as well. Moreover, the critical case (i.e. when the gradient of the implicit function is null in the initial point) is also investigated.

As in the previous section, we are interested in the implicit equation (in dimension two, here):

$$g(x,y) = 0, \quad (x,y) \in D \subset R^2. \tag{12}$$

We assume now that $g \in C^1(D)$ and $g(x_0, y_0) = 0$, for some given (x_0, y_0) in D. We associate to (12) the Hamiltonian system:

$$x'(t) = -\frac{\partial g}{\partial y}(x(t), y(t)), \tag{13}$$

$$y'(t) = \frac{\partial g}{\partial x}(x(t), y(t)), \tag{14}$$

$$x(0) = x_0, \; y(0) = y_0. \tag{15}$$

By the Peano theorem, the system (13) - (15) has at least one local solution around (x_0, y_0), for $t \in I_{\max} = (T_-, T_+)$, the maximal existence interval.

Theorem 3.1. *We have*

$$g(x(t), y(t)) = 0, \quad \forall\, t \in I_{\max}.$$

Remark 3.1. This is in fact the conservation property of Hamiltonian systems. In[18],[19] a *generalized solution* of (12) is constructed as a Hausdorff-Pompeiu limit in the case $\nabla g(x_0, y_0) = 0$, in higher dimension as well.

Remark 3.2. The geometric property behind the system (13) - (15) is that $\nabla g(x, y)$ is the normal vector to the level curve (12), if nonzero, and $\left(-\frac{\partial g}{\partial y}, \frac{\partial g}{\partial x}\right)$ is the tangent vector, in dimension two. In higher dimension, the construction is more involved,[18],[19].

Remark 3.3. If $\nabla g(x, y) \neq 0$, by combining the Peano theorem with the implicit function theorem, we get that the solution of (13) - (15) is unique in the class of equivalent parametrizations.

Remark 3.4. We have a local solution for (13) - (15) on $t \in I_{\max}$, in the usual sense for differential equations. We underline that this sense is different from that of local solution as appearing in the implicit function theorem. Namely, the requirement to obtain locally a function is essentially influenced by the choice of the axes, which is completely arbitrary and not a part of the problem. See[16], Ch. 1, where this influence is put clearly into evidence. The system (12) - (15) provides in fact a maximal solution, which is an advantage comparing with implicit functions.

We indicate now some continuity and differentiability properties for the implicit equation (12), again using (13) - (15) and following[18]. We consider the perturbed implicit equation

$$\widetilde{g}(x, y, \lambda) = 0, \quad (x, y) \in D,\ \lambda \in U, \tag{16}$$

where $U \subset R^m$ is open, $0 \in U$ and $\widetilde{g}(x, y, 0) = g(x, y)$ in D. We assume that $\widetilde{g} \in C^1(D \times U)$ with locally Lipschitzian gradient. We denote by x_λ, y_λ the solution of (13)-(14) associated to (16) and initial conditions x_0^λ, y_0^λ.

Theorem 3.2. *For any $\eta > 0$, there is $\varepsilon > 0$, such that for any λ with $|\lambda| < \varepsilon$ and any x_0^λ, y_0^λ with $|x_0 - x_0^\lambda| + |y_0 - y_0^\lambda| < \varepsilon$, the solution (x_λ, y_λ) is defined on $[T_- + \eta, T_+ - \eta]$ and the corresponding application $(x_0^\lambda, y_0^\lambda, \lambda) \to (x_\lambda, y_\lambda)$ is continuous from a ball around $(x_0, y_0, 0)$ to $C^1([T_- + \eta, T_+ - \eta])^2$.*

Remark 3.5. One may take η small such that $0 \in (T_- + \eta, T_+ - \eta)$.

For differentiability with respect to g in (12), we consider functional variations that are a special case of (16):

$$g(x, y) + \lambda h(x, y) = 0, \quad \text{in } D, \tag{17}$$

$\lambda \in R$ and $h \in C^1(D)$ with $g(x_0, y_0) = h(x_0, y_0) = 0$.

We associate to (17) the Hamiltonian system:

$$x'_\lambda = -\frac{\partial g}{\partial y}(x_\lambda, y_\lambda) - \lambda \frac{\partial h}{\partial y}(x_\lambda, y_\lambda), \tag{18}$$

$$y'_\lambda = \frac{\partial g}{\partial x}(x_\lambda, y_\lambda) + \lambda \frac{\partial h}{\partial y}(x_\lambda, y_\lambda), \tag{19}$$

$$x_\lambda(0) = x_0,\ y_\lambda(0) = y_0. \tag{20}$$

By the previous Theorem 3.2, the solutions $(x_\lambda(t), y_\lambda(t))$ exist for $|\lambda|$ small on a common compact interval I, with $0 \in \text{int}\, I$. We denote by:

$$z_\lambda = \frac{x_\lambda - x}{\lambda},\ w_\lambda = \frac{y_\lambda - y}{\lambda},\ t \in I, \lambda \neq 0. \tag{21}$$

Theorem 3.3. *Assume that $g \in C^2(D)$, $h \in C^1(D)$, with locally Lipschitzian derivatives of the highest order. We have $z_\lambda \to z$, $w_\lambda \to w$ in $C^1(I)$ and z, w satisfy the system in variations:*

$$z' = -\nabla \left[\frac{\partial g}{\partial y}(x, y)\right] \cdot (z, w) - \frac{\partial h}{\partial y}(x, y),\ in\ I, \tag{22}$$

$$w' = \nabla \left[\frac{\partial g}{\partial x}(x, y)\right] \cdot (z, w) + \frac{\partial h}{\partial x}(x, y),\ in\ I, \tag{23}$$

$$z(0) = w(0) = 0. \tag{24}$$

Remark 3.6. The existence interval for (22) - (24) is the same as for (13) - (15). The limit in (21) may be defined on the same interval via some extension procedure.

4. Directional Derivatives

We fix now the problem to be studied, as follows:

$$\underset{\Omega \in \mathcal{O}}{\text{Min}} \int_\Lambda \left(\frac{\partial y_\Omega}{\partial n}\right)^2 d\sigma, \tag{25}$$

$$-\Delta y_\Omega = f \quad \text{in } \Omega, \tag{26}$$

$$y_\Omega = 0 \quad \text{on } \partial\Omega, \tag{27}$$

$$E \subset \Omega \subset D,\ \forall\, \Omega \in \mathcal{O}, \tag{28}$$

where $f \in L^2(D)$, Λ is defined below, $E \subset\subset D \subset R^2$ are given bounded domains and the class $\mathcal{O}$ of admissible domains is given by

$$\mathcal{O} = \{\Omega_g \text{ connected}; g \in G_{\text{ad}}\}, \tag{29}$$

$$\Omega_g = \text{int}\{(x, y) \in D; g(x, y) \geq 0\} \tag{30}$$

and $G_{\text{ad}} \subset C^2(\overline{D})$ satisfies

$$g|_E \geq 0, \ \forall \, g \in G_{\text{ad}}, \tag{31}$$

$$\nabla g(x_0, y_0) \neq 0, \ g(x_0, y_0) = 0, \ \forall \, g \in G_{\text{ad}}, \tag{32}$$

where $(x_0, y_0) \in D \setminus \overline{E}$ is given.

Notice that the family G_{ad} defined by (31), (32) and, correspondingly, the family $\mathcal{O}$ defined in (29), (30) are very rich.

The condition that Ω_g is connected, i.e. a domain, does not follow from (30) and has to be imposed separately in (29). That is, in (30), one has to choose the connected component that satisfies (28).

In order that the cost functional (25) makes sense, we impose first that

$$\nabla g(x, y) \neq 0, \ \forall \ (x, y) \in D \text{ with } g(x, y) = 0. \tag{33}$$

Then, it is known that the equation $g(x, y) = 0$ defines C^2 curves in R^2, by the implicit functions theorem, and, $\nabla g(x, y)$ gives the normal to these manifolds, in each of their points (x, y). We may take $n(x, y) = \dfrac{\nabla g(x, y)}{|\nabla g(x, y)|_{R^2}}$ and $\dfrac{\partial y_\Omega}{\partial n} = n \cdot \nabla y_\Omega$ is well defined. Then (26), (27) has strong solution $y_\Omega \in H^2(\Omega)$ under the above assumptions and $\dfrac{\partial y_\Omega}{\partial n} \in H^{1/2}(\partial\Omega)$, that is the cost (25) makes sense.

Remark 4.1. While the above considerations are valid on $\partial\Omega$, the cost functional in (25) is defined in fact on Λ, which is the connected component of the boundary $\partial\Omega$ passing through (x_0, y_0). In principle, the admissible domain Ω defined above may not be simply connected and its boundary may have several disjoint components.

We also impose the hypothesis

$$g(x, y) < 0, \ \forall \ (x, y) \in \partial D, \forall \, g \in G_{ad}. \tag{34}$$

Remark 4.2. Under conditions (33), (34) the level lines $g(x, y) = 0$ contain no critical points for g and are "far" from the boundary ∂D. Hypothesis (33) also excludes self intersections. Each level line is a compact in D (assumed bounded). The structure theorem in[18], for the solution of the Hamiltonian system (13)-(15), shows that it describes equivalently the level line through (x_0, y_0). The fact that the solution of the Hamiltonian system is on the level line is obvious. Conversely, one has to use as well the implicit function theorem and a reparametrization of the solution trajectory to obtain the Hamiltonian system from the differentiability formula for implicit functions. If $g(x, y(x)) = 0$ and $g_y(x_0, y_0) \neq 0$, then $y'(x) = -g_x(x, y(x))/g_y(x, y(x))$ and $y(x_0) = y_0$. We define $\tilde{y} = y(x)$ and introduce $x' = -g_x(x, y(x(t)), x(0) = x_0$ and we get that $\tilde{y}(t) = y(x(t))$ satisfies (13)-(15) by differentiation in t.

It is of interest to obtain periodic solutions for the Hamiltonian system in dimension two (then the corresponding domain is well defined). However, the Poincaré-Bendixson theorem[3] allows examples of the type (in polar coordinates): $[(1 - 1/\theta)(cos\theta, sin\theta)], \theta > 0.5$. It is unclear how to remove such situations.

We consider functional variations around g of the form

$$g + \lambda h, \quad \lambda \in R, \tag{35}$$

with $h \in G_{ad} \subset C^2(\overline{D})$, that is $h(x_0, y_0) = 0$, $h|_E \geq 0$. We denote shortly by Ω_λ the domain defined by (30), associated to (35) and by y_λ the corresponding solution of (26), (27).

All the above regularity considerations remain valid for Ω_λ, y_λ, for λ small, since (33), (34) are satisfied by $g + \lambda h$, due to $g, h \in C^2(\overline{D})$.

The aim of this section is to study the directional differentiability properties of (25) - (27) with respect to g, if $\Omega = \Omega_g$. The objective is to compute the limit:

$$\lim_{\lambda \to 0} \frac{1}{\lambda} \left[\int_{\Lambda_\lambda} \left(\frac{\partial y_\lambda}{\partial n} \right)^2 d\sigma - \int_{\Lambda_g} \left(\frac{\partial y_g}{\partial n} \right)^2 d\sigma \right], \tag{36}$$

where y_g is a short notation for y_{Ω_g} and the notations $\Lambda_\lambda, \Lambda_g$ are obvious.

We shall study (36) in an approximate sense. Namely, taking into account the results from Section 2, we approximate (26), (27) by (9), (10), that is by extending the elliptic equation to the whole domain D. We may assume that D is at least of class $C^{1,1}$ and the solution of (9), (10) is in $H^2(D)$ since the coefficient in (9) is in $C^2(\overline{D})$ if $g \in C^2(\overline{D})$.

The cost functional (25) makes sense, due to the above discussion and we don't modify it. Instead of (36), we shall study

$$\lim_{\lambda \to 0} \frac{1}{\lambda} \left[\int\limits_{\Lambda_\lambda} \left(\frac{\partial y_{\varepsilon\lambda}}{\partial n} \right)^2 d\sigma - \int\limits_{\Lambda_g} \left(\frac{\partial y_\varepsilon}{\partial n} \right)^2 d\sigma \right], \tag{37}$$

where y_ε, $y_{\varepsilon\lambda} \in H^2(D)$ satisfy respectively

$$-\Delta y_\varepsilon + \frac{1}{\varepsilon}(1 - H^\varepsilon(g))y_\varepsilon = f \text{ in } D, \tag{38}$$

$$y_\varepsilon = 0 \quad \text{on } \partial D; \tag{39}$$

$$-\Delta y_{\varepsilon\lambda} + \frac{1}{\varepsilon}(1 - H^\varepsilon(g + \lambda h))y_{\varepsilon\lambda} = f \text{ in } D, \tag{40}$$

$$y_{\varepsilon\lambda} = 0 \quad \text{on } \partial D. \tag{41}$$

Subtracting the above boundary value problems in D, we get

$$\begin{aligned} &-\Delta \frac{y_{\varepsilon\lambda} - y_\varepsilon}{\lambda} + \frac{1}{\varepsilon}(1 - H^\varepsilon(g)) \frac{y_{\varepsilon\lambda} - y_\varepsilon}{\lambda} = \\ &= \frac{1}{\varepsilon} \frac{H^\varepsilon(g + \lambda h) - H^\varepsilon(g)}{\lambda} y_{\varepsilon\lambda} \text{ in } D, \end{aligned} \tag{42}$$

$$\frac{y_{\varepsilon\lambda} - y_\varepsilon}{\lambda} = 0 \quad \text{on } \partial D. \tag{43}$$

Here, $\varepsilon > 0$ is fixed and we take $\lambda \to 0$. We notice first that $H^\varepsilon(g + \lambda h) \to H^\varepsilon(g)$ uniformly in D and, consequently, $y_{\varepsilon\lambda} \to y_\varepsilon$ strongly in $H^2(D) \cap H_0^1(D)$, as $\lambda \to 0$, due to (38) - (41).

Moreover, $\frac{1}{\lambda}[H^\varepsilon(g + \lambda h) - H^\varepsilon(g)] \to \dot{H}^\varepsilon(g)h$, for $\lambda \to 0$, uniformly in D, where $\dot{H}^\varepsilon(\cdot)$ denotes the derivative of the function $H^\varepsilon(\cdot)$. Some convergence properties are valid in stronger topologies, but we shall use just the above mentioned convergences.

From (42), (43) we get that $\left\{ \frac{y_{\varepsilon\lambda} - y_\varepsilon}{\lambda} \right\}_\lambda$ is compact in $H^2(D) \cap H_0^1(D)$.

If $z_\varepsilon = \lim_{\lambda \to 0} \frac{y_{\varepsilon\lambda} - y_\varepsilon}{\lambda}$, then it satisfies

$$-\Delta z_\varepsilon + \frac{1}{\varepsilon}(1 - H^\varepsilon(g))z_\varepsilon = \frac{1}{\varepsilon} \dot{H}^\varepsilon(g) y_\varepsilon \text{ in } D, \tag{44}$$

$$z_\varepsilon = 0 \quad \text{on } \partial D. \tag{45}$$

Such arguments are explained in more detail in[8]. We obtain the following proposition:

Proposition 4.1. *The mapping* $g \to y_\varepsilon$*, from* $C^2(\overline{D})$ *to* $H^2(D) \cap H_0^1(D)$*, defined by* (38), (39) *is Gateaux differentiable and the corresponding equation in variations is given by* (44), (45).

One can compute now the limit in (37):

$$
\begin{aligned}
&\lim_{\lambda \to 0} \frac{1}{\lambda} \left[\int_{\Lambda_\lambda} \left(\frac{\partial y_{\varepsilon\lambda}}{\partial n} \right)^2 d\sigma - \int_{\Lambda_g} \left(\frac{\partial y_\varepsilon}{\partial n} \right)^2 d\sigma \right] \\
&= \lim_{\lambda \to 0} \frac{1}{\lambda} \left[\int_{\Lambda_g} \left[\left(\frac{\partial y_{\varepsilon\lambda}}{\partial n} \right)^2 - \left(\frac{\partial y_\varepsilon}{\partial n} \right)^2 \right] d\sigma + \int_{\Lambda_\lambda} \left(\frac{\partial y_{\varepsilon\lambda}}{\partial n} \right)^2 d\sigma - \int_{\Lambda_g} \left(\frac{\partial y_{\varepsilon\lambda}}{\partial n} \right)^2 d\sigma \right] \\
&= 2 \int_{\Lambda_g} \frac{\partial z_\varepsilon}{\partial n} \frac{\partial y_\varepsilon}{\partial n} d\sigma + \lim_{\lambda \to 0} \frac{1}{\lambda} \left[\int_{\Lambda_\lambda} \left(\frac{\partial y_{\varepsilon\lambda}}{\partial n} \right)^2 d\sigma - \int_{\Lambda_g} \left(\frac{\partial y_{\varepsilon\lambda}}{\partial n} \right)^2 d\sigma \right].
\end{aligned} \tag{46}
$$

Notice that all the integrals make sense since $y_{\varepsilon\lambda}, y_\varepsilon, z_\varepsilon$ are regular enough in D.

For the last term in (46) we use the representation of Λ_λ, Λ_g via the implicit parametrization theorem, as explained in §3. We have:

$$
\int_{\Lambda_g} \left(\frac{\partial y_{\varepsilon\lambda}}{\partial n} \right)^2 d\sigma = \int_{I_g} \frac{\partial y_{\varepsilon\lambda}^2}{\partial n} (x_g(t), y_g(t)) \sqrt{(\dot{x}_g(t))^2 + (\dot{y}_g(t))^2 dt}, \tag{47}
$$

$$
\int_{\Lambda_\lambda} \left(\frac{\partial y_{\varepsilon\lambda}}{\partial n} \right)^2 d\sigma = \int_{I_\lambda} \frac{\partial y_{\varepsilon\lambda}^2}{\partial n} (x_\lambda(t), y_\lambda(t)) \sqrt{\dot{x}_\lambda(t)^2 + \dot{y}_\lambda(t)^2 dt}, \tag{48}
$$

where I_g, I_λ are the maximal existence intervals of the solutions for the systems (13) - (15), respectively (18) - (20), used in (47), (48). By Rem.4.2, these systems provide a complete description of the boundary components Λ_g, respectively Λ_λ.

We know that $\{\nabla y_{\varepsilon\lambda}\}_\lambda$ is bounded uniformly in D since $D \subset R^2$ and $\{y_{\varepsilon\lambda}\}$ is bounded in $W^{2,p}(D)$, $p > 2$ if we assume $f \in L^p(D)$. Then $\left\{ \dfrac{\partial y_{\varepsilon\lambda}}{\partial n} \right\}$ is uniformly bounded on any regular curve included in D since the normal is of norm one in any point.

Assume for simplicity that I_g is finite and, for λ "small", there is $I \subset I_g \cap I_\lambda$ compact, including the origin in the interior, such that the length $|I_\lambda - I| + |I_g - I|$ is of order λ^2. This is motivated by Theorem 3.2 and excludes situations as in the end of Remark 4.2.

Due to the above boundedness, we may analyze just the limit of

$$\frac{1}{\lambda}\int_I \left[\frac{\partial y_{\varepsilon\lambda}^2}{\partial n_\lambda}(x_\lambda(t), y_\lambda(t))\sqrt{\dot{x}_\lambda(t)^2 + \dot{y}_\lambda(t)^2}\right.$$
$$\left. -\frac{\partial y_{\varepsilon\lambda}^2}{\partial n_g}(x_g(t), y_g(t))\sqrt{(\dot{x}_g(t))^2 + (\dot{y}_g(t))^2}\right] dt. \tag{49}$$

In (49) we have $\dfrac{\partial y_{\varepsilon\lambda}}{\partial n_g} = \nabla y_{\varepsilon\lambda} \cdot \dfrac{\nabla g}{|\nabla g|_{R^2}}$ and $\dfrac{\partial y_{\varepsilon\lambda}}{\partial n_\lambda} = \nabla y_{\varepsilon\lambda} \cdot \dfrac{\nabla(g+\lambda h)}{|\nabla(g+\lambda h)|_{R^2}}$, for λ small.

They are defined on the curves $(x_\lambda(t), y_\lambda(t))$ or $(x_g(t), y_g(t))$ respectively, but clearly they can be extended to the whole domain D. We use Thm. 3.3 and Prop. 4.1 to compute the limit in (49). The existence of the limit follows due to our regularity assumptions and several applications of the chain rule. We denote it by L and we get

Proposition 4.2. *If $f \in L^p(D), p > 2$, the limit in (49) is given by $L = \int_I [A + B + C]$, where*

$$A = 2\frac{\partial y_\varepsilon}{\partial n_g}(x_g(t), y_g(t))\{\frac{\partial}{\partial n_g}[\nabla y_\varepsilon(x_g(t), y_g(t))].(z(t), w(t))$$

$$+[\nabla y_\varepsilon(x_g(t), y_g(t))].[\nabla n_g(x_g(t), y_g(t)).(z(t), w(t))]\}\sqrt{(\dot{x}_g(t))^2 + (\dot{y}_g(t))^2},$$

$$B = 2\frac{\partial y_\varepsilon}{\partial n_g}(x_g(t), y_g(t))\{[\nabla y_\varepsilon.\nabla h/|\nabla g|](x_g(t),$$

$$y_g(t)) - [\frac{\partial y_\varepsilon}{\partial n_g}\nabla g.\nabla h/|\nabla g|^2](x_g(t), y_g(t))\}\sqrt{(\dot{x}_g(t))^2 + (\dot{y}_g(t))^2},$$

$$C = 2\{[\frac{\partial y_\varepsilon}{\partial n_g}^2 |\nabla g|^{-2}](x_g(t), y_g(t))\}(\dot{x}_g(t), \dot{y}_g(t)).(z(t), w(t)).$$

The integration interval in the limit (37) *is I_g and (z, w) are defined in Thm.* 3.3.

Proof. This is standard calculation, quite tedious, and we indicate just a sketch. By simply adding and subtracting terms, the integrand in (49) may be rewritten as

$$\begin{aligned} &\frac{\partial y^2_{\varepsilon\lambda}}{\partial n_\lambda}(x_\lambda(t), y_\lambda(t))\sqrt{\dot{x}_\lambda(t)^2 + \dot{y}_\lambda(t)^2} \\ &-\frac{\partial y^2_{\varepsilon\lambda}}{\partial n_\lambda}(x_g(t), y_g(t))\sqrt{\dot{x}_\lambda(t)^2 + \dot{y}_\lambda(t)^2} \end{aligned} \tag{50}$$

$$\begin{aligned} &+\frac{\partial y^2_{\varepsilon\lambda}}{\partial n_\lambda}(x_g(t), y_g(t))\sqrt{\dot{x}_\lambda(t)^2 + \dot{y}_\lambda(t)^2} \\ &-\frac{\partial y^2_{\varepsilon\lambda}}{\partial n_g}(x_g(t), y_g(t))\sqrt{\dot{x}_\lambda(t)^2 + \dot{y}_\lambda(t)^2} \end{aligned} \tag{51}$$

$$\begin{aligned} &+\frac{\partial y^2_{\varepsilon\lambda}}{\partial n_g}(x_g(t), y_g(t))\sqrt{\dot{x}_\lambda(t)^2 + \dot{y}_\lambda(t)^2} \\ &-\frac{\partial y^2_{\varepsilon\lambda}}{\partial n_g}(x_g(t), y_g(t))\sqrt{(\dot{x}_g(t))^2 + (\dot{y}_g(t))^2} \end{aligned} \tag{52}$$

We use the previously explained regularity and boundedness properties and the fact that the normal derivatives can be extended to the whole domain due to their expression. Then, it is possible to pass to the limit in each pair in (50)-(52) and obtain the terms A, B, C via classical chain rules. □

Remark 4.3. The approximating directional derivative of the cost (25), defined in (36), (37), is computed by combining Prop. 4.2 with (44), (45). While the obtained result is rather involved, the advantage is the application of general functional variations, of no prescribed form, enlarging significantly the class of admissible geometries. We underline that we compute here a usual analytic directional derivative. By the geometric interpretation of the mapping g in (30), one can speak about a geometric directional derivative in shape optimization - but this is a language abuse since the parametrization of a domain by a function g is not unique and the obtained limits depend on the representation mapping. However, the case of distributed problems[8] shows that one can solve numerically shape optimization problems in this way (their solution is not unique as well). The introduction of a corresponding adjoint system for boundary cost functionals and related algorithms will be the object of future work.

Acknowledgement

This work was supported by Grant with contract 145/2011 of CNCS Romania.

References

1. D. Bucur, G. Buttazzo, *Variational methods in shape optimization problems*, Springer-Verlag, Berlin (2006).
2. T.F. Chan, L. Vese, *Variational image restoration and segmentation models and approximations*, VCLA Computational and Applied Mathematics Reports, vol. 97, Issue 47 (1997).
3. Ph. Hartman, *Ordinary differential equations*, SIAM, Philadelphia (2002).
4. J. Haslinger, P. Neittaanmaki, *Finite element approximation of optimal shape design*, J. Wiley and Sons, New York (1988).
5. A. Henrot, M. Pierre, *Variation et optimization de formes: une analyse geometrique*, Springer Verlag, Berlin (2005).
6. J.L. Lions, *Controle optimal de systemes gouvernés par des equations aux derivees partielle*, Paris, Dunod (1968).
7. R. Makinen, P. Neittaanmäki, D. Tiba, *On a fixed domain approach for a shape optimization problem*, In: Computational and applied mathematics II (W.F. Ames, P.J. van der Houwen Eds.), Amsterdam, North-Holland, pp. 317-326 (1992).
8. P. Neittaanmäki, A. Pennanen, D. Tiba, *Fixed domain approaches in shape optimization problem with Dirichlet boundary conditins*, Inverse Problems 25 (5), pp. 1-18 (2009).
9. P. Neittaanmäki, J. Sprekels, D. Tiba, *Optimization of elliptic systems. Theory and applications*, Springer, Berlin (2006).
10. P. Neittaanmäki, D. Tiba, *Fixed domain approaches in shape optimization problems*, Inverse Problems 28 (9) (2012), pp. 1-35.
11. M. R. Nicolai, D. Tiba. *Implicit functions and parametrizations in dimension three: generalized solution*, DCDS-A vol. 35, no. 6, pp. 2701-2710 (2015). doi:10.3934/dcds.2015.35.2701.
12. S. Osher, J.A. Sethian, *Fronts propagating with curvature-dependent speed: algorithms based on Hamilton-Jacobi formulations*, J. Comp. Phys. 74, pp. 12-44 (1988).
13. P. Philip, D. Tiba, *A penalization and regularization technique in shape optimization problems*, SIAM J. Control and Optim., vol. 51, no. 6, pp. 4295-4317 (2013).
14. P. Philip, D. Tiba, *Shape optimization via control of a shape function on a fixed domain: theory and numerical results*, Comp. Meth. in Appl. Sc. 27, Springer, pp. 305-320 (2013).
15. O. Pironneau, *Optimal shape design for elliptic systems*, Springer

Verlag, Berlin (1984).
16. H. Schlichtkrull, *Curves and surfaces*, Lect. Notes for Geom. 1, Univ. of Copenhagen (2011).
17. J. Sokolowski, J.-P. Zolesio, *Introduction to shape optimization. Shape sensitivity analysis*, Springer Verlag, Berlin (1992).
18. D. Tiba, *The implicit function theorem and implicit parametrizations*, Ann. Acad. Rom. Sci. Ser. Math. Appl., vol. 5 (1-2), p. 193-208 (2013).
19. D. Tiba. *Implicit systems in arbitrary dimension: constructive local parametrization and generalized solution*, (2015), `http://arxiv.org/pdf/1408.6726v2.pdf`.

Recent progress on steady gravity water waves

Eugen Vărvărucă
Faculty of Mathematics, "Alexandru Ioan Cuza" University,
Bdul. Carol I, nr. 11, 700506 Iaşi, Romania
E-mail: eugen.varvaruca@uaic.ro

We present some recent results on the existence of two-dimensional travelling free surface gravity water waves of large amplitude that may have critical layers or overhanging profiles.

Keywords: steady water waves; global bifurcation

1. Introduction

We consider the classical problem of two-dimensional waves propagating under the influence of gravity at the surface of a layer of an incompressible, inviscid and homogeneous fluid situated above a flat bed. This problem has a rich history spanning more than two centuries, during which time it provided a motivation for some fundamental developments in mathematics, such as Cauchy's theory of holomorphic functions of one complex variable, see Strauss[1] and the references therein. While recent years have seen great progress in the study of the well-posedness question for the associated initial-value problem, see Lannes[2], in the present paper we restrict attention to travelling waves[a], assuming in addition that they are spatially periodic.

This problem may be formulated mathematically as that of finding a domain Ω in the upper half of the (X,Y)-plane, whose boundary consists of the real axis $\mathcal{B}=\{(X,0): X\in\mathbb{R}\}$, which represents the flat impermeable water bed, and an a priori unknown curve expressed in parametric form as $\mathcal{S}=\{(\widetilde{X}(s),\widetilde{Y}(s)): s\in\mathbb{R}\}$, where the map $s\mapsto(\widetilde{X}(s)-s,\widetilde{Y}(s))$ is periodic of period L, which represents the free surface of the water, and a function $\psi:\Omega\to\mathbb{R}$ which represents the *stream function* giving the velocity field $(\psi_Y,-\psi_X)$ in a frame moving at the constant wave speed, which satisfies

[a]Solutions for which the flow is stationary in a frame of reference moving at some constant speed.

$\psi(X+L,Y)=\psi(X,Y)$ for all $(X,Y)\in\Omega$, and the following equations and boundary conditions:

$$\Delta\psi = \Upsilon(\psi) \quad \text{in } \Omega, \tag{1a}$$

$$\psi = 0 \quad \text{on } \mathcal{S}, \tag{1b}$$

$$\psi = -m \quad \text{on } \mathcal{B}, \tag{1c}$$

$$|\nabla\psi|^2 + 2gY = Q \quad \text{on } \mathcal{S}. \tag{1d}$$

Here g is the gravitational constant of acceleration, the positive constant L is the spatial period of the wave, the constant m is the relative mass flux, while the constant Q is related to the hydraulic head (see the discussion in Ref. 3). The function $\Upsilon : I \to \mathbb{R}$, where I is an interval of $\mathbb{R}$ containing both 0 and $-m$, is the *vorticity function*, which expresses the functional dependence between the *vorticity* $\omega := \Delta\psi$ of the flow and the stream function ψ. A nontrivial vorticity is required in order to model the presence of waves on non-uniform running streams. The connection between the formulation (1) of the steady water wave problem[b] and the Euler equations for an incompressible inviscid fluid with a free surface is explained in Section 2 of the paper.

Problem (1) has been the subject of extensive investigations, both analytical and computational. General overviews of the mathematical theory can be found in Refs. 1,4–6 and of the numerics in Refs. 7–11. The aim of this paper is to survey some of the main existence results for this problem, with emphasis on the results dealing with waves having critical layers[c] and whose profile may be an overhanging curve (rather than the graph of a single-valued function).

Problem (1) admits many solutions, the simplest of which are those for which $\mathcal{S}$ is a horizontal line, say $\{Y = h\}$ for some $h > 0$, and ψ is a function of the vertical variable Y only. In this situation, (1a) is a second-order ordinary differential equations that can be integrated explicitly, and taking also into account the boundary conditions (1b)–(1d) leads to a one-parameter family of solutions. For example, in the case when the vorticity function is a constant Υ, the problem has, for each $m \in \mathbb{R}$, in the domain

[b]This is a *free-boundary problem*, in the sense that the domain Ω occupied by the fluid is not prescribed a priori, but has to be determined as part of the solution.

[c]Recall that the level sets of ψ are called *streamlines*, a point where the gradient of ψ vanishes is called a *stagnation point*, while a *critical layer* is a curve along which $\psi_Y = 0$. Such curves arise in the context of flow reversal, where a fluid region in which the flow is oriented towards the direction of propagation of the wave is adjacent to a fluid region in which the flow is adverse to it.

bounded below by the rigid bed $\mathcal{B}$ and above by the free surface $\{Y = h\}$, the solution

$$\psi(X,Y) = \frac{\Upsilon}{2}Y^2 + \left(\frac{m}{h} - \frac{\Upsilon h}{2}\right)Y - m, \qquad X \in \mathbb{R}, 0 \leq Y \leq h, \qquad (2)$$

where Q is determined uniquely in terms of m. For each of these flows, all particles move on horizontal lines, with a velocity depending on the height of the respective line above the bed (the velocity being the same at all heights if and only if $\Upsilon = 0$). Such parallel flows may be constructed for any vorticity function Υ, see Refs. 3,12. Another family of explicit solutions of (1) was described in 1802 by Gerstner[13], for the related problem of waves of infinite depth (in which the fluid domain consists of the region below a curve $\mathcal{S}$ and extending to $-\infty$ in the vertical direction) for a special choice of vorticity function. However, no other explicit solutions of (1) are known, and the existence of other steady waves is a question that has required the tools of mathematical analysis for its solvability.

One of the pioneering researchers on steady water waves was Stokes[14,15], who concerned himself with *irrotational* waves[d]. He argued formally that there exists nontrivial waves which are small perturbations of some of the trivial flows given by (2) with $\Upsilon = 0$ [e]. Moreover, he conjectured that there exists a family of waves, parametrized by amplitude, 'connecting' a trivial wave with a so-called 'extreme wave', the latter being distinguished by the fact that it has a stagnation point at the crest, where the wave profile has lateral tangents making an angle of 120°. Stokes thus recognized in (1) the structure of a bifurcation problem, and, as we shall see in what follows, subsequent research efforts on the existence question for the steady water wave problem, with or without vorticity, have been largely devoted to giving rigorous proofs to the results he had predicted.

After early results in the 1920s on the existence of waves of small amplitude, see Strauss[1] for references, existence results for waves of large amplitude started appearing in the 1960s, based on methods of nonlinear functional analysis. Major existence results for irrotational waves are due to Krasovskii[16], Keady and Norbury[17], and Amick and Toland[18]. A key step in these works is that of equivalently reformulating the free-boundary problem (1) over a fixed domain, the new problem being then regarded as one of finding the zeros of a nonlinear mapping between suitable Banach

[d]The situation when the vorticity is identically 0.

[e]The specific flows that can be perturbed in such a way depend on the spatial period of the waves sought.

spaces. For irrotational waves, given that the stream function is harmonic, a natural approach has been the use of complex analysis. Indeed, by means of a conformal mapping the problem is transformed into one over a fixed domain, which may then be reduced to an equation on the boundary of that domain for a function of one variable. In particular, an elegant modern theory of global and subharmonic bifurcation for irrotational waves of infinite depth has been given by Buffoni, Dancer and Toland[19,20], making use of a reformulation of the problem originally due to Babenko[21].

The first existence result for waves of large amplitude with general vorticity is due to Constantin and Strauss[3]. They used a change of variables originally due to Dubreil-Jacotin[22] to transform (1) into a boundary-value problem for a quasilinear elliptic equation in a strip, and proved the existence of solutions corresponding to waves of large amplitude by using a theory of global bifurcation based on a generalized topological degree. The Dubreil-Jacotin transformation is applicable, however, only if the stream function is strictly monotone in the vertical direction, a condition which excludes the presence of critical layers or overhanging profiles.

While in the irrotational case it is known that any wave profile is necessarily the graph of a single-valued function, and the stream function is necessarily monotone in the vertical direction, see Refs. 23–26, a cursory inspection of (2) reveals that even the problem of waves with constant vorticity admits solutions (parallel flows) with critical layers. Trivial flows with critical layers can also be easily constructed for large classes of vorticity distributions, see Ref. 12. Moreover, numerical work, see Refs. 7,10,11, strongly indicates that the problem of waves with constant vorticity admits solutions with critical layers[f], as well as solutions with overhanging profiles.

The first rigorous result on waves with critical layers is due to Wahlén[28], who proved the existence of waves of small amplitude with constant vorticity. It is based on a different reformulation of problem (1) over a fixed domain. Namely, by means of a flattening transformation, one is led to solving in strip a problem involving a function of two variables and one of one variable. This approach is not restricted to waves of constant vorticity, and has indeed been used subsequently in more general settings in Refs. 29–31. More precisely, in Refs. 30,31 small amplitude waves with multiple critical layers are constructed for affine vorticity functions, by making use of multiparameter bifurcation theory, while local bifurcation results are

[f]A typical feature of the dynamics of such flows is the 'cat's eye' pattern of Thomson (Lord Kelvin)[27].

proven in Ref. 29 for general vorticity functions. However, the flattening transformation approach requires that the wave profile is the graph of a function. In addition, the resulting fixed-domain formulation appears rather complicated, and so far it has not led to existence results for waves of large amplitude.

The only currently known fixed-domain reformulation of problem (1) that is capable of describing steady waves with no restrictions on the geometry of the wave profile or on the streamline pattern, and that has been successfully employed in a global bifurcation theory, is that very recently uncovered by Constantin, Strauss and the present author in Ref. 36. This reformulation is valid, however, only in the case when the vorticity is a constant, and is an extension of the finite-depth analogue of Babenko's formulation for irrotational waves of infinite depth. The paper Ref. 36 substantially develops the earlier work in Ref. 35, where a conformal mapping approach was introduced in the problem of steady waves with constant vorticity and of finite depth, leading to a nonlinear pseudo-differential equation for a function of one variable describing the wave elevation in the conformal parametrization of the free surface. In Ref. 36 this equation is further transformed, by an application of Riemann–Hilbert theory, into an equivalent one, which similarly involves a periodic Hilbert transform on the boundary of a strip, but has much more convenient structural properties. These properties are instrumental in proving the existence of a countable family of curves of solutions of global extent by employing the real-analytic bifurcation theory due to Dancer[38], Buffoni and Toland[39]. Qualitative features of the waves corresponding to points on the two curves that contain the 'least oscillatory' solutions are also studied in some detail in Ref. 36. Namely, it is proven there that the vertical coordinate of the free surface of such waves strictly decreases from crest to trough even as the wave profile is allowed to overturn. Moreover, the solution curve either goes to infinity in the function space, or the velocity at the wave crest approaches 0, or there exists a point on the curve corresponding to a limiting configuration where the wave profile has a self-intersection above the trough line.

The remainder of the paper is organized as follows. After discussing in Section 2 the equivalence of (1) with the free surface Euler equations, in Section 3 we present the Dubreil-Jacotin transformation and the flattening transformation that lead to equivalent reformulations of (1) under suitable conditions, and provide some additional details on the existence results that have been proved using these reformulations. In Section 4 we present the conformal mapping approach in Ref. 35 that leads to the generalized

Babenko equation for waves with constant vorticity, and in Section 5 we present the main existence results in Ref. 36. We conclude the paper with Section 6, where some open problems concerning steady water waves with vorticity are briefly mentioned.

2. The governing equations for free surface water waves

Consider the flow of a two-dimensional incompressible inviscid fluid of unit density, acted on by gravity, over a flat, horizontal, impermeable bed and with a free surface. Let $\Omega(t) \subset \mathbb{R}^2$ be the domain in the (X, Y)-plane occupied by the fluid at time t, where the boundary of $\Omega(t)$ consists of the real axis $\mathcal{B}$, which represents the flat impermeable bed, and a curve $\mathcal{S}(t)$ in the upper half-plane and of unbounded horizontal extent in both directions, which represents the free surface. Then the dynamics of the fluid is described by the Euler equations for the vector velocity field $(U(t,\cdot), V(t,\cdot)) : \Omega(t) \to \mathbb{R}^2$ and the scalar pressure field $\mathcal{P}(t,\cdot) : \Omega(t) \to \mathbb{R}$:

$$U_t + UU_X + VU_Y = -\mathcal{P}_X \qquad \text{in } \Omega(t), \tag{3a}$$

$$V_t + UV_X + VV_Y = -\mathcal{P}_Y - g \qquad \text{in } \Omega(t), \tag{3b}$$

$$U_X + V_Y = 0 \qquad \text{in } \Omega(t), \tag{3c}$$

where subscripts denote partial derivatives and g is the gravitational constant of acceleration. The equations are supplemented by standard kinematic and dynamic boundary conditions at the boundary of the fluid:

$$V = 0 \qquad \text{on } \mathcal{B}, \tag{3d}$$

$$(U, V) \cdot \nu = W \qquad \text{on } \mathcal{S}(t), \tag{3e}$$

$$\mathcal{P} = \mathcal{P}_{\mathrm{atm}} \qquad \text{on } \mathcal{S}(t), \tag{3f}$$

where W is the normal velocity of the free surface, ν is the outer normal vector, and $\mathcal{P}_{\mathrm{atm}}$ is the constant atmospheric pressure.

We consider in all that follows only travelling-wave solutions of the above problem, that is, those solutions for which there exist a domain $\Omega \subset \mathbb{R}^2$ whose boundary consists of the real axis $\mathcal{B}$ and a curve $\mathcal{S}$ in the upper half-plane, and $c \in \mathbb{R}$, $(\widetilde{U}, \widetilde{V}) : \Omega \to \mathbb{R}^2$ and $\widetilde{\mathcal{P}} : \Omega \to \mathbb{R}$ such that $\Omega(t) = \Omega + ct(1, 0)$ for all $t \in \mathbb{R}$, and, for all $t \in \mathbb{R}$ and $(X, Y) \in \Omega(t)$,

$$U(X, Y, t) = \widetilde{U}(X - ct, Y), \quad V(X, Y, t) = \widetilde{V}(X - ct, Y),$$

and $\mathcal{P}(X, Y, t) = \widetilde{\mathcal{P}}(X - ct, Y)$. The corresponding stationary system satisfied by $\widetilde{U}, \widetilde{V}$ and $\widetilde{\mathcal{P}}$ in Ω is easily written down. Now the incompressibility condition (3c) ensures the existence of a *stream function* ψ in Ω such that

$$\psi_X = -\widetilde{V}, \quad \psi_Y = \widetilde{U} - c \qquad \text{in } \Omega. \tag{4}$$

The equations corresponding to (3d) and (3e) yield that ψ is constant on each of $\mathcal{S}$ and $\mathcal{B}$. Since ψ is only determined up to an additive constant, we assume with no loss of generality that (1b) holds. If m denotes the relative mass flux, that is the difference between the values of ψ on $\mathcal{S}$ and on $\mathcal{B}$, it follows that (1c) holds. Moreover, eliminating $\widetilde{\mathcal{P}}$ from the equations corresponding to (3a) and (3b) yields

$$(\Delta\psi)_X\psi_Y = (\Delta\psi)_Y\psi_X \qquad \text{in } \Omega, \tag{5}$$

which means that the vorticity $\omega := \Delta\psi$ of the flow is constant on each level set of ψ. Equation (5) is satisfied whenever there exists a function $\Upsilon : I \to \mathbb{R}$, where $I \subset \mathbb{R}$ is an interval containing all the values of ψ in $\overline{\Omega}$, such that (1a) holds. (The existence of such a function Υ can also be proved to be a necessary condition for (5) under additional assumptions on the velocity field.) We assume in what follows that the function $\Upsilon : I \to \mathbb{R}$ is prescribed. Let $\Xi : I \to \mathbb{R}$ be given by $\Xi(r) = \int_0^r \Upsilon(s)\,ds$ for all $r \in I$. It then follows that

$$\widetilde{\mathcal{P}} + \frac{1}{2}|\nabla\psi|^2 + gY + \Xi(\psi) = \text{constant} \qquad \text{in } \Omega, \tag{6}$$

and then (3f) leads to the existence of a constant Q such that (1d) holds.

Conversely, it is easy to check that, for any solution of (1), the functions $(\widetilde{U}, \widetilde{V})$ given by (4) and $\widetilde{\mathcal{P}}$ given by (6) satisfy the travelling-wave form of the Euler equations (3).

3. Some fixed-domain reformulations

Compared to the travelling-wave form of (3), the system (1) has the advantage of involving a single function ψ. The remaining difficulty that (1) is still a free-boundary problem can be circumvented by transforming it into an equivalent problem over a fixed domain. There are several ways to achieve this, which will be presented in what follows together with some of the results to which they lead.

3.1. *The Dubreil-Jacotin formulation*

The earliest such approach is that of Dubreil-Jacotin[22], which uses in an essential way the fact that ψ is constant on the top and the bottom of Ω. It does however require that ψ_Y does not change sign in $\overline{\Omega}$. Let us assume for definiteness that $\psi_Y < 0$ in $\overline{\Omega}$. (A similar analysis can be carried out in the other case.) This assumption implies in particular that $\mathcal{S}$ is the graph of a function. Let us consider the partial hodograph (or semi-Lagrangian) mapping $(X, Y) \mapsto (q, p) = (X, -\psi(X, Y))$, which is a bijection between $\overline{\Omega}$ and the closure of the strip $\mathcal{R} = \{(q, p) : q \in \mathbb{R}, p \in (m, 0)\}$. Then the inverse mapping, from $\overline{\mathcal{R}}$ to $\overline{\Omega}$, necessarily has the form $(q, p) \mapsto (X, Y) = (q, h(q, p))$, for some function $h : \overline{\mathcal{R}} \to \mathbb{R}$ with $h_p > 0$ in $\overline{\mathcal{R}}$. Then (1) may be reformulated as the following nonlinear elliptic boundary-value problem for the function h defined above:

$$(1 + h_q^2)h_{pp} - 2h_q h_p h_{qp} + h_p^2 h_{qq} - \Upsilon(-p)h_p^3 = 0 \qquad \text{in } \mathcal{R}, \tag{7a}$$

$$h = 0 \qquad \text{on } p = m, \tag{7b}$$

$$1 + h_q^2 + (2gh - Q)h_p^2 = 0 \qquad \text{on } p = 0. \tag{7c}$$

It is also easy to see that, conversely, if h satisfies (7) with $h_p > 0$ in $\overline{\mathcal{R}}$, then a solution of (1) may be constructed.

In the analysis of Constantin and Strauss[3], the left-hand sides of (7a) and (7c) are regarded as defining a nonlinear mapping between suitable Hölder spaces, while condition (7b) is incorporated in the definition of the domain space. The system (7) has a one-parameter family of trivial solutions, depending on the vertical coordinate p only, and, with Q being regarded as a parameter, local bifurcation of nontrivial solutions is obtained as a consequence of the Crandall–Rabinowitz Theorem[32]. As long as h_p and $Q - 2gh$ are bounded away from 0, problem (7) may be regarded as a quasilinear elliptic equation with an oblique boundary condition. Since there is no obvious way in which (7) may be rewritten as a compact perturbation of the identity, the Leray–Schauder degree, which is the key tool in the global bifurcation theory of Rabinowitz[33], is not available. However, global bifurcation is proved in Ref. 3 by an application of a generalized topological degree theory due to Healey and Simpson[34].

3.2. *The flattening transformation formulation*

Let us now assume that $\mathcal{S}$ is a graph, namely $\mathcal{S} = \{(X, \eta(X)) : X \in \mathbb{R}\}$, where $\eta : \mathbb{R} \to \mathbb{R}$ is an L-periodic function. It is then possible to consider the vertical scaling mapping $(X, Y) \mapsto (X, S) = \left(X, \frac{Y}{\eta(X)}\right)$, which is, obviously, a bijection between $\overline{\Omega}$ and the closure of the strip $\mathcal{R} = \{(X, S) : X \in \mathbb{R}, S \in (0, 1)\}$. Then the inverse mapping, from $\overline{\mathcal{R}}$ to $\overline{\Omega}$, has the form $(X, S) \mapsto (X, Y) = (X, S\eta(X))$. For any function ψ on Ω, one can associate the function w on $\mathcal{R}$ given by $w(X, S) = \psi(X, S\eta(X))$ for all $(X, S) \in \mathcal{R}$.

If ψ is a solution of (1) in Ω, it follows that the function w on $\mathcal{R}$ and the function η of one variable that defines $\mathcal{S}$ satisfy the system

$$\begin{aligned} \left(w_X - \frac{S\eta'}{\eta}w_S\right)_X - \frac{S\eta'}{\eta}\left(w_X - \frac{S\eta'}{\eta}w_S\right)_S & \\ + \frac{1}{\eta'}w_{SS} - \Upsilon(w) = 0 & \qquad \text{in } \mathcal{R}, \qquad (8a) \\ w = -m & \qquad \text{on } S = 0, \qquad (8b) \\ w = 0 & \qquad \text{on } S = 1, \qquad (8c) \\ \left(w_X - \frac{S\eta'}{\eta}w_S\right)^2 + \frac{1}{\eta^2}w_S^2 + 2g\eta = Q & \qquad \text{on } S = 1. \qquad (8d) \end{aligned}$$

Conversely, if a periodic function $\eta : \mathbb{R} \to \mathbb{R}$ and a function $w : \mathcal{R} \to \mathbb{R}$ periodic in the horizontal direction, satisfy (8), then a solution of (1) may be easily constructed.

In the analysis of Wahlén[28], the left-hand sides of (8a) and (8d) are regarded as defining a nonlinear mapping between suitable Hölder spaces, conditions (8b) and (8c) are incorporated in the definition of the domain space, and a suitable application of the Crandall–Rabinowitz Theorem leads to the existence of waves of small amplitude (with constant vorticity), some of which have critical layers. A similar argument is employed by Kozlov and Kuznetsov[29] for more general vorticity distributions, while in Refs. 30,31, a Lyapunov–Schmidt reduction method is used for a bifurcation problem with more than one parameter, leading in some situations to the existence of higher-dimensional sheets of solutions of small amplitude.

4. The conformal mapping approach

We consider in what follows the case when the vorticity is a constant Υ, and present the approach in Ref. 35 that leads to the reformulation of problem (1) as the quasilinear pseudodifferential equation (16) for a periodic function of a single variable, expressing the elevation of the wave profile in the conformal parametrization of the free surface. We emphasize that no assumptions are made on the geometry of the free surface $\mathcal{S}$ or on the topology of the streamlines in the interior of the fluid domain.

The approach in Ref. 35 is based on the idea that any domain Ω contained in the upper half of the (X, Y)-plane, whose boundary consists of the real axis $\mathcal{B}$ and a curve $\mathcal{S}$ in parametric form that is periodic in the horizontal direction, may be regarded as the conformal image of a strip. Given any $L > 0$, let $k > 0$ be such that $L = 2\pi/k$ (so that, for a steady wave of spatial period L, k is the wave number) and, for any $d > 0$, let us consider the strip $\mathcal{R}_d = \{(x, y) \in \mathbb{R}^2 : \ -d < y < 0\}$. Then, as shown in Ref. 35, there exist a unique constant $h > 0$, called the *conformal mean depth* of Ω, and a conformal map $\widetilde{X} + i\widetilde{Y}$ from $\mathcal{R}_{kh}$ onto Ω that admits an extension as a homeomorphism between the closures of these domains, such that, for all $(x, y) \in \mathcal{R}_{kh}$,

$$\widetilde{X}(x + 2\pi, y) = \widetilde{X}(x, y) + \frac{2\pi}{k}, \quad \widetilde{Y}(x + 2\pi, y) = \widetilde{Y}(x, y), \tag{9}$$

with the line $\{y = -kh\}$ being mapped onto $\mathcal{B}$ and the line $\{y = 0\}$ being mapped onto $\mathcal{S}$. We define

$$v(x) = \widetilde{Y}(x, 0) \quad \text{for all } x \in \mathbb{R}. \tag{10}$$

It then follows that necessarily $[v] = h$, where, throughout the paper, square brackets $[\cdot]$ are used to denote the mean over one period of 2π-periodic functions. Note that $\widetilde{Y}$ may be recovered uniquely from v as the harmonic function in $\mathcal{R}_{kh}$ that satisfies

$$\widetilde{Y}(x, 0) = v(x), \quad x \in \mathbb{R}, \qquad \widetilde{Y}(x, -kh) = 0, \quad x \in \mathbb{R}, \tag{11}$$

while $\widetilde{X}$ is, of course, a harmonic conjugate of $-\widetilde{Y}$ on $\mathcal{R}_{kh}$. Note also that, for any such domain Ω, classical elliptic theory ensures that (1a)–(1c) has a unique solution. Thus, (1) may be regarded as the problem of finding a domain Ω for which the corresponding solution of (1a)–(1c) satisfies also (1d). In turn, this problem may be recasted in terms of the function v of one variable defined above, using the fact that, if ψ is the unique solution of (1a)–(1c), then $(X, Y) \mapsto \psi(X, Y) - \frac{\Upsilon}{2} Y^2$ is a harmonic function in Ω, and therefore its composition with the conformal mapping $\widetilde{X} + i\widetilde{Y}$ is a harmonic function in $\mathcal{R}_{kh}$. The validity of (1d) may then be expressed as a condition on the top boundary $\{y = 0\}$ of $\mathcal{R}_{kh}$ involving the Dirichlet–Neumann operator, and using its connection with the periodic Hilbert transform for a strip, one obtains for the function v with $[v] = h$ the equation

$$\begin{aligned}\left\{\frac{m}{kh} - \frac{\Upsilon}{2kh}[v^2] - \Upsilon \mathcal{C}_{kh}(vv') + \Upsilon v\left(\frac{1}{k} + \mathcal{C}_{kh}(v')\right)\right\}^2 \\ -(Q - 2gv)\left\{(v')^2 + \left(\frac{1}{k} + \mathcal{C}_{kh}(v')\right)^2\right\} = 0,\end{aligned} \tag{12}$$

where $\mathcal{C}_{kh}$ denotes the periodic Hilbert transform associated to the strip $\mathcal{R}_{kh}$, whose definition is given below.

For any $d > 0$, the *conjugation operator* (also called *Hilbert transform*) $\mathcal{C}_d$ relates the boundary values on the real axis of the imaginary part and the real part of any holomorphic function in the strip $\mathcal{R}_d$ which is 2π-periodic in the x variable and real-valued on the line $y = -d$, see Ref. 36. It is defined, for any 2π-periodic function w of zero mean, $[w] = 0$, having the Fourier series expansion $w(x) = \sum_{n=1}^{\infty} a_n \cos(nx) + \sum_{n=1}^{\infty} b_n \sin(nx)$, $x \in \mathbb{R}$, by

$$\big(\mathcal{C}_d(w)\big)(x) = \sum_{n=1}^{\infty} a_n \coth(nd)\sin(nx) - \sum_{n=1}^{\infty} b_n \coth(nd)\cos(nx), \qquad x \in \mathbb{R}.$$

The mapping $w \mapsto \mathcal{C}_d(w)$ is a bounded linear operator from $C^{p,\alpha}_{2\pi,\circ}$ into itself, where, for any nonnegative integer p and any $\alpha \in (0,1)$, $C^{p,\alpha}_{2\pi,\circ}$ denotes the space of 2π-periodic functions of zero mean which are p times differentiable and whose p^{th} derivative is locally Hölder continuous with exponent α. The linear operator $\mathbb{C}_d$ also has the pointwise representation as a principal value integral[g]

$$\big(\mathcal{C}_d(w)\big)(x) = \frac{1}{2\pi}\, PV \int_{-\pi}^{\pi} \beta_d(x-s)\, w(s)\, ds, \qquad x \in \mathbb{R}, \tag{13}$$

for any $w \in C^{0,\alpha}_{2\pi,\circ}$, where the function $\beta_d : \mathbb{R} \setminus 2\pi\mathbb{Z} \to \mathbb{R}$ is given by

$$\beta_d(s) = -\frac{s}{d} + \frac{\pi}{d}\coth\left(\frac{\pi s}{2d}\right) + \frac{\pi}{d}\sum_{k\in\mathbb{N}} \frac{2\sinh\left(\frac{\pi s}{d}\right)}{\cosh\left(\frac{\pi s}{d}\right) - \cosh\left(\frac{2\pi^2 k}{d}\right)}, \qquad s \in \mathbb{R}\setminus 2\pi\mathbb{Z}.$$

The function β_d is 2π-periodic, odd, and continuous on $\mathbb{R} \setminus 2\pi\mathbb{Z}$, while $s \mapsto \left(\beta_d(s) - \frac{\pi}{d}\coth\left(\frac{\pi s}{2d}\right)\right)$ is continuous at $s = 0$.

Returning now to the water wave problem, a solution of (12) with $[v] = h$ gives rise to a solution of (1) if and only if it satisfies in addition

$$v(x) > 0 \qquad \text{for all } x \in \mathbb{R}, \tag{14}$$

$$\text{the mapping } x \mapsto \left(\frac{x}{k} + \big(\mathcal{C}_{kh}(v-h)\big)(x),\, v(x)\right) \text{ is injective on } \mathbb{R}. \tag{15}$$

Indeed, if $\widetilde{X}$ is the harmonic conjugate in $\mathcal{R}_{kh}$ of $-\widetilde{Y}$, where $\widetilde{Y}$ is obtained from v as the harmonic function in $\mathcal{R}_{kh}$ satisfying (11), then the

[g]As $d \to \infty$, the family of linear operators $\mathcal{C}_d$ approaches (at least formally) the standard periodic Hilbert transform

$$\big(\mathcal{C}(w)\big)(x) = \frac{1}{2\pi}\, PV \int_{-\pi}^{\pi} \cot\left(\frac{x-s}{2}\right) w(s)\, ds, \qquad \text{for all } x \in \mathbb{R},$$

that occurs in an analogous equation for irrotational (with $\Upsilon = 0$) waves of infinite depth, see Ref. 39.

validity of (14)–(15) enables us to define a non-self-intersecting curve $\mathcal{S} = \{(\frac{x}{k} + (\mathcal{C}_{kh}(v-h))(x), v(x)) : x \in \mathbb{R}\}$ in the upper half-plane so that $\widetilde{X} + i\widetilde{Y}$ is a conformal bijection between $\mathcal{R}_{kh}$ and the $2\pi/k$-periodic domain Ω whose boundary consists of $\mathcal{S}$ and $\mathcal{B}$, in which the existence of a solution ψ of (1) is then a consequence of (12).

While (12) has a one-parameter family of trivial solutions, for which $v \equiv h$ and which correspond to the parallel flows given by (2), and a construction of waves of small amplitude by means of the Crandall–Rabinowitz Local Bifurcation Theorem[32] applied to (12) was carried out in Ref. 35, the applicability of *global bifurcation theory* requires certain compactness properties of the nonlinear operators in question, which do not seem to be available for (12). Motivated by the derivation in Ref. 37 of Babenko's equation[21] in the theory of irrotational waves of infinite depth, an analogous reformulation of (12) has been very recently uncovered in Ref. 36: the *generalized Babenko equation* is the system

$$\mathcal{C}_{kh}\Big(\big(Q - 2gv - \Upsilon^2 v^2\big)\, v'\Big) + (Q - 2gv - \Upsilon^2 v^2)\Big(\frac{1}{k} + \mathcal{C}_{kh}(v')\Big) \tag{16a}$$
$$- 2\Upsilon v\Big(\frac{m}{kh} - \frac{\Upsilon}{2kh}\,[v^2] - \Upsilon\mathcal{C}_{kh}(vv')\Big) - \frac{Q - 2\Upsilon m - 2gh}{k} + 2g\,[v\,\mathcal{C}_{kh}(v')] = 0,$$

$$\left[\left\{\frac{m}{kh} - \frac{\Upsilon}{2kh}[v^2] - \Upsilon\mathcal{C}_{kh}(vv') + \Upsilon v\left(\frac{1}{k} + \mathcal{C}_{kh}(v')\right)\right\}^2\right] \tag{16b}$$
$$- \left[(Q - 2gv)\left\{(v')^2 + \left(\frac{1}{k} + \mathcal{C}_{kh}(v')\right)^2\right\}\right] = 0,$$

for unknowns $(m, Q, v) \in \mathbb{R} \times \mathbb{R} \times C^{1,\alpha}_{2\pi}(\mathbb{R})$ which have the same meaning as in (12), and where $[v] = h$. The square bracket $[\cdot]$ means the average over a period, so that (16b) is a scalar equation. The equivalence between (12) and (16) has been proved by means of Riemann–Hilbert theory on the strip. When restricted to the special case $\Upsilon = 0$ (irrotational flows), Babenko's equation can be obtained formally as the limit of the above equation as $d \to \infty$.

5. Global bifurcation

We now present, following Ref. 36, the main existence results for the problem of steady gravity waves with constant vorticity Υ, based on the reformulation (16). We assume that Υ, k and the conformal mean depth h are fixed (while g is always fixed), and regard (16) as a nonlinear equation with unknowns (m, Q, v) in the space $\mathbb{R} \times \mathbb{R} \times C^{2,\alpha}_{2\pi,e}(\mathbb{R})$ and such that $[v] = h$, where $C^{2,\alpha}_{2\pi,e}(\mathbb{R}) = \{f \in C^{2,\alpha}_{2\pi}(\mathbb{R}) :\ f(x) = f(-x) \text{ for all } x \in \mathbb{R}\}$, for some constant

$\alpha \in (0,1)$. The requirement that v is an even function reflects the symmetry of the corresponding wave profile. The family of trivial solutions of (1) described by (2) corresponds to the solutions of (16) for which $v \equiv h$, while Q and m are related by $Q = 2gh + \left(\frac{m}{h} + \frac{\Upsilon h}{2}\right)^2$, where $m \in \mathbb{R}$ is arbitrary; this family represents a curve $\mathcal{K}_{\text{triv}} = \left\{\left(m, 2gh + \left(\frac{m}{h} + \frac{\Upsilon h}{2}\right)^2, h\right) : m \in \mathbb{R}\right\}$ in the space $\mathbb{R} \times \mathbb{R} \times C^{2,\alpha}_{2\pi,e}(\mathbb{R})$. The main result of existence of nontrivial solutions is the following.

Theorem 5.1 (Ref. 36, Theorem 3). *Let h, $k > 0$ and $\Upsilon \in \mathbb{R}$ be given. For each $n \in \mathbb{N}$, let*

$$m^*_{n,\pm} = -\frac{\Upsilon h^2}{2} + \frac{\Upsilon h \tanh(nkh)}{2nk} \pm h\sqrt{\frac{\gamma^2 \tanh^2(nkh)}{4n^2k^2} + g\,\frac{\tanh(nkh)}{nk}}$$

and $Q^*_{n,\pm} = 2gh + \left(\frac{m^*_{n,\pm}}{h} + \frac{\Upsilon h}{2}\right)^2$.

First, for any $m \in \mathbb{R}$ with $m \notin \{m^*_{n,\pm} : n \in \mathbb{N}\}$, there exists a neighbourhood in $\mathbb{R}\times\mathbb{R}\times C^{2,\alpha}_{2\pi,e}(\mathbb{R})$ of the point (m,Q,h) on $\mathcal{K}_{\text{triv}}$, within which all the solutions of (16) are those on $\mathcal{K}_{\text{triv}}$. Secondly, consider the points $m^*_{n,\pm}$. For each integer $n \geq 1$ and each choice of sign in $\pm$, there exists in the space $\mathbb{R} \times \mathbb{R} \times C^{2,\alpha}_{2\pi,e}(\mathbb{R})$ a continuous curve $\mathcal{K}_{n,\pm} = \{(m(s), Q(s), v_s) : s \in \mathbb{R}\}$ of solutions of (16) such that the following properties hold:

(i) $(m(0), Q(0), v_0) = (m^*_{n,\pm}, Q^*_{n,\pm}, h)$;
(ii) $v_s(x) = h + s\cos(nx) + o(s)$ *in* $C^{2,\alpha}_{2\pi,e}(\mathbb{R})$ *if* $0 < |s| < \varepsilon$, *for some* $\varepsilon > 0$ *sufficiently small;*
(iii) *there exist a neighbourhood $\mathcal{W}_{n,\pm}$ of $(m^*_{n,\pm}, Q^*_{n,\pm}, h)$ in $\mathbb{R} \times \mathbb{R} \times C^{2,\alpha}_{2\pi,e}(\mathbb{R})$ and $\varepsilon > 0$ sufficiently small such that*

$$\{(m,Q,v) \in \mathcal{W}_{n,\pm} :\ v \not\equiv h, (16) \text{ holds}\} = \{(m(s), Q(s), v_s) :\ 0 < |s| < \varepsilon\};$$

(iv) $Q(s) - 2gv_s(x) > 0$ *for all* $s, x \in \mathbb{R}$;
(v) *$\mathcal{K}_{n,\pm}$ has a real-analytic reparametrization locally around each of its points;*
(vi) *One of the following alternatives occurs:*
   (α) *either*

$$\min\left\{\frac{1}{1 + ||(m(s), Q(s), v_s)||_{\mathbb{R}\times\mathbb{R}\times C^{2,\alpha}_{2\pi,e}(\mathbb{R})}},\ \min_{x\in\mathbb{R}}\left\{Q(s) - 2gv_s(x)\right\}\right\} \to 0 \qquad \text{as } s \to \pm\infty;$$

(β) *or there exists* $T > 0$ *such that*

$$(m(s+T), Q(s+T), v_{s+T}) = (m(s), Q(s), v_s) \qquad \textit{for all } s \in \mathbb{R}.$$

Moreover, for each integer $n \geq 1$ *and each choice of sign in* $\pm$*, such a curve of solutions of (16) with the properties* (i)–(vi) *is unique (up to reparametrization).*

In this theorem, (i) identifies the local bifurcation points along the trivial solution curve $\mathcal{K}_{\text{triv}}$, (ii) and (iii) describe the local behavior of the bifurcating curve of nontrivial solutions, and (iv)–(vi) describe the global behavior. The alternative (α) means that either the curve is unbounded in the function space or the fluid velocity at the crest of the corresponding waves approaches 0, while the alternative (β) is that the curve forms a loop. The proof of Theorem 5.1 is based on a slightly corrected version of the global bifurcation theorem for real-analytic operators due to Dancer[38] and improved by Buffoni and Toland (see Theorem 9.1.1 in Ref. 39). The required compactness properties of the operators involved are rather subtle and depend on *commutator estimates* for the periodic Hilbert transform for a strip[h] [i].

Let us now describe some qualitative features of the waves whose existence is ensured by Theorem 5.1. Note that the parallel flow corresponding to $m = m^*_{1,-}$ admits critical points if and only if $\Upsilon < 0$ and $\frac{\tanh(kh)}{kh} \leq \frac{\Upsilon^2 h}{g+\Upsilon^2 h}$, while the parallel flow corresponding to $m = m^*_{1,+}$ admits critical points if and only if $\Upsilon > 0$ and the same inequality holds. Such a parallel flow having a horizontal critical layer consisting of stagnation points that mark the flow reversal, in the case $\Upsilon < 0$, is depicted in the left half of Figure 1. The information provided by part (ii) in Theorem 5.1 is sufficient to derive some properties of the waves located on the bifurcation curves $\mathcal{K}_{1,\pm}$ close enough to the bifurcation points. For $s \neq 0$ small enough, the wave profiles obtained thereby are graphs that are strictly

[h]A typical such estimate is the following: if $f \in C^{j,\alpha}_{2\pi,\circ}(\mathbb{R})$ and $g \in C^{j-1,\alpha}_{2\pi,\circ}(\mathbb{R})$ with $j \in \mathbb{N}$ and $\alpha \in (0,1)$ are such that $[fg] = 0$, then $\big(f\,\mathcal{C}_d(g) - \mathcal{C}_d(fg)\big) \in C^{j,\delta}_{2\pi}(\mathbb{R})$ for all $\delta \in (0,\alpha)$ (with the inequality $\delta < \alpha$ being sharp), and there exists a constant $C = C(j,\alpha,\delta)$ such that

$$||f\,\mathcal{C}_d(g) - \mathcal{C}_d(fg)||_{j,\delta} \leq C||f||_{j,\alpha}||g||_{j-1,\alpha}.$$

[i]Such commutator estimates also lead, in combination with bootstrap arguments, to the result that, for any solution (m, Q, v) of (16), the function v is necessarily of class C^∞ on $\mathbb{R}$.

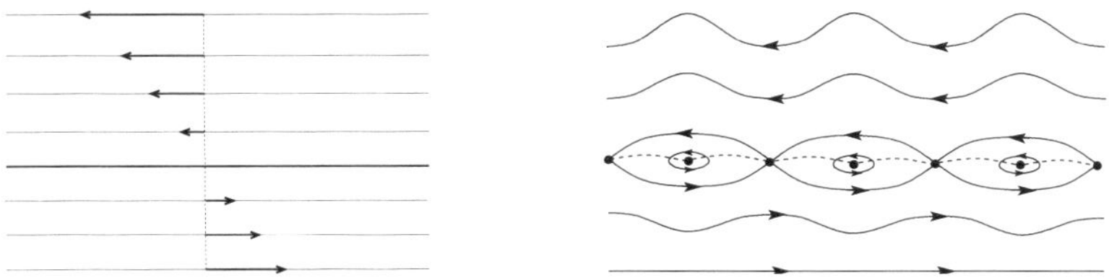

Fig. 1.

monotone between crests and troughs and are symmetric about the crest line; for $s > 0$ the crest is located at $x = 0$, while for $s < 0$ the trough is at $x = 0$. Moreover, if the parallel flow at the bifurcation point has a critical line *strictly above* the flat bed, then an analysis of the leading-order behavior with respect to the small parameter s of the velocity field reveals that the waves of small amplitude on the bifurcation curve have a 'cat's eye' flow pattern[j] as depicted, in the case $\Upsilon < 0$ and $s > 0$, in the right half of Figure 1.

The next result addresses the questions, which turn out to be related, of which of the alternatives (α) and (β) occurs for the curves $\mathcal{K}_{1,\pm}$, of whether all points on those curves gives rise to solutions of (1) (a matter which depends on the validity of (14)–(15)), and of the validity for waves of large amplitude of the monotonicity properties of the waves of small amplitude. Let us to consider the following properties[k] of a function $v \in C^{2,\alpha}_{2\pi,e}$:

$$v'(x) < 0 \quad \text{for all } x \in (0, \pi)\,, \tag{17}$$

$$v(x) > 0 \qquad \text{for all } x \in \mathbb{R}, \tag{18}$$

$$0 < \frac{x}{k} + \big(\mathcal{C}_{kh}(v - [v])\big)(x) < \frac{\pi}{k} \qquad \text{for all } x \in (0, \pi)\,. \tag{19}$$

Then the following result holds.

Theorem 5.2 (Ref. 36, Theorem 5). *For either choice of sign in* $\pm$, *let* $\mathcal{K}^{>}_{1,\pm} = \{(m(s), Q(s), v_s) : s \in (0, \infty)\}$ *be the curve of solution of (16). Then one of the following alternatives occurs:*

(A_1) v_s *satisfies* (17)–(19) *for any* $s \in (0, \infty)$, *and alternative* (α) *in Theorem 5.1 occurs;*

[j] Results of this type were first obtained by Wahlén[28] by working with the flattening transformation formulation.

[k] Note that conditions (17) and (19) are sufficient to ensure the validity of the injectivity condition (15). In fact, in the presence of (17), (19) is both necessary and sufficient for (15).

(A_2) *there exists some* $s^* \in (0,\infty)$ *such that* v_s *satisfies* (17) – (19) *for any* $s \in (0,s^*)$, *while* v_{s^*} *satisfies* (17) – (18), *and instead of* (19) *it satisfies*

$$0 < \frac{x}{k} + \big(\mathcal{C}_{kh}(v-h)\big)(x) \le \frac{\pi}{k} \qquad \text{for all } x \in (0,\pi), \tag{20a}$$

$$\frac{x_0}{k} + \big(\mathcal{C}_{kh}(v-h)\big)(x_0) = \frac{\pi}{k} \qquad \text{for some } x_0 \in (0,\pi). \tag{20b}$$

Roughly speaking, alternative (A_1) means that all the solutions on $\mathcal{K}^{>}_{1,\pm}$ correspond to physical water waves that are symmetric and whose vertical coordinate strictly decreases between each of its consecutive global maxima and minima, which are unique per minimal period. However, since no claim is being made that the horizontal coordinate is strictly monotone, such waves could have overhanging profiles. Alternative (A_2) means that solutions on $\mathcal{K}^{>}_{1,\pm}$ that correspond to physical water waves with qualitative properties as described above do exist until a limiting configuration with a profile that self-intersects on the line strictly above the trough[1] is reached at $s = s^*$. Thus in particular the alternative (β) of a loop cannot occur in Theorem 5.1 unless such a limiting configuration as above is first reached along the curve. This intriguing possibility is supported by the numerical computations in Refs. 7,11. The proof of Theorem 5.2 is based on an intricate series of arguments based on repeated use of the strong maximum principle together with the Hopf and Serrin inequalities at the boundary.

6. Conclusions and some open problems

There are several ways, each valid under certain assumptions, in which the steady water wave problem (1) can be equivalently rewritten as a problem over a fixed domain. The resulting problem is then regarded as a bifurcation problem in suitable Banach spaces, and existence of small-amplitude solutions is typically achieved by an application of the Crandall–Rabinowitz local bifurcation theorem. To extend the local bifurcation curve, there are essentially two versions of global bifurcation theory that can be used, depending on the structure of the problem at hand.

The first, which in an abstract setting goes back to Rabinowitz[33], is based on topological methods, namely the Leray–Schauder degree or more general versions of the degree, such as that of Healey and Simpson, and provides existence of a connected set of solutions. The second, which is due

[1]Indeed, if (20b) holds, then the points corresponding to the values x_0 and $2\pi - x_0$ of the parameter x coincide, and they have the same horizontal coordinate as the point corresponding to $x = \pi$.

to Dancer[38], is applicable for equations involving real-analytic operators, is based on the structure theorem of real-analytic varieties, and provides existence of a curve of solutions. Both abstract approaches provide several alternatives for the behavior in-the-large of the family of solutions so constructed. One possibility is that this family 'returns' to the curve of trivial solutions. This has been completely ruled out for the Dubreil-Jacotin formulation in Ref. 3, while Theorem 5.2 shows that, for the generalized Babenko formulation, it must necessarily be preceded by the occurrence of a limiting wave with a self-intersecting profile. Another alternative in the abstract theory is that the solution set contains a sequence that either goes to infinity in the function space in which the problem is formulated, or approaches the 'boundary' of the domain in which the equations are well behaved.

The situation is much better understood in the irrotational problem. There it has been proved that the solution set remains bounded in suitable spaces of functions of weaker regularity, and contains in its closure a solution with stagnation points at its crests, where it has lateral tangents making an angle of 120°, just as Stokes had predicted, see Refs. 1,4 for the relevant references. A new proof, based on a modern geometric analysis approach, of the fact that any irrotational wave with stagnation points on the free surface must have symmetric corners of 120° at any such point has recently been given in Ref. 40.

It is widely believed that a similar scenario happens in many circumstances, but not always, for the problem of steady waves with vorticity as well. The fact that any wave with vorticity having a stagnation point at the crest must (except in some degenerate cases) have a corner of 120° there has been proved in Ref. 41. However, there are only some partial results on the existence of such waves. Namely, in Ref. 42 it is proved that if certain simple a priori bounds are valid along a family of smooth waves (of the type constructed in Ref. 3), then a subsequence can be extracted that converges in weak sense to a solution with a stagnation point at the crest. However, the validity of such a priori bounds has not yet been proved, except in limited circumstances; proving that such bounds are valid in more general situations is a problem currently under investigation.

Another important open problem is to determine which of the alternatives (A_1) and (A_2) occurs in Theorem 5.2. Alternative (A_1) is most likely to occur when the vorticity is small (compared to some other parameters in the problem). In such a case, it would be important to determine a priori bounds on the set of solutions, that could lead to the existence of an 'extreme wave', whose profile might be either a graph or an overhanging curve. It would also be interesting to confirm rigorously that alternative (A_2) is indeed possible in certain situations.

In the case when the vorticity is not a constant, the most general fixed-domain reformulation currently known is that provided by the flattening transformation, which is capable of describing waves with critical layers. It would be of interest to develop a global bifurcation theory for the flattening transformation formulation.

Finally, it is unsatisfactory that, apart from the generalized Babenko formulation (which is valid only for waves with constant vorticity) no fixed-domain reformulation is known for the steady water wave problem that would impose no restriction on the geometry of the free surface or the topological structures of the streamlines. Exhibiting such a formulation and developing a global bifurcation theory for it are important open problems.

References

1. W. Strauss, Steady water waves, *Bull. Amer. Math. Soc.* **47** (2010), 671–694.
2. D. Lannes, *The water waves problem. Mathematical analysis and asymptotics.* Mathematical Surveys and Monographs, 188. American Mathematical Society, Providence, RI, 2013.
3. A. Constantin and W. Strauss, Exact steady periodic water waves with vorticity, *Comm. Pure Appl. Math.* **57** (2004), 481–527.
4. J. F. Toland, Stokes waves, *Topol. Methods Nonlinear Anal.* **7** (1996), 1–48.
5. M. D. Groves, Steady water waves, *J. Nonl. Math. Physics* **11** (2004), 435–460.
6. A. Constantin, *Nonlinear water waves with applications to wave-current interactions and tsunamis*, CBMS-NSF Conf. Ser. Appl. Math., Vol. 81, SIAM, Philadelphia, 2011.
7. A. F. T. da Silva and D. H. Peregrine, Steep, steady surface waves on water of finite depth with constant vorticity, *J. Fluid Mech.* **195** (1988), 281–302.
8. J. Ko and W. Strauss, Effect of vorticity on steady water waves, *J. Fluid Mech.* **608** (2008), 197–215.
9. J. Ko and W. Strauss, Large-amplitude steady rotational water waves, *Eur. J. Mech. B Fluids* **27** (2008), 96–109.
10. H. Okamoto and M. Shoji, *The mathematical theory of permanent pro-*

gressive water waves, World Scientific, New Jersey-London-Singapore-Hong Kong, 2001.

11. J.-M. Vanden-Broeck, Some new gravity waves in water of finite depth, *Phys. Fluids* **26** (1983), 2385–2387.
12. V. Kozlov and N. Kuznetsov, Steady free-surface vortical flows parallel to the horizontal bottom, *Quart. J. Mech. Appl. Math.* **64** (2011), 371–399.
13. F. Gerstner, Theorie der Wellen, *Abhand. Koen. Boehmischen Gesel. Wiss.*, Prague, 1802.
14. G. G. Stokes, On the theory of oscillatory waves, *Trans. Camb. Phil. Soc.* **8** (1847), 441–455.
15. G. G. Stokes, Considerations relative to the greatest height of oscillatory irrotational waves which can be propagated without change of form, Math. and Phys. Papers, vol. I, 225–228, Cambridge University Press, Cambridge, 1880.
16. Yu. P. Krasovskii, On the theory of steady-state waves of large amplitude, *USSR Comput. Math. and Math. Phys.* **1** (1961), 996–1018.
17. G. Keady, J. Norbury, On the existence theory for irrotational water waves, *Math. Proc. Camb. Phil. Soc.* **83** (1978), 137–157.
18. C. J. Amick and J. F. Toland, On periodic water-waves and their convergence to solitary waves in the long-wave limit, *Phil. Trans. Roy. Soc. London Ser. A* **303** (1981), 633–669.
19. B. Buffoni, E. N. Dancer and J. F. Toland, The regularity and local bifurcation of steady periodic water waves, *Arch. Ration. Mech. Anal.* **152** (2000), 207–240.
20. B. Buffoni, E. N. Dancer and J. F. Toland, The sub-harmonic bifurcation of Stokes waves, *Arch. Ration. Mech. Anal.* **152** (2000), 241–271.
21. K. I. Babenko, Some remarks on the theory of surface waves of finite amplitude, *Dokl. Akad. Nauk* **294** (1987), 1033–1037.
22. M.-L. Dubreil-Jacotin, Sur la détermination rigoureuse des ondes permanentes périodiques d'ampleur finie, *J. Math. Pures Appl.* **13** (1934), 217–291.
23. E. R. Spielvogel, A variational principle for waves of infinite depth, *Arch. Ration. Mech. Anal.* **39** (1970), 189–205.
24. J. F. Toland, On a pseudo-differential equation for Stokes waves, *Arch. Ration. Mech. Anal.* **162** (2002), 179–189.
25. E. Varvaruca, Some geometric and analytic properties of solutions of Bernoulli free-boundary problems, *Interfaces Free Bound.* **9** (2007), 367–381.

26. E. Varvaruca, Bernoulli free-boundary problems in strip-like domains and a property of permanent waves on water of finite depth, *Proc. Roy. Soc. Edinburgh A* **138** (2008), 1345–1362.
27. W. Thomson (Lord Kelvin), On a disturbing infinity in Lord Rayleigh's solution for waves in a plane vortex stratum, *Nature* **23** (1880), 45–46.
28. E. Wahlén, Steady water waves with a critical layer, *J. Differential Equations* **246** (2009), 2468–2483.
29. V. Kozlov and N. Kuznetsov, Dispersion equation for water waves with vorticity and Stokes waves on flows with counter-currents, *Arch. Ration. Mech. Anal.* **214** (2014), 971–1018
30. M. Ehrnström, J. Escher and E. Wahlén, Steady water waves with multiple critical layers, *SIAM J. Math. Anal.* **43** (2011), 1436–1456.
31. M. Ehrnström and E. Wahlén, Trimodal steady water waves, *Arch. Ration. Mech. Anal.* **216** (2015), 449–471.
32. M. G. Crandall and P. H. Rabinowitz, Bifurcation from simple eigenvalues, *J. Funct. Anal.* **8** (1971), 321–340.
33. P. H. Rabinowitz, Some global results for nonlinear eigenvalue problems, *J. Funct. Anal.* **7** (1971), 487–513.
34. T. Healey and H. Simpson, Global continuation in nonlinear elasticity, *Arch. Rat. Mech. Anal.* **143** (1998), 1–28.
35. A. Constantin and E. Varvaruca, Steady periodic water waves with constant vorticity: regularity and local bifurcation, *Arch. Ration. Mech. Anal.* **199** (2011), 33–67.
36. A. Constantin, W. Strauss and E. Varvaruca, Global bifurcation of steady gravity water waves with critical layers, *Acta Math.*, in press.
37. E. Shargorodsky and J. F. Toland, Bernoulli free-boundary problems, *Mem. Amer. Math. Soc.* **196** (2008), no. 914.
38. E. N. Dancer, Bifurcation theory for analytic operators, *Proc. London Math. Soc.* **26** (1973), 359–384.
39. B. Buffoni and J. F. Toland, *Analytic theory of global bifurcation*, Princeton University Press, Princeton and Oxford, 2003.
40. E. Varvaruca and G. S. Weiss, A geometric approach to generalized Stokes conjectures, *Acta Math.* **206** (2011), 363–340.
41. E. Varvaruca and G. S. Weiss, The Stokes conjecture for waves with vorticity, *Ann. Inst. H. Poincaré Anal. Non Linéaire* **29** (2012), 861–885.
42. E. Varvaruca, On the existence of extreme waves and the Stokes conjecture with vorticity, *J. Differential Equations* **246** (2009), 4043–4076.

Author index